# 2nd International Conference on Energy and Environmental Science 2018



IOP Conference Series: Earth and Environmental Science
Volume 164

Kuala Lumpur, Malaysia
16 – 18 January 2018

ISBN: 978-1-5108-6581-5
ISSN: 1755-1307

**Printed from e-media with permission by:**

Curran Associates, Inc.
57 Morehouse Lane
Red Hook, NY 12571

proceedings.com

**Some format issues inherent in the e-media version may also appear in this print version.**



Printed by Curran Associates, Inc. (2018)

For permission requests, please contact the Institute of Physics
at the address below.

Institute of Physics
Dirac House, Temple Back
Bristol BS1 6BE UK

Phone: 44 1 17 929 7481
Fax: 44 1 17 920 0979

techtracking@iop.org

**Additional copies of this publication are available from:**

Curran Associates, Inc.
57 Morehouse Lane
Red Hook, NY 12571 USA
Phone: 845-758-0400
Fax: 845-758-2633
Email: curran@proceedings.com
Web: www.proceedings.com

# Table of contents

## Volume 164

**2018 2nd International Conference on Energy and Environmental Science**

**16–18 January 2018, Kuala Lumpur, Malaysia**

**Accepted papers received: 12 June 2018**
**Published online: 21 June 2018**

### Preface

2018 2nd International Conference on Energy and Environmental Science

Committees

Peer review statement

### Papers

#### Energy Development and Utilization

Mapping And Development Of Geographic Information System Spreading And Biomass Potential In Java And Sumatra

R. A. E. Virgana and Parlindungan Harahap.....1

Spatial Differentiation of Energy Poverty in Eu Countries

M Zawada, M Szajt, J Kowalik and R Kucęba.....7

Prosumer energy as a stimulator of micro- Smart Grids development - on the consumer side

R Kucęba, M Zawada, M Szajt and J Kowalik.....13

Implementation of Energy Policy in the European Union Countries in the Aspect of their Sustainable Development

J Kowalik, M Zawada, M Szajt and R Kucęba.....18

A Study on Leaching Potential of Bio-gasification Tar in Water

Hemantajeet Medhi, M Ronghang, B Chakrabarty and A K Baruwa.....23

Comparative Analysis of Corrosion Inhibition: Between Jatrophacurcas, Palm and Diesel Oil based Emulsified Acids for Acid Stimulation Operations

Muhammad Mohsin Yousufi, Mysara Eissa Mohyaldinn Elhaj and Muhammad Moniruzzaman.....28

Kinetic study on oil palm biomass conversion to levulinic acid via indium trichloride-ionic liquids

Y W Tiong, C L Yap, S Gan and W S P Yap.....34

Numerical simulations for biomass burning aerosol transport in northern Thailand

Nurzahziani and C. Surussavadee.....39

An investigation of the vertical diffuse efficacy of solar irradiance for Thailand's climate

Sutiwat Umar and Pipat Chaiwiwatworakul.....45

CFD and experimental analysis of solar crop dryer with waste heat recovery system of exhaust gas from diesel engine

R Meenakshi Reddy, E Siva Reddy, C Uma Maheswari and K Krishna Reddy.....51

**Power System and Industrial Engineering**

Multiple Distribution Generation Location in Reconfigured Radial Distribution System Distributed generation in Distribution System

Smieee Ashwani Kumar and Ravi Teja Bhimarasetti.....57

Electricity Consumption and Associated Risks in European Countries - Convergence Analysis

M Szajt, R Kuczba, M Zawada and J Kowalik.....66

Performance Analysis of Multi-Photovoltaic (PV)-Grid Tied Plant in Malaysia

Laith M Halabi and Saad Mekhilef.....72

Control and energy management of grid-connected pv system with battery-supercapacitor hybrid energy storage

A Guichi, A Talha, El Berkouk and N Sabeur.....77

A New Under-Frequency Load Shedding Scheme Based on Adaptive Neuro-Fuzzy Inference System and Evolutionary Programming Shedding Priority

Mohammad Dridy, Hazlie Mokhlis and Saad Mekhilef.....82

Effect Method Evaluation of Forecasting Scour by Propeller in Harbours

P T Ghazvinei, H H Darvishi and S Soon.....87

Enhanced wind-driven infiltration model by incorporating wind direction for design purpose

L G Perdamaian and Z J Zhai.....93

The Mediating Role of Organizational Learning Capability On Green Distribution and Green Packaging Towards Sustainability Performance as A Function Environmental Dynamism: Indonesia and Malaysia Fishery Industries

A Hutomo, M Haizam and O Sinaga.....106

Increasing public awareness of consuming Indonesian coffee through electronic marketing

A. Lisdayanti and T.A Anwar.....117

**Environmental Science and Ecological Protection**

Environmental Management in Central Siberia (Based On Analysis of the Krasnoyarsk Territory)

N P Koptseva and A A Sitnikova.....123

A Study on the Evaluation Indicators of the Regeneration of Brownfields in Taiwan - A Case Study of Formosa Chemicals & Fibre Corporation (Changhua Plant)

S J Feng, N C Chao and P T Li.....128

Predicting Fish Ecological As Indicator of River Pollution Using Decision Tree Technique

Che-Yu Hsu, Sheng-Jung Ou and Wei-Fan Hsieh.....137

Investigation of Haloacetic acids (HAAs) levels in water supply and its correlated HAAs formation

P Saengchut, S Karuchit and P Pentamwa.....143

Estimation of Urban Air Pollutant Levels using AERMOD: A Case Study in Nakhon Ratchasima, Thailand.

K Katika and S Karuchit.....149

Treatment of $H_2S$ and $NH_3$ from wastewater using a pilot-scale tray biofilter

S Karuchit and A Rakthaisong.....154

The effect of profitability and leverage to the carbon emission disclosure on companies that registered consecutively in sustainability reporting award period 2014-2016

Khairun Nisak and Rita Yuniarti.....159

Removal of Pesticides Using Aquatic Plants in Water Resources: A Review

Premla Devi Chander, Chow Ming Fai and Chai Mee Kin.....164

Corrosion Study of Carbon Steel in $CO_2$ Loaded Amine-Amino Acid Solutions-Case of Mixtures of NMethyldiethanolamine and L-Arginine

A G Talkhan, A Benamor and M S Nasser.....169

Effects of slopes on the stormwater attenuation performance in extensive green roof

M F Chow, M F Abu Bakar and M H Mohd Razali.....175

Numerical Groundwater Modeling of an Eogenetic Karst Catchment Using Analytical Head-Guided Zonation Method: A New Analytical Approach

D K S Y Klaas, M A Imteaz, A Arulrajah, I Sudiayem, E M E Klaas and E C M Klaas.....181

The Influence of shales characteristics on $CO_2$ adsorption behaviour under sub-critical conditions

Ahmed Mohammed Al-Mutarreb, Shiferaw Regassa Jufar, Hesham Abdulelah, Eswaran Padmanabhan, Ghareb Mostafa and Maqsood Ahmed.....186

**Urban Infrastructure Construction and Planning**

The Environmental Load Characteristic Analysis of LCA-based IPC Girder Bridge

G C Choi, J S Kim, S Sackey and B S Kim.....193

Analysis of Construction-specific Environmental Load Characteristics in the Road Paving Work Maintenance & Management Phase

J G Im, D H Kim, J N Liu, J Y Park and B S Kim.....198

The Impact of Taiwan's Urban-renewal Policies, Laws and Decrees on Urban Redevelopment and Urban Disaster Prevention

Jen-Chih Chao and Kuo-Wei Hsu.....203

Design and Implementation of Earthquake Emergency Response System Based on Grid Data

J Zhao and Z H Ma.....210

Research Direction in Realizing Sustainable IoT Based Smart City Ecosystem

Iqbal Yulizar Mukti and Yudha Prambudia.....217

IOP Conf. Series: Earth and Environmental Science **164** (2018) 011001 doi:10.1088/1755-1315/164/1/011001

# Preface

2018 2nd International Conference on Energy and Environmental Science (ICEES 2018), was successfully held in University of Malaya, Kuala Lumpur, Malaysia on January 16-18, 2018.

The proceedings are the result of ICEES 2018, which presents 36 accepted articles selected from 115 papers submitted to the conference from 25 countries and districts including USA, Japan, Australia, Poland, Canada, Germany, China, India, Pakistan, Iran, South Korea, Russia, Indonesia, Taiwan, Nigeria, Brazil, Qatar, Egypt, Ecuador, Malaysia, Thailand, Philippines, Turkey, Algeria and Jordan. All of the papers were subjected to peer-review by the program technical committee members and international reviewers. The papers are presented orally or with posters in 8 parallel sessions, and all the presenters have done very well during the presentation.

The volume contains 4 chapters, which tends to present to the readers the recent advances in the field of Energy and Environmental Science including Energy Development and Utilization, Power System and Industrial Engineering and Environmental Science and Ecological Protection, as well as Urban Infrastructure Construction and Planning.

On behalf of the organizing committee, I would like to express my sincere gratitude to guest editors, program technical committee members and international reviewers, who provided constructive criticism and stimulated comments and suggestions to the authors. It's a tough work to select papers from the submissions, and without their hard work the proceedings will not be possible to come into being. Besides, we also want to thank the chairs, who are dedicated to make sure to run the conference smoothly, the keynote speakers and invited speakers who shared with us their precious research findings. Moreover, our appreciation also goes to the staff in University of Malaya, who gave the delegates very vivid and professional demonstration during the lab visit. Last but not at least, I would like thank University of Malaya, Malaysia; Widyatama University, Indonesia and ASEPTEC SDN BHD Company for their financial and technical sponsorship.

Finally, I hope the readers will find the proceedings interesting and worthwhile. ICEES 2019 conference will be held in Seoul, South Korea, and we look forward to seeing you next year in Seoul.

Yours Sincerely,

Prof. Saad Mekhilef
Conference Chair of ICEES 2018
University of Malaya, Malaysia

IOP Conf. Series: Earth and Environmental Science **164** (2018) 011002 doi:10.1088/1755-1315/164/1/011002

# Committees

**Conference Chairs**

Prof. Dr. Saad Mekhilef, University of Malaya, Malaysia

Prof. Chan Siew Hwa, Nanyang Technology University, Singapore

**Program Chairs**

Prof. Hazlie Mokhlis, University of Malaya, Malaysia

Prof. Chul H. Jo, Inha University, South Korea

Prof. JB Superales, JH Cerilles State College, Philippines

**Local Organizing Committee**

Mr. Leong Wen Chek, University of Malaya, Malaysia

Mr. Nasser Sabeur, University of Malaya, Malaysia

**Technical Committee**

Asst. Prof. Jennifer L. Santos, Centro Escolar University, Philippines

Assoc. Prof. Sandro Nizetic, FESB Faculty, University of Split, Croatia

Assoc. Prof. Thanakom Soontornchainacksaeng, King Mongkut's University of Technology, Thailand

Prof. Yee-Chang Feng, National Cheng Kung University, Taiwan

Prof. Yogesh Pahariya, Lakshmi Narain College of Technology, India

Asst. Prof. Varun Goel, National Institute of Technology, India

Asst. Prof. Kumar K, SV College of Engineering, India

Dr. Tsung-Mou Huang, Taiwan power company, Taiwan

Dr. Jiahao Liu, University of Science and Technology of China, China

Dr. Lau Woei Jye, Universiti Teknologi Malaysia, Malaysia

Lecturer Brabha Hari Nagaratnam, Universiti Utara Malaysia, Malaysia

Dr. Umarin Sangpanich, Kasetsart University, Thailand

Lecturer Unggul Wibawa, Universitas Brawijaya, Indonesia

Prof. Ali CHEKNANE, Amar Telidji University of Laghouat, Algeria

Dr. Alireza Ghasempour, University of Applied Science and Technology, Iran

Dr. Song He, Xiamen University, China

Assoc. Prof. R. K. Saket, Indian Institute of Technology (Banaras Hindu University), India

Dr. G. Kumaresan, Anna University, India

Dr. Sandip Singh Bhatti, Guru Nanak Dev University, India

Dr. Lau Woei Jye, Universiti Teknologi Malaysia, Malaysia

Dr. Umarin Sangpanich, Kasetsart University,Thailand

Prof. Manish Bhardwaj, SRM University, India

Asst. Prof. Sohail Ahmed, University of Haripur, Pakistan

# Peer review statement

All papers published in this volume of *IOP Conference Series: Earth and Environmental Science* have been peer reviewed through processes administered by the proceedings Editors. Reviews were conducted by expert referees to the professional and scientific standards expected of a proceedings journal published by IOP Publishing.

IOP Conf. Series: Earth and Environmental Science **164** (2018) 012001 doi:10.1088/1755-1315/164/1/012001

# Mapping And Development Of Geographic Information System Spreading And Biomass Potential In Java And Sumatra

**R. A. E. Virgana[1], Parlindungan Harahap[2]**

[1,2] Information System Department , Faculty of Engineering, University of Widyatama, Jl. Cikutra No. 204, Bandung, Indonesia

rae.virgana@widyatama.ac.id; parlindungan.mt@widyatama.ac.id

**Abstract.** Indonesia is one of the countries on the equator. This geographical position makes the soil of Indonesia become fertile and have high rainfall. It also makes forests and plantations grow well. The main product of forest and plantation products in Indonesia gives positive and economical value to society and country. However, it also gives negative value, one of them is the biomass waste that it produces. If not managed properly, then this will be bad for the cleanliness and aesthetic environment. Through the bioenergy program, it can be done how to utilize biomass waste into Biomass Power Plant (PLTBM). In this research, mapping and development of geographical information distribution system and bioenergy potential by classifying the biomass sources used by the plant and also the analysis of land area to potential energy that can be used.

## 1. Introduction

Identification of Potential and Study of Biomass Utilization, Municipal Rubbish and Biogas for Electrical Energy in Sumatera and Kalimantan Island, so that biomass potential can be obtained, for electricity on the island of Sumatra of 15,588 MWe and potential for biomass for electricity in Borneo island of 5.062 Mwe. The biomass potential for electricity is unevenly distributed and requires a power grid to be connected to the bioenergy plant that has been built. The current state of electricity network development by the State Electricity Company is still constrained by geographical aspects, and economics so that the available power distribution network is not yet connected to the state electricity company network. Forest Energy is a program of the energy sector and the forestry sector, to encourage the involvement of all parties in campaigning for the potential of forests to prosper the surrounding communities and create national energy independence and resilience. Energy Plantation Forest is the allocation of productive forest land and growing crops above it as biomass feedstock devoted entirely to biomass-based bio-based fuel or electricity production for domestic purposes. The process of energy formation from this biomass alone, kind of a lot to be able to generate energy, from the start of the combustion system (boiler) to the fermentation system, and so on. [1] In this research will map out the scattered biomass resources especially in Sumatera and Kalimantan Island, where in these two islands, biomass resources are very large, especially the remaining resources of palm oil waste and waste woodchip waste. [2][3]

This research hopes to describe the map of the distribution of biomass-based energy, so it can be concluded and analyzed the benefits for the people and prosperity of the Unitary State of the Republic of Indonesia.

## 2. Method

The methodology used in this study is how data are collected, including non spatial data (biomass power generation points, industrial forests, oil palm plantations), and spatial data on village, sub-district, district and provincial districts [4][5]. Then the data is processed, there are processes of segregation / merging / intersection data, then data in overview or visualized in GIS tools [6] [7], as shown in Figure 1. below :

**Figure 1.** General Process Methodology Mapping and Development of Geographic Information System Distribution and Biomass Potential in Java and Sumatera

Data input can be tables, reports, maps, satellite photos, remote sensing, field measurements etc. This process is to collect, prepare and store spatial data, as well as convert to a format that can be used by GIS devices. Then Data management, in this process is to organize spatial data and related attribute tables into a database system so easily recalled or retrieve, updated and edited. Then manipulation and data analysis, this process to determine the information that can be generated by GIS. In addition, it also manipulates and modeling to produce the expected information. Finally output / analysis results, shaped map display results include exporting it to the desired format as in the form of tables, graphics, reports, maps and so on [8] [9][10].
The steps are as shown in Figure 2. as follows:

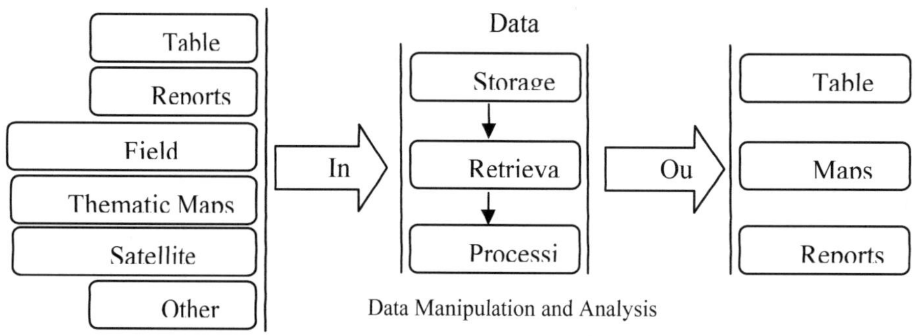


**Figure 2**. Special Process Methodology Mapping and Geographic Information System Development Analysis of Biomass Distribution and Potential in Java and Sumatera

## 3. Data Analysis

Plantations are all natural resource management, human resources, production, tools and machinery, cultivation, harvesting, processing and marketing activities related to Plantations. In the framework of energy security, plantation waste can be a source of raw materials of bio-energy such as palm oil, coconut, sugar cane, sago, kemiri sunan and jatropha. The yield of the oil palm area is a source available for bioenergy feedstock, of the total area Total Area 10.9 Million Ha, smallholdings: 4.6 M ha, large plantation of the country: 0.7 M Ha, large private plantation: 5, 6 Million Ha, can be classified as follows: People's Plantation, Large State Plantations, Large Private Plantations, Plants have not yet produced, Plants Produce, and Plant Damaged / Old. Oil palm plantations have a high CO2 absorption capacity (251.9 tons / ha / yr) and are very useful in reducing CO2 concentrations in the air due to increased greenhouse gases causing climate change on Earth as well as O2 production capability of O2 (183, 2 ton / ha / yr), and high biomass (C). Biomass production of oil palm plantations is higher than that of tropical forests. Oil palm waste, both trees, midribs, empty fruit bunches and shells are a considerable source of energy that can be utilized for biofuels and suppress

the use of fossil fuels, thereby significantly reducing emissions. Oil palm plantations have potential for biofuels, data from the Ministry of Forestry show that, area of oil palm plantations in Indonesia: 10,465 million hectares, CPO production (crude palm oil): 27.78 million tons (Exports of 20.57 million tons, domestic demand 7.21 million tons). Cultivation:
Farmers: 44%, and Large State / Private Plantations: 56%. Palm Oil byproducts: Pome: 73,294,200 tons (58%), Shell: 17,102,248 tons (7%), Empty bunches: 28,096,386 tons (23%), Fiber: 14,650,984 tons (14%), Farmers: 2.01 million, and Labor: 1.8 million

**Table 1.** Recapitulation of Oil Palm Companies in Indonesia

| No | Province | Number of Companies | Area (Ha) | Production | Number of Oil Palm Plant | Capacity (Ton / Hour) |
|---|---|---|---|---|---|---|
| 1 | Aceh | 117 | 366.343,54 | 828.256,81 | 26 | 863,00 |
| 2 | Sumatera Utara | 68 | 260.149,53 | 793.176,28 | 88 | 3.552,00 |
| 3 | Bengkulu | 42 | 129.165,00 | 171.232,76 | 26 | 1.090,00 |
| 4 | Riau | 310 | 1.462.118,03 | 4.237.736,70 | 215 | 9.610,00 |
| 5 | Jambi | 133 | 646.996,40 | 1.060.985,05 | 44 | 1.965,50 |
| 6 | Sumatera Selatan | 48 | 42.047,49 | 158.813,33 | 45 | 345,00 |
| 7 | Lampung | 25 | 58.353,89 | 166.217,61 | 12 | 592,00 |
| 8 | Bangka Belitung | 32 | 151.398,63 | 510.717,48 | 12 | 780,00 |
| 9 | Banten | 4 | 4.057,14 | 9.440,96 | 1 | 30,00 |
| 10 | Jawa Barat | 1 | 7.768,30 | 34.147,00 | 1 | 20,00 |
| 11 | Kalimantan Barat | 104 | 590.345,70 | 861.048,30 | 58 | 2.370,00 |
| 12 | Kalimantan Tengah | 87 | 642.383,81 | 1.563.091,97 | 64 | 4.347,00 |
| 13 | Kalimantan Selatan | 84 | 650.082,87 | 394.066,09 | 33 | 1.532,30 |
| 14 | Kalimantan Timur | 96 | 942.143,05 | 491.103,21 | 65 | 3.075,00 |
| 15 | Sulawesi Selatan | 4 | 49.492,74 | 54.689,48 | 2 | 60,00 |
| 16 | Sulawesi Tengah | 10 | 140.806,63 | 108.980,37 | 3 | 210,00 |
| 17 | Sulawesi Barat | 8 | 48.036,85 | 152.467,65 | - | - |
| 18 | Papua | 8 | 99.041,60 | 93.070,80 | 7 | 390,00 |
| | Total | 1181 | 6.290.731,20 | 11.689.241,85 | 702,00 | 30.831,80 |

**Table 2.** Number of Palm Oil Mill in Sumatera and Kalimantan

| Territory | Area (Ha) | Production (Ton) | Number of Oil Palm Companies | Capacity (Ton / Hour) |
|---|---|---|---|---|
| Sumatera | 3.116.572,51 | 7.927.136,02 | 468 | 18.797,5 |
| Kalimantan | 2.824.955,43 | 3.309.309,57 | 220 | 3.075,0 |
| Total | 5.941.527,94 | 11.236.445,59 | 688 | 30.121,8 |

**Table 3.** Sample of Data, Distribution of potential sources and utilization of Bioenergy In plantation sector, in Aceh Province, Babel, Bengkulu & Jambi, as follows:

| City or District | Type of Industrial Area | Potential (MWe) | | Utilized (MWe) | |
|---|---|---|---|---|---|
| | | Technical | Optimization | on-grid | off-grid |
| Aceh | shell of desicated coconut, Sawit | 0 | 0 | 1.2 | 0 |
| Aceh Barat | Oil Palm Factory (Fiber, Shell, Empty Bunch, Liquid Waste) | 20.14 | 23.66 | 0 | 4.5 |
| Aceh Besar | Oil Palm Factory (Fiber, Shell, Empty Bunch, Liquid Waste) | 4.24 | 4.92 | 0 | 1 |
| Aceh Singkil | Oil Palm Factory (Fiber, Shell, Empty Bunch, Liquid Waste) | 2.28 | 2.68 | 0 | 0.5 |
| Aceh Tamiang | Oil Palm Factory (Fiber, Shell, Empty Bunch, Liquid Waste) | 11.99 | 14.77 | 0 | 2 |
| | Pabrik Sawit, Tg Seumantoh, POM Residue, Sawit | 0 | 0 | 10 | 0 |
| Aceh Timur | Oil Palm Factory (Fiber, Shell, Empty Bunch, Liquid Waste) | 27.11 | 32.31 | 0 | 5.58 |
| Aceh Utara | Oil Palm Factory (Fiber, Shell, Empty Bunch, Liquid Waste) | 2.12 | 2.46 | 0 | 0.5 |
| Langsa | POME, Oil | 0 | 0 | 2 | 0 |
| | POM Residue, Oil | 0 | 0 | 1.5 | 0 |
| Nagan Raya | Oil Palm Factory (Fiber, Shell, Empty Bunch, Liquid Waste) | 5.07 | 6.09 | 0 | 1 |

| | | | | | |
|---|---|---|---|---|---|
| Simeulue | Oil Palm Factory (Fiber, Shell, Empty Bunch, Liquid Waste) | 3.44 | 4.06 | 0 | 0.75 |
| Bangka | Oil Palm Factory (Fiber, Shell, Empty Bunch, Liquid Waste) | 6.57 | 7.44 | 0 | 1.75 |
| Bangka Barat | Oil Palm Factory (Fiber, Shell, Empty Bunch, Liquid Waste) | 13.52 | 14.4 | 0 | 4.5 |
| Belitung | Oil Palm Factory (Fiber, Shell, Empty Bunch, Liquid Waste) | 4.36 | 5.09 | 0 | 1 |
| Belitung Timur | Oil Palm Factory (Fiber, Shell, Empty Bunch, Liquid Waste) | 21.85 | 25.3 | 0 | 6.75 |
| Bengkulu Selatan | Oil Palm Factory (Fiber, Shell, Empty Bunch, Liquid Waste) | 10.43 | 13.08 | 0 | 1.5 |
| Bengkulu Utara | Oil Palm Factory (Fiber, Shell, Empty Bunch, Liquid Waste) | 40.15 | 50.12 | 0 | 6 |
| Muko Muko | Oil Palm Factory (Fiber, Shell, Empty Bunch, Liquid Waste) | 28.25 | 35.49 | 0 | 4 |
| Seluma | Oil Palm Factory (Fiber, Shell, Empty Bunch, Liquid Waste) | 8.36 | 10.44 | 0 | 1.25 |
| Batang Hari | Oil Palm Factory (Fiber, Shell, Empty Bunch, Liquid Waste) | 30.11 | 37.59 | 0 | 4.5 |
| Bungo | Oil Palm Factory (Fiber, Shell, Empty Bunch, Liquid Waste) | 14.13 | 17.75 | 0 | 2 |
| Jambi | POM Residue, Sawit | 0 | 0 | 10 | 0 |
| Merangin | Oil Palm Factory (Fiber, Shell, Empty Bunch, Liquid Waste) | 40.15 | 50.12 | 0 | 6 |
| Muaro Jambi | Oil Palm Factory (Fiber, Shell, Empty Bunch, Liquid Waste) | 33.45 | 41.77 | 0 | 5 |
| Sarolangun | Oil Palm Factory (Fiber, Shell, Empty Bunch, Liquid Waste) | 10.04 | 12.53 | 0 | 1.5 |
| Tanjung Jabung Barat | Oil Palm Factory (Fiber, Shell, Empty Bunch, Liquid Waste) | 31.78 | 39.93 | 0 | 4.5 |
| Tebo | Oil Palm Factory (Fiber, Shell, Empty Bunch, Liquid Waste) | 10.04 | 12.53 | 0 | 1.5 |

## 4. Results

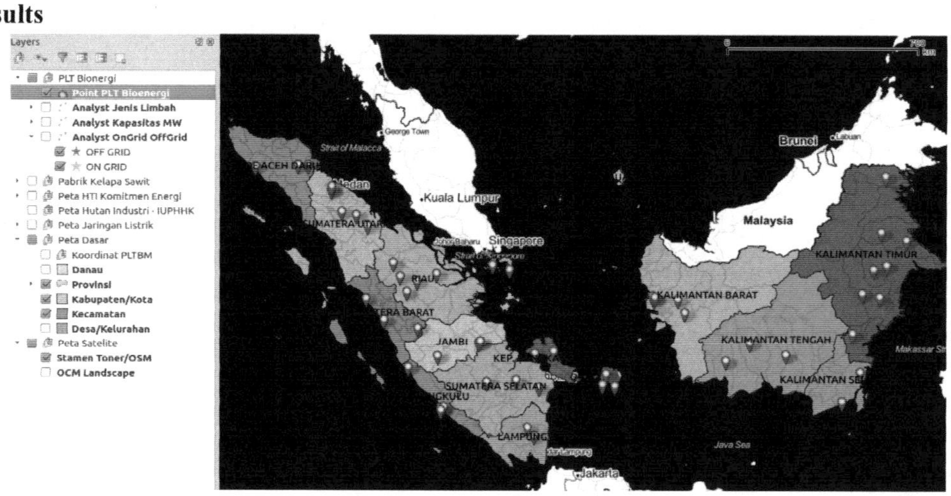


**Figure 3.** Composite Layer, Red Polygon is a spread of plantation / natural forest that has IUPHHK, and red dot is the distribution of Bioenergy power plants in Sumatra Island and Kalimantan

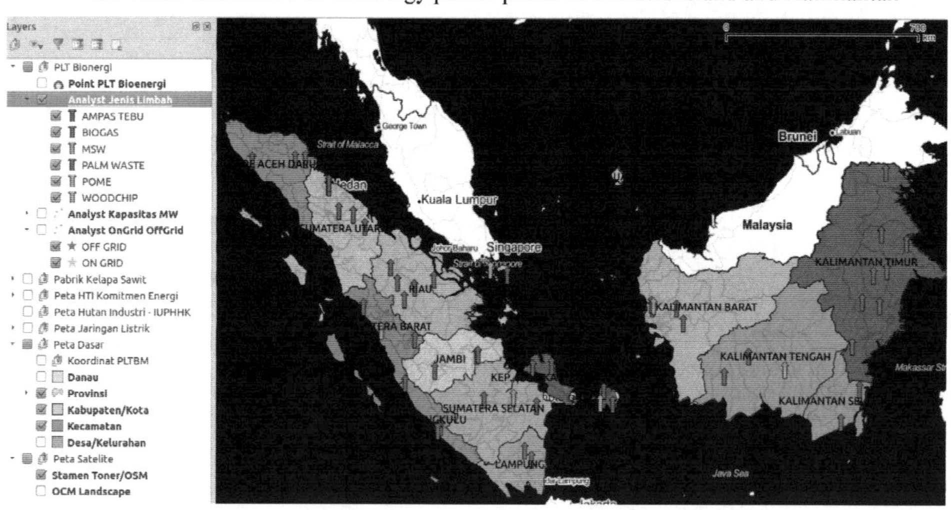

**Figure 4.** The following thematic maps appear on the map, the Bioenergy power plant is classified into: Sugar Cane, Biogas, MSW, Palm Waste, Pome and Woodchip. Palm waste appears to dominate the distribution of waste types for Bioenergy power plants.

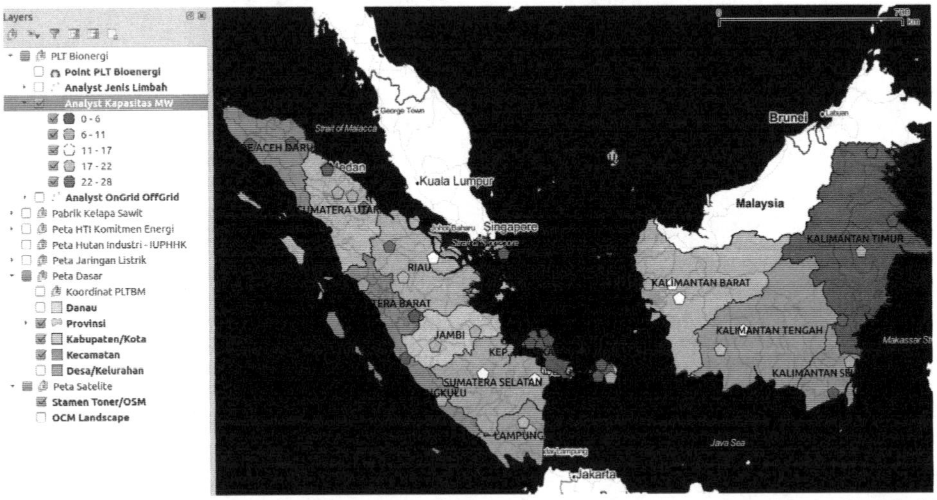


**Figure 5.** In the following thematic maps shown on the map, the Bioenergy power plant is classified into: 0-6 MW, 6-11 MW, 11-17 MW, 17-22 MW, and 22-28 MW. Looks a capacity of 0-6 MW dominates the distribution for Bioenergy power plants

## 5. Conclusion

These forests and plantations give positive and economic value to the people and the state. But giving a negative value, especially the waste of biomass that it produces, if not managed properly will adversely affect the cleanliness and environmental aesthetics. So with the bioenergy program how to utilize biomass waste into Biomass Power Plant (PLTBM). In this research, mapping and development of geographical information distribution system and bioenergy potential by classifying biomass sources used by plant and also analysis of land area to potential energy that can be used. So the conclusions of this research have been analyzed and described the potential of biomass energy, especially on the island of Sumatra and Kalimantan.

## References

[1] IRENA. Renewable Energy Technologies: Cost Analysis Series, Biomass For Power Generation. The International Renewable Energy Agency (IRENA). 2012;

[2] Nisha Sriram, Mohammad Shahidehpour. Renewable Biomass Energy. International Journal of Business, Economics and Management. IEEE. 2005;

[3] Donald L. Klass. Biomass for Renewable Energy And Fuels. Encyclopedia of Energy, published by Elsevier, Inc. r 2004;

[4] Abdeen Mustafa Omer. Biomass Energy Resources Utilisation And Waste Management. Journal of Agricultural Biotechnology and Sustainable Development Vol. 3(8), pp. 149 -170, October 2011;

[5] Sumit Sharma, Rajendra Meena, Amit Sharma, Pawan Kumar Goyal. Biomass Conversion Technologies for Renewable Energy and Fuels. IOSR Journal of Mechanical and Civil Engineering (IOSR-JMCE). 2014;

[6] Jonathan Campbell, Michael Shin. Essentials to Geographic Information Systems. Flat World Knowledge, Inc. 2011;

[7] Burrough P. Principle of Geographical Information System for Land Resources Assesment. Oxford, Claredon Press. 1986;

[8] Stephanie Rogers, Patricia Vivas. A study on the use of Geographical Information Systems (GIS) for the creation of addressing systems. Universal Postal Union. 2014;

[9] Tomislav Hengl. A Practical Guide to Geostatistical Mapping. Office for Official Publications of the European Communities, Luxembourg. 2009;
[10] Retno Astrini, Patrick Oswald. Modul Pelatihan Quantum GIS. GIZ Decentralization as Contribution to Good Governance (DeCGG). 2012;

**Acknowledgments**
Thanks to fellow Widyatama University lecturers, State Electricity Company, Palm Oil Company, Biomass Power Plants Company, Ministry of energy and mineral resources of the Republic of Indonesia.

IOP Conf. Series: Earth and Environmental Science **164** (2018) 012002 doi:10.1088/1755-1315/164/1/012002

# SPATIAL DIFFERENTIATION OF ENERGY POVERTY IN EU COUNTRIES

**M Zawada[1a], M Szajt[1], J Kowalik[1] and R Kucęba[2b]**

[1] Technical University of Czestochowa, Faculty of Management, Department of Econometrics and Statistics, Al. Armii Krajowej 39b 42-200 Częstochowa, Poland
[2] Technical University of Czestochowa, Faculty of Management, Institute of Information Management Systems, Al. Armii Krajowej 19b 42-200 Częstochowa, Poland

[a]marcinzawada04@gmail.com; [b]robertkuceba@wp.pl

**Abstract:** The provision of heat and electricity in the household is one of the basic human needs. However, many households are experiencing energy poverty, so it is difficult to satisfy basic energy needs. These households have to choose between drastic energy savings, which results in experiencing lack of thermal comfort and energy expenditure at the expense of other basic needs, such as food or health. The main aim of this article is to analyse the state of energy poverty in the European Union countries by comparing the basic characteristics of its measurement such as: income, energy consumption, energy prices in the years 2006-2015. This allowed to indicate the existence of spatial variations between members of the European Union.

## 1. Introduction

Low incomes force people to limit their expenditures and more Modest existence. However, there are so called basic goods, resignation from which is difficult. It has been assumed that energy belongs to such goods. It is an essential element of our lives. It makes our lives easier, creates comfortable working and relaxation conditions. It is also indispensable to illuminate the households, prepare meals, prepare hot domestic water and also to heat our houses. The necessity to decrease the total energy consumption has been recognized to be a priority at the EU level, however, it needs to be stressed that energy is required in everyday lives of everybody, regardless of whether they are rich or poor. Therefore, it is this context in which the term of energy poverty should be analysed, which consists in, among others, lack of equal access of households to energy and problems that follow as a result of that situation [1].

A uniform definition of energy poverty does not exist. One of the first to propose such a definition was B. Boardman [2] in 1991, who stated that „energy poverty occurs if a household spends more that 10% of its income on all types of energy, in this energy required to heat the household at a satisfactory level". The 10-percent-treshold accompanied this definition of energy poverty, which is the most frequently applied index used to identify this phenomenon. In the course of years defining energy poverty has been undergoing permanent changes depending on the researcher of this phenomenon and the country of their origin. In 2011 the European Economic and Social Committee in its opinion suggested that the general definition of energy poverty should be each time adapted to the local situation of each of the EU's member states [3].

Spatial differentiation of particular dimensions of energy poverty is vital for constructing effective instruments of social policy. Efficiency of poverty counteracting instruments depends not only on proper identification of the group in need, but also on adjusting the type of support to the characteristics of the poor in the given region.

## 2. Characteristics of the European Union countries in the scope of energy consumption and economic development

Energy consumption and economic development remain in a close relationship [4]. Basic macroeconomic indexes that characterize particular countries usually comprise (per capita): GDP, energy consumption (in this electricity) and recently also CO2 emission. They are frequently supplemented by the value of Human Development Index (HDI) and average life expectancy of inhabitants. All these values for 28 countries of the European Union have been presented in Table 1.

In 2015 compared to 2006 the average life expectancy in each of the countries increased (most in Estonia), and the level of HDI likewise (most in Poland). In case of GDP its decrease was recorded only in two countries (Cyprus and Greece). The level of electricity consumption per capita in 10 countries increased (most in Estonia), in the remaining 18 counties it decreased, while the largest decrease was recorded in Luxembourg, Finland and Sweden. Electricity prices in comparative periods grew in all the countries except the Netherlands. The largest growth was recorded in Spain, the lowest one in Hungary and Slovakia. In 27 countries the emission of CO2 significantly decreased (most in Luxembourg and Denmark) The only country where the emission grew was Lithuania.

**Table 1.** Energy and development indicators in 2006, 2014 and 2015

| **Country** | HDI | | GDP per capita $ PPP | | Life expectancy (years) | | Electricity consumption per capita (kW h) | | Energy consumption per capita (toe) | | Electricity prices (EUR per kWh) | | CO2 per capita (t) | |
|---|---|---|---|---|---|---|---|---|---|---|---|---|---|---|
| | 2006 | 2015 | 2006 | 2015 | 2006 | 2015 | 2006 | 2014 | 2006 | 2014 | 2006 | 2015 | 2006 | 2014 |
| Austria | 0,860 | 0,893 | 41907 | 50078 | 79,9 | 81,8 | 8244 | 8361 | 4,1 | 3,8 | 0,134 | 0,201 | 8,7 | 6,9 |
| Belgium | 0,871 | 0,896 | 35406 | 46283 | 79,4 | 81,3 | 8684 | 7709 | 5,5 | 4,7 | 0,144 | 0,213 | 10,2 | 8,3 |
| Bulgaria | 0,755 | 0,794 | 14963 | 19199 | 72,6 | 74,5 | 4367 | 4709 | 2,7 | 2,5 | 0,066 | 0,094 | 6,4 | 5,9 |
| Croatia | 0,793 | 0,827 | 19205 | 23596 | 75,8 | 77,3 | 3635 | 3714 | 2,2 | 1,9 | 0,092 | 0,132 | 5,1 | 4,0 |
| Cyprus | 0,836 | 0,856 | 33295 | 32580 | 78,7 | 80,3 | 4238 | 3625 | 2,2 | 1,7 | 0,143 | 0,196 | 7,4 | 5,3 |
| Czech Republic | 0,851 | 0,878 | 27659 | 34711 | 76,5 | 79,5 | 6529 | 6259 | 4,5 | 3,9 | 0,099 | 0,139 | 11,9 | 9,2 |
| Denmark | 0,904 | 0,925 | 43083 | 49696 | 78,1 | 81,1 | 6825 | 5859 | 3,7 | 2,9 | 0,236 | 0,307 | 10,1 | 5,9 |
| Estonia | 0,829 | 0,865 | 21623 | 29365 | 72,7 | 77,1 | 5869 | 6732 | 3,8 | 4,6 | 0,073 | 0,130 | 12,0 | 14,8 |
| Finland | 0,873 | 0,895 | 38812 | 43053 | 79,2 | 81,4 | 17215 | 15249 | 7,1 | 6,2 | 0,108 | 0,155 | 12,6 | 8,7 |
| France | 0,873 | 0,897 | 36027 | 41466 | 80,8 | 82,7 | 7540 | 6938 | 4,2 | 3,7 | 0,119 | 0,168 | 5,9 | 4,6 |
| Germany | 0,898 | 0,926 | 39263 | 48730 | 79,1 | 81,1 | 7212 | 7035 | 4,2 | 3,8 | 0,183 | 0,295 | 9,9 | 8,9 |
| Greece | 0,855 | 0,866 | 28203 | 26783 | 79,4 | 81,6 | 5435 | 5063 | 2,7 | 2,1 | 0,070 | 0,177 | 8,8 | 6,2 |
| Hungary | 0,809 | 0,836 | 21467 | 26681 | 73,1 | 76,0 | 3882 | 3966 | 2,7 | 2,3 | 0,108 | 0,113 | 5,7 | 4,3 |
| Ireland | 0,902 | 0,923 | 43221 | 68883 | 79,2 | 81,5 | 6351 | 5722 | 3,4 | 2,8 | 0,149 | 0,243 | 10,1 | 7,4 |
| Italy | 0,862 | 0,887 | 35076 | 38161 | 81,3 | 83,5 | 5833 | 5002 | 3,2 | 2,4 | 0,211 | 0,245 | 8,1 | 5,3 |
| Latvia | 0,814 | 0,830 | 17593 | 26031 | 70,9 | 74,1 | 2967 | 3507 | 2,1 | 2,2 | 0,083 | 0,164 | 3,6 | 3,5 |
| Lithuania | 0,812 | 0,848 | 20130 | 29966 | 71,1 | 75,1 | 3353 | 3821 | 2,7 | 2,4 | 0,072 | 0,126 | 4,3 | 4,4 |
| Luxemburg | 0,877 | 0,898 | 85779 | 105882 | 79,3 | 82,2 | 16457 | 13915 | 9,2 | 6,9 | 0,160 | 0,177 | 24,0 | 17,4 |
| Malta | 0,808 | 0,858 | 27863 | 37899 | 79,4 | 81,9 | 4898 | 5007 | 2,1 | 1,8 | 0,095 | 0,126 | 6,4 | 5,5 |
| Netherlands | 0,899 | 0,924 | 44586 | 50898 | 79,7 | 81,7 | 7026 | 6713 | 4,9 | 4,3 | 0,209 | 0,199 | 11,0 | 9,9 |
| Poland | 0,808 | 0,855 | 21089 | 27811 | 75,1 | 78,2 | 3585 | 3972 | 2,5 | 2,5 | 0,119 | 0,144 | 8,4 | 7,5 |
| Portugal | 0,797 | 0,843 | 27361 | 30624 | 78,4 | 81,5 | 4827 | 4663 | 2,4 | 2,0 | 0,141 | 0,228 | 5,7 | 4,3 |
| Romania | 0,766 | 0,802 | 17181 | 23626 | 72,2 | 75,0 | 2446 | 2584 | 1,9 | 1,6 | 0,094 | 0,130 | 4,9 | 3,5 |
| Slovenia | 0,865 | 0,890 | 27793 | 32885 | 78,1 | 81,1 | 7124 | 6728 | 3,6 | 3,2 | 0,105 | 0,159 | 8,1 | 6,2 |
| Slovakia | 0,802 | 0,845 | 25011 | 30632 | 74,2 | 77,2 | 5153 | 5137 | 3,5 | 2,9 | 0,145 | 0,151 | 7,3 | 5,7 |
| Spain | 0,849 | 0,884 | 31984 | 36310 | 80,8 | 83,4 | 6105 | 5356 | 3,2 | 2,5 | 0,115 | 0,231 | 7,9 | 5,0 |
| Sweden | 0,895 | 0,913 | 41668 | 49175 | 80,7 | 82,6 | 15263 | 13480 | 5,5 | 5,0 | 0,144 | 0,185 | 5,5 | 4,5 |
| United Kingdom | 0,889 | 0,910 | 35741 | 42609 | 79,2 | 81,6 | 6201 | 5130 | 3,6 | 2,8 | 0,102 | 0,213 | 8,9 | 6,5 |

*Source:* World Bank and Eurostat

IOP Conf. Series: Earth and Environmental Science **164** (2018) 012002 doi:10.1088/1755-1315/164/1/012002

In most cases moderate linear dependencies occur among analyses values. A strong correlation can be observed in case of CO2 emission and energy consumption and life as well as life expectancy and HDI. Statistically insignificant proved to be the correlation between life expectancy and CO2 emission. Weak dependencies occur between CO2 and HDI and energy consumption and life expectancy (Table 2). Electricity price is negatively correlated with all the other variables. An exception is life expectancy. Unfortunately, these dependencies are weak or very weak and simultaneously statistically insignificant.

**Table 2.** Correlation coefficients among values characterizing development and energy consumption (2014)

| | HDI | GDP | Life expec | Electricity | Energy | Prices | CO2 |
|---|---|---|---|---|---|---|---|
| HDI | 1 | 0,6650 | 0,7793 | 0,5390 | 0,5492 | -0,0298 | 0,3812 |
| GDP | | 1 | 0,5433 | 0,6473 | 0,6748 | -0,1766 | 0,6352 |
| Life expec | | | 1 | 0,4787 | 0,3512 | 0,1222 | 0,2053 |
| Electricity | | | | 1 | 0,9149 | -0,0652 | 0,5362 |
| Energy | | | | | 1 | -0,2444 | 0,7488 |
| Prices | | | | | | 1 | -0,2361 |
| CO2 | | | | | | | 1 |

Source: own calculations on the basis of data by Eurostat and World Bank

To sum up, it can be concluded that energy consumption is necessary, but it alone is not sufficient for economic development. Moreover, from a certain level of development it is implemented political strategies that decide whether it is possible to improve or maintain the standards of welfare without an increase in energy consumption. Undoubtedly, the best way to understand the importance of energy is to analyse the effects of lack of access to it – for various reasons – that is the phenomenon of energy poverty.

## 3. Energy poverty measurement

Main measures that allow to assess the level of energy poverty most frequently applied in the research on this phenomenon include [5]:

- The Ten-Percent-Rule (TPR)
- Double Median or Mean indicator
- Low Income, High Cost (LIHC) indicator
- Minimal Income Standard (MIS) indicator

The first one of them has already been mentioned while defining the energy poverty phenomenon. This definition was accepted as a starting point to analyse energy poverty. In Great Britain, the country that as the first one took interest in this problem and possesses the greatest experience in measuring and limiting it, active works aimed at changing it for the one that better classifies the energy poor are ongoing. One of the most widely discussed definitions is the one coming from the breakthrough report of J. Hills [6]. The Ten-Percent-Rule (TPR) and Double Median indicators share a common history. However, I argue that they can and should be differentiated. Both indicators define energy poverty as excess spending on energy beyond a certain threshold, most prominently, a ten percent share or double median share of energy expenditure for all households relative to net income.

Low Income, High Cost LIHC is a relative measure. It is based on the concept of low incomes and high expenditures (model ones) on energy. In both dimensions the terms "high" and "low" are dependent on the distribution of incomes and expenditures of the population. Technically, they consist with reference to the level of median: households that achieve equivalent incomes lower than 60% of the median and expenditure on energy above the expense of the median household.

These minimum income standards (MIS) can be used with data from the Housing Survey to estimate the number of households in fuel poverty. In this context, households are deemed to be in fuel poverty if, after deducting their actual housing costs, they have insufficient residual net income to meet their total required fuel costs after all other minimum living costs (as defined by the MIS) have been met. Conversely, a household is in MIS based fuel poverty if Fuel costs (HS) > Net household income (HS) – housing costs (HS) – minimum living costs (MIS).

Figure 1 presents information concerning the percentage of population in particular EU countries who are in arrears with utility bills during 2015. The largest arrears occur in Greece (42%), Bulgaria (31,4%) and Croatia (28,7%), the lowest ones in Luxembourg (2,4%), the Netherlands (2,7%) and the Czech Republic (3%).

The relationship between arrears in utility bills payments and the average share of energy products in total consumption expenditures of households seems to be weak. However, it needs to be stressed that in case of some social and economic groups, whose share of energy products in total consumption expenditures is larger compared to the average national, the probability of arrears in utility bills is higher.

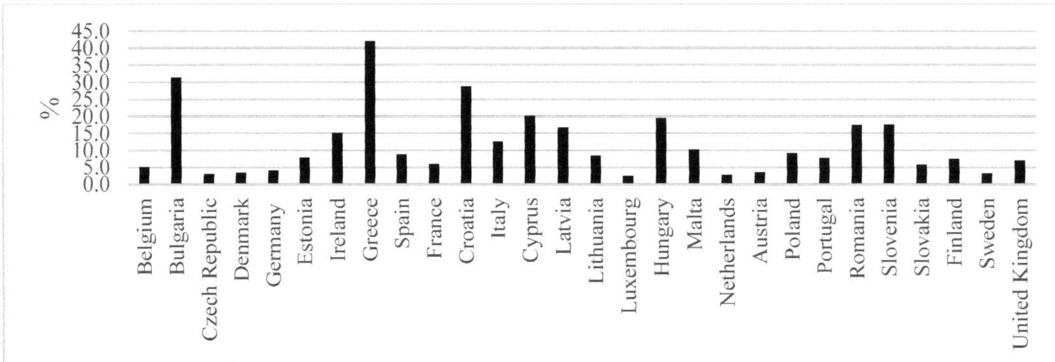


**Figure 1.** Percentage of population in arrears with utility bills in 2015
Source: own elaboration on the basis of data by Eurostat [ilc_mdes07]

Among the inhabitants whose incomes are below 60% of the median the largest percentage of those in arrears with utility bills can be observed in Greece, Bulgaria and Hungary. Among those whose incomes are above 60% of the median the largest arrears can be noticed in case of Greece, Bulgaria and Croatia (Figure 2).

Another picture (Figure 3) presents information concerning the percentage of expenditures incurred by households for electricity, gas and other fuels used to heat the accommodation and domestic hot water preparation. This information comes from the survey on consumption expenditures that was conducted by Eurostat in 5-year time intervals. The latest data from this survey comes from the year 2010. Another edition of this survey was conducted in 2015, but the data was not available yet at the time when the present paper was prepared. The largest expenditures are incurred by households in Slovakia, Hungary and the Czech Republic, the lowest ones in Malta, Cyprus and Spain. The largest growth of expenditures compared to the year 2005 concerns three countries: Latvia, Estonia and Great Britain, the largest decrease was recorded in Romania.

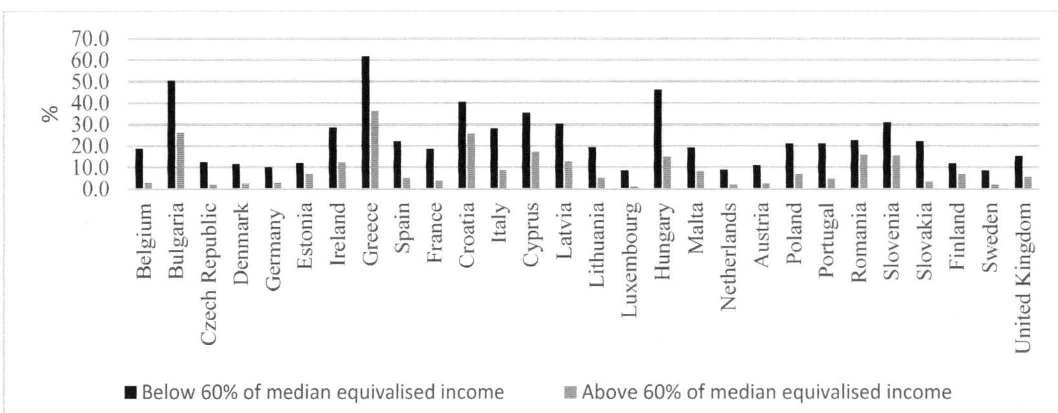

**Figure 2.** Percentage of population in arrears with utility bills: populations above and below 60% of median equivalised income in each Member State in 2015
Source: own elaboration on the basis of data by Eurostat [ilc_mdes07]

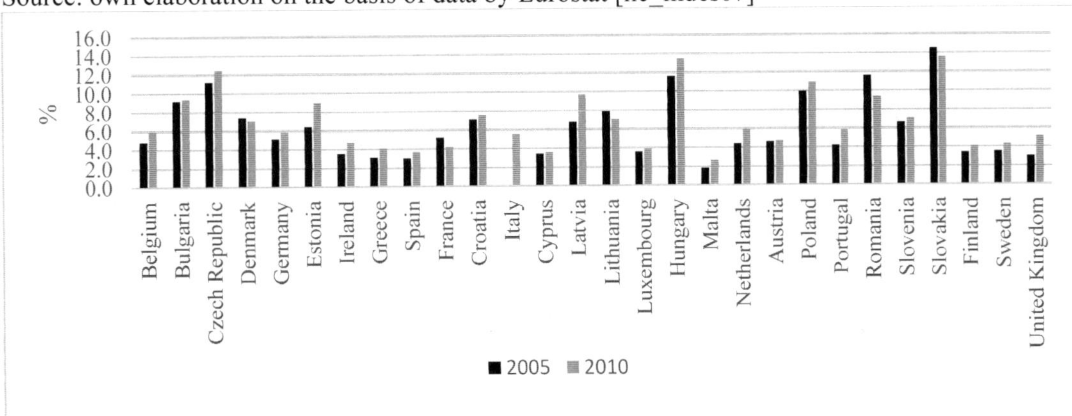


**Figure 3.** Percentage of expenditures incurred by households for electricity, gas and other fuels used to heat the accommodation and prepare hot domestic water
Source: own elaboration on the basis of data by Eurostat [hbs_str_t211].

## 4. Conclusions

In search for new measures that allow for better description of the problem of energy poverty the Authors suggest that the ones should be constructed that will consider [7]:

- Information on the choice and availability of energy carriers. It is important to note that this should not be limited to mains gas and electricity only – rather it should incorporate all potential sources, including, for example, self-collected firewood and peat. This would enable examination of 'energy degradation' issues.
- Information on householder flexibility to move to new energy services, to understand infrastructural and built environment contexts.
- Technical energy efficiency and housing quality data, to allow estimation of theoretical energy costs
- Information on the ways in which households may have rationed their provision of energy services in the home – for example by restricting heating, lighting and/or usage of appliances.

The EU has adopted numerous initiatives in order to evaluate the problem connected with energy poverty at the local, regional and national level and also to determine the most effective palliative actions. However, the most effective measures to prevent energy poverty in households or mitigate it are actions directed at the level of households.

## 5. References

[1] Szamrej-Baran I 2017 Ranking krajów UE ze względu na ubóstwo energetyczne. *Gospodarka w Praktyce i Teorii 43*, pp. 79-91

[2] Boardman B 1991 Fuel Poverty: From Cold Homes to Affordable Warmth, Belhaven Press London, pp. 2-224

[3] European Economic and Social Committee 2011 Opinion on Energy Poverty in the Context of Liberalisation and the Economic Crisis 2011/C 44/09, pp.53-56

[4] Zawada M Włodarczyk A Piłatowska M 2015 *CO2 Emissions, Energy Consumption and Economic growth in the EU Countries: Evidence from Threshold Cointegration Analysis*, 12th International Conference on the European Energy Market (EEM), Lisbon, Portugal, IEEE, pp.1-5

[5] Schuessler R 2014 Energy Poverty Indicators: Conceptual Issues-Part I: The Ten-Percent-Rule and Double Median/Mean Indicatorspp, Center for European Economic Research, pp.3-23

[6] Hills J 2012 *Getting the measure of fuel poverty. Final report of the fuel poverty review*, Raport CASE nr 72, London

[7] Thomson H Bouzarovski S Snell C 2017 Rethinking the measurement of energy poverty in Europe: A critical analysis of indicators and data, *Indoor and Built Environment*, 1420326X17699260

IOP Conf. Series: Earth and Environmental Science **164** (2018) 012003 doi:10.1088/1755-1315/164/1/012003

# Prosumer energy as a stimulator of micro- Smart Grids development - on the consumer side

**R Kucęba[1a], M Zawada[2b], M Szajt[2] and J Kowalik[2]**

[1] Technical University of Czestochowa, Faculty of Management, Institute of Information Management Systems, Al. Armii Krajowej 19b 42-200 Częstochowa, Poland
[2] Technical University of Czestochowa, Faculty of Management, Department of Econometrics and Statistics, Al. Armii Krajowej 39b 42-200 Częstochowa, Poland

[a] robertkuceba@wp.pl; [b] marcinzawada04@gmail.com

**Abstract.** In the present paper its Author introduces the taxonomy of energy prosumer in the sustainable development context. This concerns in particular micro-installations of Renewable Energy Sources dedicated directly to final energy customers. The four-layer structure of micro-Smart Grids has been indicated in the paper, which includes: the consumer layer (RES micro-installations and intelligent energy receivers), electricity networks layer, Smart Metering layer and the Decentralized Energy Management Systems. The Author has aggregated the binary associations between prosumer energy diffusion and micro- Smart Grid development – which favour creation of zero-energy micro-regions.

## 1. Introduction

Prosumer energy constitutes a stimulant of new and simultaneously promising, social „green initiatives" for sustainable region development. Prosumer energy is a distinguished segment of electricity sector often defined as the third production pillar – frequently isolated from Centralized Energy Systems (after large-scale/corporate energy – first pillar and dispersed and distributed energy – second pillar). Prosumer energy due to its decentralized and sustainable nature is simultaneously a stimulant of micro-regions sustainable development. Certainly, this is justified in case if it is in line with the third wave prosumer, according to A. Toffler [1]. It needs to be stressed here that A. Toffler rejected the traditional concept of a passive consumer, indicating the development of an active one, involved in the search for a product that will meet best their expectations, at economic, ecological and social justification. In his grasp prosumers are specific market entities that produce energy for own needs. In this understanding production and consumptions are the elements of the same cycle. The value is created when consumers make effort and proper meaning into the products they generate. While prosumption is a continuous process, not a single act [2]. Prosumers who participate in this process contribute their money, time, efforts and skills expecting specific benefits. A reference should be made here to postulates of D. Tapsott and A. Williams who believe that prosumption is a process connected with the will to possess a particular consumer good compliant with own imagination of the co-creator (prosumer). With this reference the activeness of consumers who adapt qualities of prosumers is demonstrated in co-creation of the desired good. Activities of prosumers defined as prosumption concern basically self-adjustment of products to specific own needs, and frequently also the needs of the immediate environment. Consumers become co-producers who play an active role in production of goods, creating the value, brand and service, frequently being elementary socially responsible market entities [3].

On the basis of the approach by A. Toffler a modified concept of the prosumer as a new market entity has also been introduced by S. Kirsner. He proposes a term of a „professional prosumer" – the one who possesses an advanced knowledge that allows them to participate in a constructive dialogue with the producer and the environment of co-operating prosumers – which is in line with the subject matter of the present paper.

## 2. Prosumer energy market

Prosumer energy should be defined as transformation from energy products acquired from fossil fuels in centralized energy sources – towards energy value chains of RES energy, co-created in economic micro-spaces that generate economic, ecological and social benefits. The new energy value chains stimulate new business models on energy markets (a new paradigm) – departing from centralized towards decentralized models. Prosumer energy development stimulants can be aggregated in a generalised set, which includes:

- forecasts of electricity demand growth,
- expectations as to the future higher electricity prices,
- high cost of system electricity distribution,
- limited access to electricity generated in conventional sources (vast territory of the state, obsolete technological infrastructure, lack of energy raw resources),
- the need to make use of new technologies, in particular RES,
- growth of ecological awareness,
- the necessity to protect the environment,
- implementation of international agreements in the scope of reducing the demand for fossil fuels, reducing the emission of greenhouse gases accompanied by a simultaneous increase of RES (e.g. according to the declaration 3x20 of 2008 the UE-28 member states are obliged to increase until 2020 the share of energy produced in RES up to 20%, simultaneously reducing resources and fossil fuels by 20% and reducing greenhouse gases emission in CO2 equivalent by 20% - in 2030 CO2 reduction at the level of 40% [4].

It is estimated that the European market of energy prosumers (in the aggregated grasp of the EU-28) is currently at the stage of growth. It is estimated that the number of RES prosumer installations in the EU-28 amounts about 10 million [5]. For instance, in 2015 the total and cumulative power of photovoltaic installations in EU-28 that can be recognised as prosumer ones amounted 94,87 GW. A growth of those installed powers by 6,12 GW occurred in 2016. Also, a large differentiation as to the pace of prosumer energy development, in particular in the scope of RES micro-installations. The European leader in this area is Germany, with the number of 2 million of functioning prosumers, who acquire electricity mainly from photovoltaic installations, and the total power of RES amounts about 36 GW. Only in 2014 alone the new installed photovoltaic power in Germany reached 1,9 GW [6].
Prosumer energy is poorly developed in the countries of East Europe, in this the EU-28 member states. For instance, in Poland the total power of RES as of the 3$^{rd}$ quarter of 2015 reached 6,5 GW, in this it is estimated that the total power of installed RES micro-installations amounts barely 0,2 GW (including photovoltaic sources) [7].

However, it should be stressed here that the prosumer energy market in the EU-28 is open to new technologies – in case of prosumer energy this concerns primarily RES micro-installations[1], but also the environment of their integration in micro- Smart Grid – Virtual Power Plants. Integration of prosumer energy sources in the Smart Grid environment increases their value and is a stimulator of their diffusion. Simultaneously, a positive feedback can be observed between prosumer energy and micro- Smart Grid development – therefore, frequently energy self-contained micro-regions development. The aggregated bidirectional associations between prosumer energy diffusion and micro- Smart Grids development have been presented in the next part of the paper.

## 3. Prosumer energy in correlation with micro- Smart Grid

Micro- Smart Grid is a convergence of low voltage networks and tele-informatic networks. Micro- Smart Grid networks integrate heterogenous links (e.g. technologies on energy prosumers side) in final energy value creation chains (not only the consumed value, but also environmental, economic and social values). Processes of energy creation and consumption frequently occur in the integrated links. Micro- Smart Grid – in accordance with the idea of computer intelligent networks – include also co-sharing this network's resources by various entities of producers/consumers/prosumers. This comprises also bidirectional information exchange among the cooperating entities. Micro- Smart Grid can be defined as intelligent technologies for electricity management, in order to balance the supply and demand.

In another grasp micro- Smart Grid are intelligent grids of energy transmission in local low-voltage networks, which generate bonds among the links and the processes of electricity value creation that occur in them.

Thus, Smart Grid networks are self-controlling „intelligent energy grids", which automatically and online, balance in time – demand and supply. In addition, it is a physical, functional, logical and environmental integration environment of dispersed electricity production sources, including prosumer Renewable Energy Sources. The integrated prosumer energy sources in Smart Grid networks are basically decentralized – isolated from the central energy system. At the same time, they are differentiated with regard to technology, production factors, in this raw energy resources and the volume of installed power.

On the basis of the analysis of the logical and functional structure of selected micro- Smart Grids, as well as created in the laboratory conditions intelligent environment of prosumer micro-installations virtual integration with energy receivers, the Authors have distinguished four basic layers in these networks: Prosumer Technology – PT, Operational Technology – OT, Smart Metering – SM and Decentralized Prosumer Energy Management Systems - DPEMS. Distinguishing these layers allows for separating the technologies that fulfil operational functions (production, consumption, transmission and distribution of energy – where the order is not a coincidence) from the technologies that fulfil analytic functions (functions that support management of the heterogeneous value created in micro- Smart Grid) [8].

The first two layers are called operational ones, where technologies that create the electricity value are introduced on the side of the prosumer. In the first layer - Prosumer Technology (PT) the following technologies can be distinguished: micro- dispersed generation, RS mini-installations, energy reservoirs, controlled electricity receivers (e.g. home appliances controlled in the IoT technology) and systems of intelligent electric vehicles charging: plug-in hybrid electric vehicle (PHEV), plug-in hybrid vehicle (PHV).

The second layer of micro- Smart Grid - Operational Technology (OT) is a platform of integrating technologies and energy receivers in decentralized low-voltage networks – technological resources aggregated in the first layer – PT. Computer networks are also implemented in this layer (e.g. wireless

[1] Micro-installation – an installation of a renewable energy source of total installed electric power not higher than 40 kW, which can operate in off-grid systems, in local decentralized networks or connected to electricity networks of nominal voltage lower than 110 kV or heat power achieved in cogeneration not higher than 120 kW.

networks – Wi-Fi), which create the logical metering and controlling infrastructure, and which integrate in the virtual environment all the links of the intelligent grid. The third layer of micro- Smart Grid is Intelligent Metering Infrastructure – Smart Metering (SM), which comprises the metering infrastructure - bidirectional intelligent meters in order to measure the amount of produced energy and the volume of its consumption in particular links of intelligent grid. Metering values are the source of information and knowledge on supply and demand volumes of prosumers. Systems that control prosumer technologies of the first layer of Smart Grid are also introduced in the third layer (e.g. with the use of IoT technology). The fourth layer of micro- Smart Grid is created by Decentralized Prosumer Energy Management Systems – DPEMS. They constitute the superior part in the processes of managing the dispersed structure of micro-installations in prosumer micro- Smart Grid networks.

In the synthetic way, the logical and functional structure of micro- Smart Grid with the consideration of the distinguished four layers – has been visualized in the form of an operational and analytic pyramid of heterogeneous energy value creation in these networks (Figure 1).

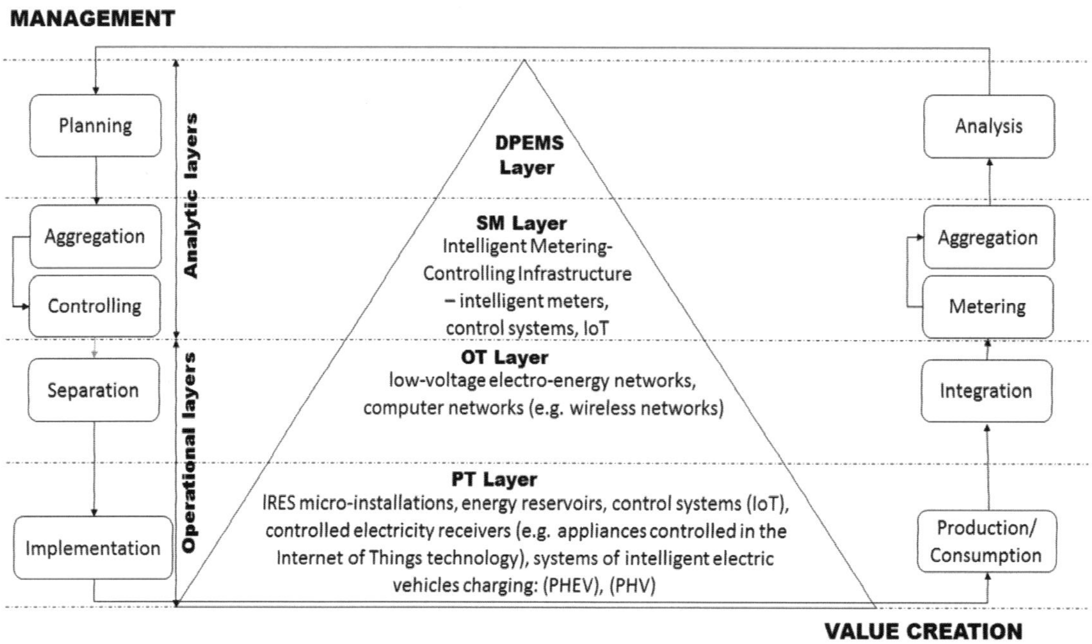


**Figure 1.** Logical and functional structure of Micro- Smart Grid – energy value creation pyramid.

Processes of generation and growth of energy value in prosumer RES micro-installations occur in particular layers of micro- Smart Grid. It is emphasized that the created values possess additional dimensions: environmental, economic and social ones [1]. In this context relevant values that result from the binary relations between prosumer energy diffusion and micro- Smart Grid development in the economic, environmental and social dimension have synthetically aggregated below.

**ECONOMIC DIMENSION-** reducing the times and costs of marginal investments of prosumer RES micro-installations through optimizing the use and supporting the management of generated energy, optimum energy use through making use of forecasting applications that support scheduling devices on the supply and demand side of the customer, reduction of exploitation costs, elimination of constant control and manual control through automatic monitoring in metering and controlling modules, supporting in the decision-making processes optimization of prosumer micro-installations operation – cost reduction.

**ENVIRONMENTAL DIMENSION** – monitoring the decrease in pollution of lithosphere, hydrosphere, aerosphere on the prosumer side, introducing a reliable, comprehensive environment evaluation through integration of dispersed prosumer RES micro-installations in the virtual environment, optimizing the processes of prosumer adjustment with reference to environmental restrictions – regional perspective.

**SOCIAL DIMENSION**- activating prosumers in the aspect of their market game in the scope of energy autonomy, stimulating the growth of the number of energy investors/producers in prosumer RES micro-installations, creating motivating systems for energy prosumers, stimulating economic activeness, transforming the energy consumption culture that stimulates growth of pro-environmental awareness.

## 4. Conclusions

To sum up, energy prosumers are essential entities that participate in energy creation in the local perspective of local multi-dimensional economic, environmental and social values. Given the high density of RES micro-installations their physical and logical integration in the micro- Smart Grid environment is justified. Micro- Smart Grid in their four-layer structure are the foundations of development for regions of full or partial energy autonomy. Autonomy being the result of prosumer energy development diffusion in correlation with micro- Smart Grid development will distinguish a given region and may confirm not only its energy but also social and cultural sovereignty.

## References

[1] Toffler A 1980 *The Third Wave.* William Collins Sons & Co. Ltd., New York

[2] Kucęba R Bylok F Pabian A Zawada M 2014 *Prosumer Energy Dimension in the Conditions of Sustainable Micro-region Development in the EU* (ICSSAM, International Conference on Social Science and Management. ISEPSS. International Symposium on Education Psychology and Social Sciences May 2014 Kyoto Japan Conference Proceedings) pp 1040–1051

[3] Prahalad CK. Ramaswamy V 2004. *The Future of Competition: Co-Creating Unique Value with Customers*, Harvard Business School Press, Boston, pp. 215-218

[4] Popczyk J 2011 *Dispersed Energy.* Warszawa: PKEOM, p. 28.

[5] United Nations Industrial Development Organizations 2015 *Industrial Prosumers of Renewable Energy. Contribution to Inclusive and Sustainable Industrial Development.* Vienna, https://www.unido.org/fileadmin/media/documents/pdf/Energy_Environment/PROSUMERS_Energy.pdf

[6] Statista - The Statistics Portal 2017 https://www.statista.com/statistics/497540/connected-and-cumulated-photovoltaic-capacity-in-the-european-union-eu/

[7] Portal of Green Energy 2017 http://gramwzielone.pl/energia-sloneczna/25391/moc-elektrowni-fotowoltaicznych-w-polsce-przekroczyla-190-mw

[8] Kucęba R Kiełtyka, L 2013 *Smart grid as a management network of distributed energy technologies.* [in:] L. Varkoly (Scientific Editors), Present Day Trends of Innovations, Issued by Dubnica Institute of Technology in Dubnica nad Vahom, pp. 306-313.

IOP Conf. Series: Earth and Environmental Science **164** (2018) 012004 doi:10.1088/1755-1315/164/1/012004

# Implementation of Energy Policy in the European Union Countries in the Aspect of their Sustainable Development

**J Kowalik[1], M Zawada[1], M Szajt[1] and R Kucęba[2]**

[1] Technical University Of Czestochowa, Faculty of Management, Department of Econometrics and Statistics, Al. Armii Krajowej 39b 42-200 Częstochowa, Poland
[2] Technical University Of Czestochowa, Faculty of Management, Institute of Information Management Systems, Al. Armii Krajowej 19b 42-200 Częstochowa, Poland

janekowalik@yahoo.pl

**Abstract.** The energy sector in the European Union is currently undergoing very deep transformations from the transition from conventional energy to new technologies and renewable energies. By defining energy policy guidelines, the European Union has followed the principle of sustainable development and has adopted the following targets for the development of the energy sector for the years 2010-2020: 20% reduction of greenhouse gas emissions, 20% increase of renewable energy share in final energy consumption, by 20% and by increasing the share of biofuels in the total consumption of transport fuels by at least 10%. The aim of the article was to analyze the current state of implementation of the basic assumptions of the Community in the field of energy policy. The ambitious energy policy objectives for 2010-2020, as outlined by the European Union, have been demonstrated to a varying extent by the Member States of the Community. This was influenced primarily by the global economic crisis, the diversification of the political and economic interests of individual member states, and the large diversification of strategic energy carriers.

## 1. Introduction

The energy sector in the European Union is currently undergoing very deep transformations from the transition from conventional energy to new technologies and renewable energies. This is the result of an energy policy in the Union. In defining the guidelines for energy policy, the European Union subordinated it to the principle of sustainable development, which was adopted as a rule of socio-economic development of the whole Union. Consequently, the Union has gone away from a traditionally understood energy policy to produce energy that meets the needs of a given country, without paying any attention to ecological effects [1].

The concept of sustainable development was first defined in 1987 by Gro Harlem Brundtland, in the report "Our Common Future" [2]. According to this definition, sustainable development is a process of development that fulfills current needs without compromising future generations ability to meet their own needs. This means that the economic development of the country and the protection of the environment and its resources should not be treated separately but jointly.

The EU's pursuit of sustainable development in energy policy has led to the following goals and values:

- counteracting climate change and addressing serious environmental problems,
- promoting energy efficiency,

– increasing the competitiveness of renewable energy sources,
– developing a common energy market based on the same ecological priorities.

The European Union's energy policy in recent years has been largely directed at supporting the development of renewable energy. There are many causes for the development of renewable energy sources, but most important are the increase in EU energy security, potential reduction of fossil fuel imports outside the Union and creation of new jobs in this sector [3].

Renewable energy and sustainable development have long been one of the most important elements of European Union policy.

The aim of this article is to analyze current state of implementation of the basic assumptions of the Community in the field of energy policy. Conducted research should demonstrate which of the Member States of the Community is implementing its energy policy and in which implementation is at risk.

## 2. Energy policy of the European Union

The EU Energy Policy is a series of directives, strategies and legal requirements for the Member States in the energy sector. The most important acts governing the activities of the energy sector and indicating aims and directions of its development in the Community and in the individual Member States are:

– European Energy Policy,
– Europe 2020 - A strategy for smart, sustainable and inclusive growth,
– Green Paper. Framework for climate and energy policy until 2030

The European Energy Policy is a climate and energy package adopted in March 2007 by the Council of the European Union, according to which, by 2020, the European Union [4]:

– will reduce by 20% greenhouse gas emissions (with a 30% reduction if other developed countries commit to comparable emission reductions, and developing countries contribute to their commitments and opportunities) relative to the 1990 level,
– will increase by 20% the share of renewable energy in final energy consumption,
– will increase by 20% energy efficiency, compared to the 2020 projections,
– will increase the share of biofuels in the total consumption of transport fuels by at least 10%.

Europe 2020 is a long-term strategy for the development of the European Union for the years 2010-2020. It was approved by the European Council on 17 June 2010. The program offers three basic, mutually reinforcing priorities [5]:

– smart growth - that is knowledge and innovation development,
– sustainable growth - a transition towards a low-carbon, resource-efficient and competitive economy,
– inclusive growth - that is, to support a high-employment economy and ensure economic, social and territorial cohesion.

Progress towards the Europe 2020 Strategy has also been defined as measurable development goals to be achieved in 2020 at the EU level. In the context of the energy sector, this is implementation of the 3x20% initiative presented in the European Energy Policy. Instruments to deliver the Europe 2020 goals are developed by the EU Member States National Reform Programs and the flagship initiatives of the European Commission. Among the latter, attention should be paid to the initiative entitled *A resource-efficient Europe*, which points to the need for a shift towards a low-carbon economy and efficient use of environmental resources.

In March 2013, the European Commission, together with the publication of the "Green Paper. Framework for climate and energy policy until 2030." launched a debate on the future energy strategy of the EU.

As a result of this debate and consultation between countries of the Community, an agreement was reached setting out the objectives of energy and climate policy by 2030, adopted by the Council of Europe in October 2014. The four main goals to be faced by the energy sector are [6]: reducing greenhouse gas emissions by at least 40% by 2020 compared to 1990 (binding target), by at least 27%

IOP Conf. Series: Earth and Environmental Science **164** (2018) 012004 doi:10.1088/1755-1315/164/1/012004

renewable energy by 2030 At least 27% improvement of energy efficiency by 2030 (indicative target), supporting the development of a full internal energy market by urgently implementing the 10% interconnection electricity target, especially for the Baltic Sea States and the Iberian Peninsula.

## 3. The implementation of the European Union's energy policy objectives

The assessment of the implementation of EU energy policy has been carried out on the basis of indicators that characterize the four main areas identified as key in the link between the energy sector and the concept of sustainable development. These are:

- energy efficiency,
- percentage of generated electricity from renewable energy sources,
- CO2 emissions in EU countries,
- share of renewable energy in fuel consumption in transport.

Energy efficiency is an economical way to reduce energy consumption while maintaining an unchanged level of economic efficiency [7]. Therefore, the energy efficiency index for EU countries was determined as a ratio of GDP to the volume of electricity generated in the economy. Energy efficiency in the EU (Figure 1) in 2008 increased by 8% compared to 2008 (this year the climate change-energy package of the European Energy Policy was adopted), thus less than the 20% growth expected by 2020.

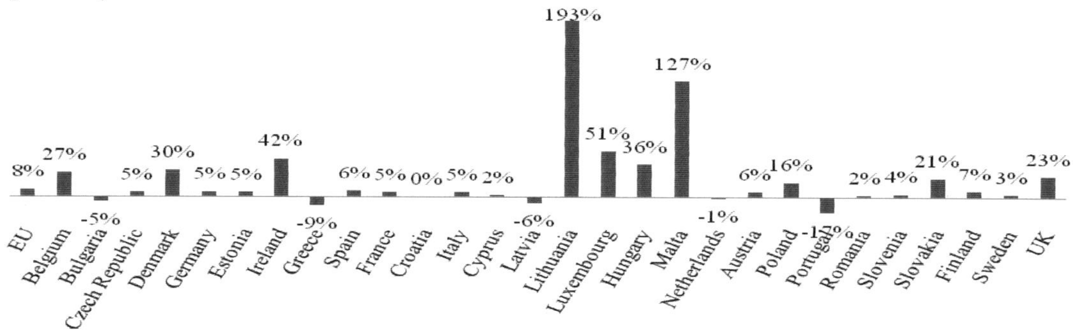


**Figure 1.** Changes in energy efficiency in EU countries in 2008-2015.

Source: Calculations and own elaboration based on Eurostat and World Bank data

The country with the highest increase in energy efficiency is Lithuania - 193% and Malta - 127%. Luxembourg, Ireland, Denmark and Belgium have also achieved significant reductions in energy consumption. During the period under review, energy efficiency declined in Portugal, Latvia and Bulgaria and Greece.

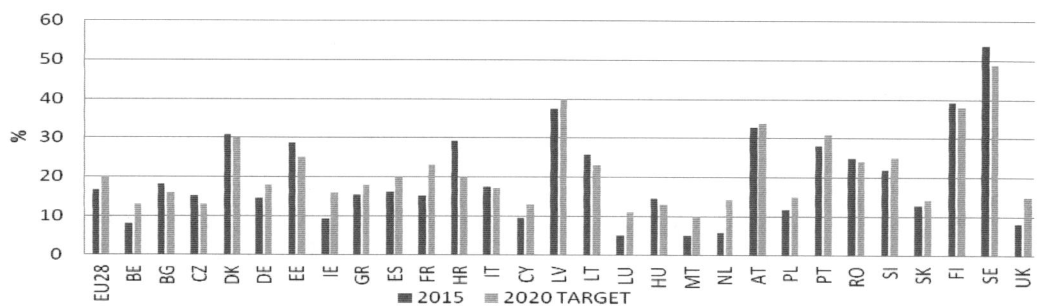


**Figure 2.** Percentage of energy from renewable energy in EU countries in 2015.

Source: Calculations and own elaboration based on Eurostat data

Taking into account the economic development of individual countries of the Community, each individual has a specific target for renewable energy. By 2015, the share of renewable energy in total energy consumption was 17%, thus the only 3% below the 20% target for 2020 (Figure 2). The group of countries that reached their "national" target for renewable energies in 2015 included Sweden (54%), Estonia (29%), Bulgaria (18%), Croatia (29%), Denmark (31%), Czech Republic (15%), Finland (39%). Moreover, in the Nordic countries, the share of renewable energy in total energy consumption is highest. The countries with the lowest levels of renewable energy use are Ireland, Malta, Cyprus, Netherlands and the United Kingdom. These countries have not yet fulfilled their targets for renewable energy use.

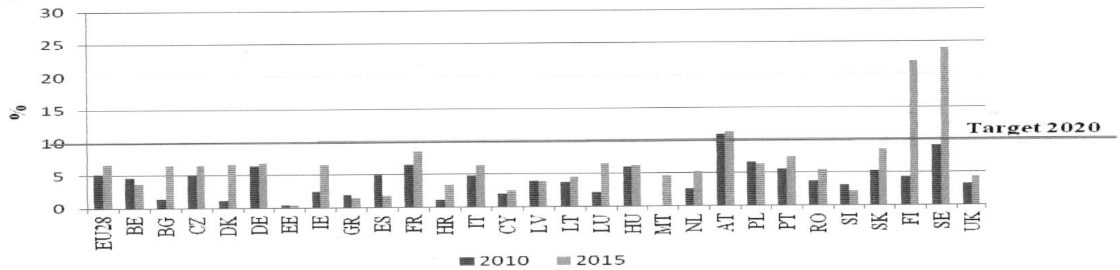


**Figure 3.** Share of renewable energy in transport fuel consumption in EU countries in 2010 and 2015. Source. Own study based on Eurostat data

According to data from the European Environment Agency, transport accounts for about a quarter of EU greenhouse gas emissions. Hence, the emphasis is on the use of renewable fuels in this sector [8]. In 2015, the share of biofuels in transport fuel consumption is growing steadily in each of the EU countries (Figure 3). For the whole Union, the share was 6.7%. Only in Finland (22%), Sweden (24%) and Austria (11.4%) the share of renewable fuels in transport fuel consumption reached the level of 10% required for 2020. The transport sector responsible for the highest CO2 emissions is road transport. The countries with the highest participation in European road transport are Poland, Spain and Germany and the Netherlands. At the same time these are countries where the use of biofuels in transport is at the lowest level among all EU countries.

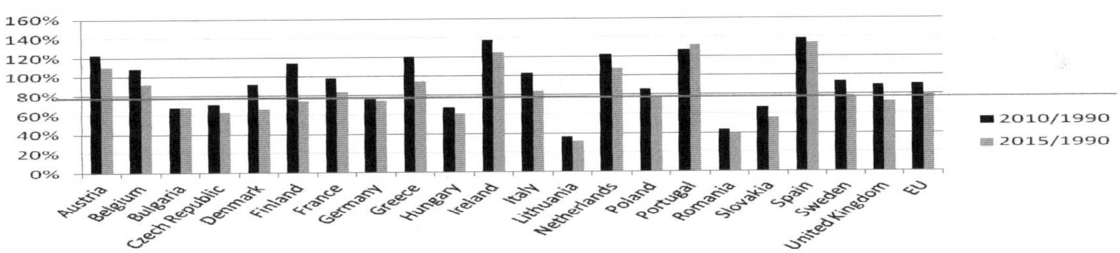


**Figure 4.** C02 emission dynamics indicators for the EU in 1990-2015.

Source: Calculations and own elaboration based on BP Statistical Review of World Energy June 2017

The climate and energy package has set a target of reducing CO2 emissions in the European Union by 20% compared to 1990, and this assumption has already been fulfilled by 2015. Countries that have reduced carbon dioxide emissions by 20% are Bulgaria, Czech Republic, Germany, Denmark, Hungary, Lithuania, Romania, Slovakia, United Kingdom and Poland. Among countries where CO2 emissions are above 100%, which means that emissions in 2015 are above 1990 levels are among others Austria, Belgium, Greece, Ireland, Portugal, Spain and the Netherlands.

## 4. Summary

The ambitious goals of the 2010-2020 energy policy outlined by the European Union are to varying degrees carried out by the member states of the Community. The EU is the best place for greenhouse gas emissions, as it achieved its target for 2020 in 2015. It is equally good in terms of renewable energy development. By far the worst target, EU member states and the whole EU is failing to reach is the share of biofuels in total fuel consumption in transport and in improving energy efficiency.

Undoubtedly, the impact on the different levels of energy policy objectives by individual EU countries has had a global economic crisis. The economic downturn has slowed down investment in the energy sector [9]. The economic situation also had a negative impact on development of renewable energy sector, whose main driving force was to subsidize investments in this area by the governments of the Member States. The crisis has caused the subsidies in many countries (Germany, Czech Republic, Italy, Bulgaria, Great Britain) to be limited.

Very slow process of communitarisation of energy policy in the European Union also adversely affects the achievement of the "3x20%" target. This is due to the diversity of political and economic interests of individual Member States, especially the largest energy consumers like Germany and France, who are eager to make the decision to cooperate with Russia on specific projects. Although a long-established fact is the significant dependence of EU countries on imports of fuels and energy from the east [10].

What is more, also the large diversity of member states in terms of country-specific energy carriers adversely affects the common goals of the EU's economic policy, which is reflected in their energy mix. This means that different countries invest in different energy-producing technologies. For example, Germany is moving away from nuclear power, while Poland and Finland have plans to develop this energy sector.

The EU's energy policy combines economic and social goals, while pursuing a philosophy of sustainable development. It must be stated that achieving the ambitious goals of energy policy will make the European Union an economy with high economic efficiency and low CO2 emissions.

**References**

[1] Włodarczyk A 2015 Evaluation of the carbon market impact on the Polish Power Exchange, *Materials, Methods and Technologies* vol. 9 pp 64-79
[2] *Our* Common *Future. Report of the World Commission on Environment and Development* 1987 Oxford University Press (New York: Oxford)
[3] Pach-Gurgul A 2014 Renewable energy in the European Union in the context of economic crisis *Prace Komisji Geografii Przemysłu Polskiego Towarzystwa Geograficznego* 27 pp 130-147
[4] *An* energy *policy for Europe* 2007 COM(2007)1 (Brussells) pp 3-28
[5] EUROPE *2020. A European strategy for smart, sustainable and inclusive growth* 2010 COM(2010) 2020, (Brusells) pp 3-29
[6] Green *paper: A 2030 framework for climate and energy policies* 2013 COM (2013) 169 (Brussells) pp 2-17
[7] Paska J and Surma T 2013 Polityka energetyczna Polski na tle polityki energetycznej Unii Europejskiej *Polityka Energetyczna* 4 T.16 pp 7-19
[8] Bunn D and Munoz J 2016 Supporting the externality of intermittency in policies for renewable energy *Energy Policy* vol 88 pp 594-602
[9] Włodarczyk A 2017 Environmental Management Instruments in the Restructuring Process of Polish Energy Enterprise [in:] ed R Borowiecki and J Kaczmarek *The Propensity to Changes in the Competitive and Innovative Economic Environment. Processes-Structures-Concepts* Publishing House: Foundation of the Cracow University of Economics (Cracow) pp 215-225
[10] Miciuła I Polityka energetyczna Unii Europejskiej do 2030 roku w ramach zrównoważonego rozwoju 2015 *Studia i Prace Wydziału Nauk Ekonomicznych i Zarządzania* 42 T.2 pp 57-67

IOP Conf. Series: Earth and Environmental Science **164** (2018) 012005 doi:10.1088/1755-1315/164/1/012005

# A Study on Leaching Potential of Bio-gasification Tar in Water

**Hemantajeet Medhi**[1], **M Ronghang**[2], **B Chakrabarty**[3] **and A K Baruwa**[4]

[1] Department of Chemical Engineering, Bineswar Brahma Engineering College, Kokrajhar, Assam 783370, India
[2] Department of Civil Engineering, Bineswar Brahma Engineering College, Kokrajhar, Assam 783370, India
[3] Department of Chemical Engineering, Assam Engineering College, Guwahati, Assam 781013, India
[4] NETES Institute of Technology & Science, Mirza, Assam 781125, India

E-mail: jeethemanta.medhi12@gmail.com

**Abstract.** Non-conventional energy sources based on biomass gasification for production of producer gas is a promising technique. However, the presence of unacceptable levels of bio-gasification tar waste puts the limit of its use. Although investigations are being carried out to study avenues for controlling the amount of tar produced during biomass gasification, the complete removal of this waste still remains a big challenge. Problems are severe when the waste comes to contact with water body leading to the leaching of components (pollutants) from tar into the environment. In this paper, we are reporting the study that has been carried out to investigate and evaluate the leaching properties of bio-gasification tar in water as a function of concentration levels, temperature and pH levels of the solvents.

## 1. Introduction

Development of non-conventional energy sources based on biomass gasification has the potential in fulfilling the requirement of modern world energy demands up to a large extent. Generated producer gas through biomass gasification has shown the potential to be used as an energy source. However, the by products, bio-gasification tar that come along with the production of producer gas have handling difficulties making the process unsuitable in commercializing the producer gas. There exist several ways to remove bio-gasification tar from the producer gas that mainly includes physical processes like filters, scrubbers, wet electrostatic precipitators and cooling tower, chemical processes like thermal cracking, catalytic cracking and catalytic reforming [1, 2, 3]. In a gasification system, bio-gasification tar draws the special attention which disposal is usually cumbersome and expensive. The amount of bio-gasification tar yield varies from 1 to 20% of the feed of biomass [4]. It mainly consists of water-soluble phenol and naphthalene that can easily pollute the water body on contact [5]. Moreover, the presence of carcinogenic compounds has to be treated properly before disposal fulfilling the health and environmental demands [6]. In recent years, there has been an increased focus on environmental issues. As all the industrial wastes eventually are dumped into the water bodies via drainage or direct dumping, there is a chance of bio-gasification tar come contact with water leading to the leaching of components (pollutants) from tar into the environment. This prompted us to study the leaching

IOP Conf. Series: Earth and Environmental Science **164** (2018) 012005 doi:10.1088/1755-1315/164/1/012005

behavior of bio-gasification tar in water at different concentration levels, at different working temperatures and at the solvents made of different pH levels.

## 2. Methodology

In the present investigation, bio-gasification tar from the rice husk based gasification unit is utilized for the evaluation of leaching property of it. The bio-gasification tar is collected from the Ankur make gasification system at NEBISCO industries Pvt. Ltd., situated in Tezpur, Assam, India that produces biscuits for consumption in the local market. In this study, the leaching property of bio-gasification tar in double distilled water (dd$H_2O$) is measured. The electrical conductivity of the leached water is measured with the help of electrical conductivity (EC) meter at different interval of time at different experimental conditions. These experimental conditions are discussed below-

### *2.1. Weight of bio-gasification tar is varied*

We take 1g, 2g, 3g, and 5g of bio-gasification tar (see figure 1 for the photograph) and 200 ml of double distilled water in beaker A, B, C & D respectively.

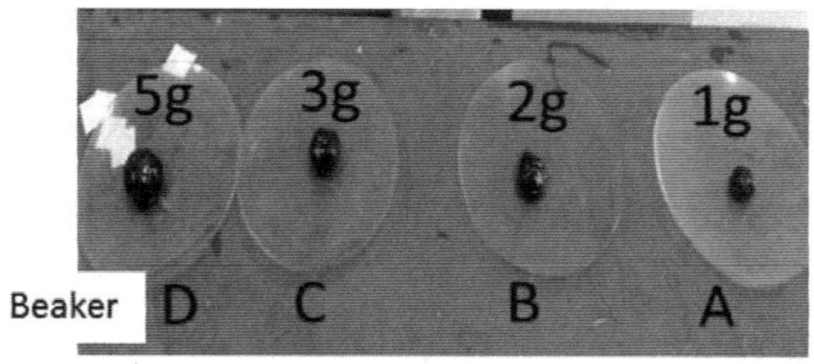


**Figure 1.** The weight of bio-gasification tar is varied. 1g, 2g, 3g, and 5g of bio-gasification tar is added in separate beakers A, B, C, and D respectively that contain 200 ml of dd$H_2O$ each.

### *2.2. Volume of distilled water is varied*

In this given set of experiment, 100 ml, 150 ml, 200 ml and 250 ml of dd$H_2O$ is used in different beakers namely A, B, C and D as shown in figure 2(a), in which 5g of bio-gasification tar is added separately to each of them. We allowed the leaching process to continue until the EC reading of the beaker solutions saturates, leading to a stage similar to the one as shown in figure 2(b).

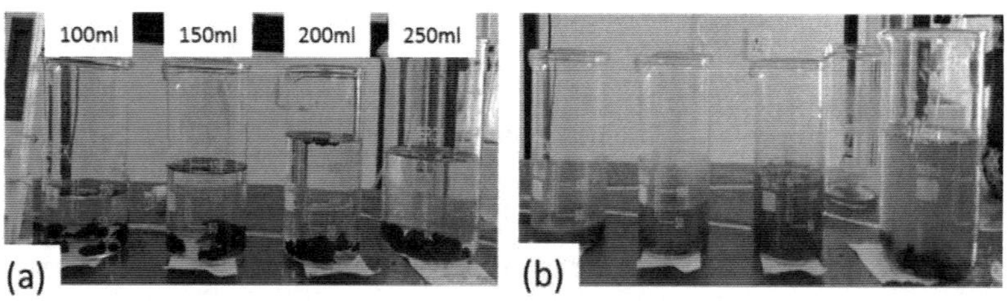


**Figure 2.** Leaching study of bio-gasification tar: 5g tar is added to each of the beakers containing ddH2O, of the volume of 100ml, 150ml, 200ml and 250ml respectively. (a) At the beginning, no leaching (b) leached, after 44hrs (see text).

### *2.3. Temperature of distilled water is varied*

We pour 300 ml of double distilled water in each beaker and place each beaker on the hot plate of the magnetic stirrer, switch on the stirrer and adjust each beaker to a different starting temperature set at 40°C for beaker A, 50°C for beaker B, 45°C for beaker C & 55°C for beaker D. We take 5g of bio-gasification tar each for all the beakers for leaching. Experimental observations for the mentioned experiments are shown in figure 4, which is also explained in section 3.

### *2.4. pH of distilled water is varied*

We place 6 beakers orderly manner and label them as A, B, C, D, E & F. In each beaker we pour 300 ml of distilled water. The pH in beaker A, B & C is maintained at pH 4, pH 5 & pH 6 by adding HCl respectively. Similarly, the pH in beaker D, E & F is maintained pH 9, pH 10 & pH 11 by adding NaOH respectively. After setting the pH levels, 3g each of bio-gasification tar is added to the specified beakers for leaching.

## 3. Results & Discussion

Leaching rate at a particular experimental condition can be obtained with the help of the electrical conductivity (EC) measurements. To achieve this, for a given set of experiment, EC data (in µS/cm) obtained at different time interval is plotted against time (in min), which are shown in figures 3-5. Calculated slope of a curve represents the leaching rate of that experiment. In another hand, biological oxygen demand (BOD) & chemical oxygen demand (COD) measurements of bio-gasification tar in distilled water are also carried out to determine its pollution in water at the District Level Water Testing Laboratory, Kokrajhar, Assam, India, which is tabulated in Table 1 and Table 2.

Experimentally measured EC data is plotted against time for measuring the leaching rate when the weight of bio-gasification tar is varied is shown in figure 3(a). It is clearly seen that the slope (hence the leaching rate) is high for the sample 5g, (represented as D, in figure 3(a)). Measured leaching rates from the graphs are 0.00499 g/cm$^2$.min for 1g, 0.00678 g/cm$^2$.min for 2g, 0.01277 g/cm$^2$.min for 3g and 0.02045 g/cm$^2$.min for 5g respectively.

In a different experimental setup as mentioned in section 2.2, the volume of ddH$_2$O is varied (100 ml, 150 ml, 200 ml and 250 ml) keeping the weight of bio-gasification tar fixed at 5g for each of the samples. For obvious reason due to the high concentration of leached tar EC measurement is high for the sample prepared with 100ml. Measured leaching rates from the graphs of figure 3(b) are 0.077 g/cm$^2$.min for 100ml, 0.0618 g/cm$^2$.min for 150ml, 0.02757 g/cm$^2$.min for 200ml and 0.02665 g/cm$^2$.min for 250ml of solution respectively.

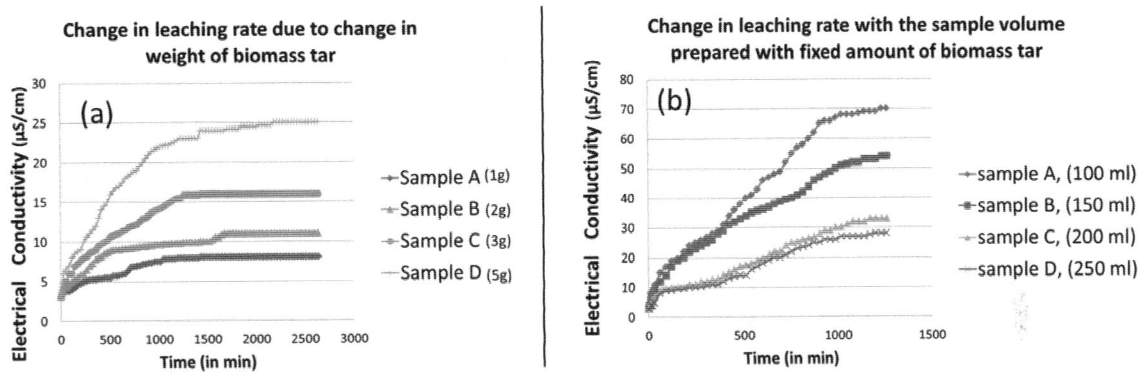


**Figure 3.** Graph representing change in EC (equivalently leaching rate) due to change in (a) weight of biomass tar vs. time and (b) volume of sample solution at fixed amount of biomass tar (5g) vs. time.

We repeat the experiments for the samples (5g biomass tar and 300ml of ddH$_2$O) prepared at different temperatures, which has been discussed in section 2.3 in details. Experimentally measured electrical conductivity is plotted against time (see figure 4). Experimentally measured leaching rates of biomass tar calculated from the slopes of the plots are 0.0483 g/cm$^2$.min for sample A at 40$^O$C, 0.0548 g/cm$^2$.min for sample B at 50$^O$C, 0.0574 g/cm$^2$.min for sample C at 55$^O$C and 0.05206 g/cm$^2$.min for sample D at 45$^O$C respectively. In this case, sample kept at higher temperature (55$^O$C) has high leaching rate.

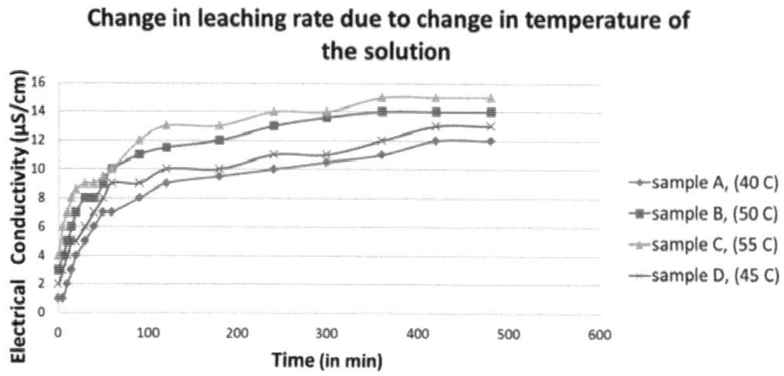


**Figure 4.** Graph representing change in EC (equivalently leaching rate) due to change in temperature of sample solution (four samples of 5g biomass tar + 300ml of ddH2O each) vs. time.

In the final set of experiment pH of the ddH$_2$O is changed which has been discussed already in section 2.4. When HCl is added, pH goes down making it acidic for which experimentally measured electrical conductivity with respect to time is shown in figure 5(a). Whereas for the case of NaOH addition to ddH$_2$O (base solution), corresponding "EC vs Time" measurements for different pH levels are shown in figure 5(b). As before, from the EC vs Time graphs, leaching rate is calculated. Considering the experiments corresponds to pH values of 4, 5 and 6, the measured leaching rates of biomass tar are 0.8065 g/cm$^2$.min, 0.648 g/cm$^2$.min and 0.1594 g/cm$^2$.min respectively. In case of experiments corresponds to pH value of 9, 10 and 11 obtained leaching rate of biomass tar are 16.15 g/cm$^2$.min, 14.661 g/cm$^2$.min and 9.37 g/cm$^2$.min respectively.

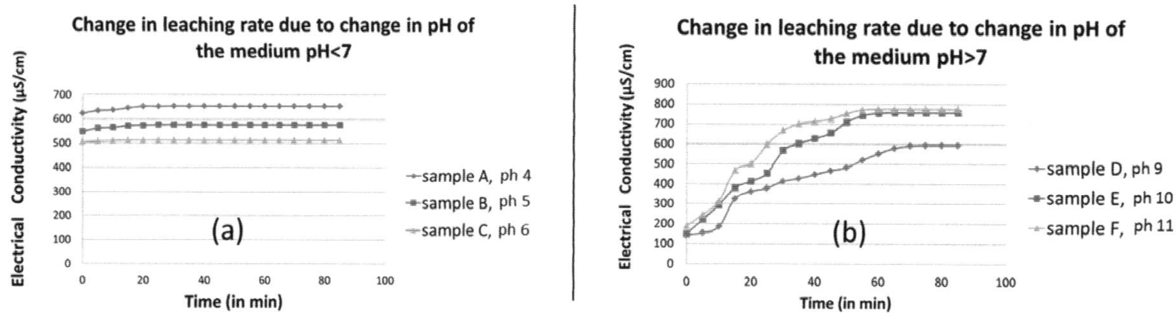


**Figure 5.** Graph representing change in EC (equivalently leaching rate) due to change in pH of sample solution vs. time (a) when pH<7 (b) when pH>7.

From the BOD and COD collected data of the experimental samples, it is found that their values are at the peak on the 2$^{nd}$ day, that starts decreasing from 3$^{rd}$ day onwards vanishing on the 4$^{th}$ day.

**Table 1.** Biological oxygen demand (BOD) measurements for the sample of section 2.3

| Sl. no. | sample | Temperature ($^0$C) | Day 1 | Day 2 | Day3 | Day4 | Day 5 |
|---|---|---|---|---|---|---|---|
| 1. | A (initial) | 40 | 1 | 4 | 1 | F | F |
| 2. | A (final) | 40 | 110 | 350 | 120 | F | F |
| 3. | B (initial) | 50 | 1 | 4 | 1 | F | F |
| 4. | B (final) | 50 | 110 | 320 | 110 | F | F |
| 5. | C (initial) | 55 | 1 | 4 | 1 | F | F |
| 6. | C (final) | 55 | 100 | 400 | 200 | F | F |
| 7. | D (initial) | 45 | 1 | 4 | 1 | F | F |
| 8. | D (final) | 45 | 100 | 200 | 100 | F | F |

IOP Conf. Series: Earth and Environmental Science **164** (2018) 012005 doi:10.1088/1755-1315/164/1/012005

**Table 2.** Chemical oxygen demand (COD) measurements for the sample of section 2.3.

| Sl. no. | sample | Temperature ($^0$C) | Day 1 | Day 2 | Day 3 | Day 4 | Day 5 |
|---|---|---|---|---|---|---|---|
| 1. | A (initial) | 40 | 0.8 | 3.2 | 0.8 | F | F |
| 2. | A (final) | 40 | 88 | 280 | 96 | F | F |
| 3. | B (initial) | 50 | 0.8 | 3.2 | 0.8 | F | F |
| 4. | B (final) | 50 | 88 | 256 | 88 | F | F |
| 5. | C (initial) | 55 | 0.8 | 3.2 | 0.8 | F | F |
| 6. | C (final) | 55 | 80 | 320 | 160 | F | F |
| 7. | D (initial) | 45 | 0.8 | 3.2 | 0.8 | F | F |
| 8. | D (final) | 45 | 80 | 160 | 80 | F | F |

Note: F represents a value which cannot be measured by the instrument.

## 4. Conclusions

The leaching behaviors (rates) of bio-gasification tar are experimentally measured for different experimental conditions. In which, the concentration of bio-gasification tar, the temperature of the medium, pH-level of the used medium are varied. The studies have shown the leaching rate of bio-gasification tar is proportional to the temperature and also increase with the decrease in pH-levels, which has been verified with the experiments conducted with $ddH_2O$ as a medium. For the used experimental conditions pH= 3 and T=55$^O$C, it is observed that the contamination rate is very rapid that exceeds the permissible level of BOD and COD suggesting not to dump such byproducts directly into the environment. Favorable conditions for higher leaching rate of bio-gasification tar can easily pollute the environment affecting the livings including humans. Therefore, it is necessary to have thorough investigations of tar before dumping it into the environment.

## 5. References


[1] Stassen H.E.M., Prins W., Swaaiji W.P.M., 2002 Thermal conversion of biomass into secondary products; the case of gasification and pyrolysis *12$^{th}$ European Biomass Conference, 17-21 June*, (Amsterdam:, The Netherlands) p 38-44

[2] Milne T.A., Evans R.J. 1998 Biomass Gasifier Tars: Their Nature, Formation and Conversion *NREL/TP-570-25357*

[3] Vivanpatarakij S., Assabumrungrat S., 2013 Thermodynamic analysis of combined unit of biomass gasifier and tar steam reformer for hydrogen production and tar removal *International Journal of Hydrogen Energy* 38, 3930- 3936

[4] Stevans D.J., 2001 Hot gas conditioning: Recent Progress with large scale biomass gasification systems, update and summary of recent progress, National renewable energy laboratory, report NREL/SR-510-29952, August

[5] Milne T.A., Evans, R.J. 1997 Developments in thermochemical biomass conversion, chemistry of tar formation and maturation in the thermochemical conversion of biomass. Blackie Academic and Professional

[6] Bhler R., Nussbaumer T., Hasler P., 1997 Evaluation of gas cleaning technologies for small scale biomass gasifiers. Swiss Federal office of Energy and Swiss Federal Office for Education and Science, August

IOP Conf. Series: Earth and Environmental Science **164** (2018) 012006 doi:10.1088/1755-1315/164/1/012006

# Comparative Analysis of Corrosion Inhibition: Between Jatrophacurcas, Palm and Diesel Oil based Emulsified Acids for Acid Stimulation Operations

**Muhammad Mohsin Yousufi [1,*], Mysara Eissa Mohyaldinn Elhaj [1], Muhammad Moniruzzaman [2]**

[1] Petroleum Engineering Department, Universiti Teknologi PETRONAS, Seri Iskandar 32610, Perak, Malaysia

[2] Chemical Engineering Department, Universiti Teknologi PETRONAS, Seri Iskandar 32610, Perak, Malaysia

E-mail: moshyousufi@hotmail.com

**Abstract.** The recent environmental concerns have raised the need for a green acidizing fluid, which is able to replace diesel oil as an oil phase in emulsion based acidizing fluids. In order to do so it must be able to provide not only retardation in acid reaction against the rock surface, but also provide suitable inhibition characteristic that can reduce the amount of corrosion which occurs due to acidizing. In this paper emulsified acids made from triglyceride oils namely; Jatropha curcas oil and Palm oil have been tested for corrosion inhibition in comparison with Diesel oil based emulsified acid and 15wt.% HCl. The emulsion made with Jatropha curcas oil showed capability to replace diesel oil as a viable and an environmental friendly substitute for oil phase in emulsified acids.

## 1. Introduction

Well acidizing is a process used for improving well productivity of a hydrocarbon reservoir suffering from lack of permeability either by restoring or increasing its permeability. This is done by injecting acidizing fluid into reservoir which reacts with the rock surface to create flow conduits and remove obstructions for formation fluid to pass through into the wellbore.

Type of acid used during any acidizing operation depends on various factors such as temperature, stability, lithology, dissolving power, reaction rate and the products that will be formed by the use of a certain acid, the acids are accordingly used either solely or by making mixtures while keeping the prior mentioned conditions under consideration, normally the mineral and organic acids utilized consist of : hydrochloric acid (HCl), hydrofluoric acid (HF), formic acid, acetic acid respectively, whereas industrial ones such chloroacetic acid and sulfamic acid have been employed [1]. The paper is limited to acid usage on carbonate reservoirs, therefore acids utilized for sandstone have been stated concisely. Generally acidizing done on carbonate rocks such as Limestone ($CaCO_3$) and Dolomite [$CaMg(CO_3)_2$] is with HCl. HF is not to be used on carbonate formations as with minerals consisting of calcium HF react in an unfavorable manner creating insoluble calcium fluoride. HCl, however, tends to react with carbonate rock in a favorable manner.

Acids although effective have a high reaction rate with the carbonate formations, especially limestone, which leads to rapid consumption of the acidizing fluid before reaching the targeted depth [2], along with creating other problems such as generating excess wide face flow paths which cave in due to closure stresses and do not cater to low permeable zones[3].

IOP Conf. Series: Earth and Environmental Science **164** (2018) 012006 doi:10.1088/1755-1315/164/1/012006

Acid are also a cause of corrosion to the metallic equipment being used, it causes weakening of the metal and propagation of $Fe^{2+}$ precipitates in spent acids which generates formation damage. Thus causing a decrease in recovery from that production zone.

Emulsified acids, which are a mixture of acid (diluted) and oil, have been used as retarding fluids which minimize the contact of acid with the rock and metal surfaces. This allows better stimulation of the rock by reducing the consumption of acidizing fluid, which is then able to create long narrow flow zones and prevents corrosion of metal to a considerable extent.

Diesel oil has been used in synthesizing emulsified acids for a number of decades. Diesel oil based emulsified acids are effective but have a high toxicity load, the use of diesel oil and related products such as xylene, benzene and kerosene have all been banned under the Energy Policy Act of 2005 (Safe Drinking Water Act, Section 322)[4]. This is because of the severe carcinogenic health issues caused by it such as, elevation in cancer rates, clotting deficiency and skin damage. During well production phase the flowback fluid containing diesel oil when introduced to local water supply is hazardous even after treatment, similarly the acidizing fluid can contaminate the groundwater as it is flushed after the acidizing operation. Due to such issues in February 2014 EPA published Underground Injection Control program to provide regulators with guidelines when giving permits to operators [5]s, in order to prevent diesel fuel usage to ensure environmental security in exploration and production operations.

In this regard, plant oils have the potential of replacing diesel oil in emulsified acids, courtesy of being more environmental friendly. The fatty acid content is suitable for providing a viscous emulsion which can retard acid reaction and provide corrosion inhibition.

Considering the use of Jatropha curcas oil and Palm oil previously in drilling [6], [7] and EOR applications [8], [9],[10]. Especially Jatropha, due to it being long chain triglyceride oil and having a high fatty acid percentage. In order to compare their potential as acidizing fluids tests were conducted on emulsified acids based on Jatropha oil and Palm oil which were compared with Diesel oil based emulsion for various properties mentioned in table 1. Based on those results and due to these presence of a sufficient amount of oleic and linoleic acids mentioned in table 2, which are able to provide corrosion inhibition property. Further study was conducted on the two plant oil based emulsions, in order to find out which one can provide corrosion inhibition comparable to diesel oil based emulsified acid.

**Table 1.** Properties of Jatropha oil, Diesel oil and Palm Oil based Emulsified Acids

| Tests | Jatropha curcas Oil based Emulsion | Diesel Oil based Emulsion | Palm Oil based Emulsion | Test Standard |
|---|---|---|---|---|
| Droplet size Distribution median size, (μm) | 2.17886 | 2.38166 | 2.24068 | ISO 13320 |
| Acid Solubility, (%) | 57.22 | 66.95 | 66.68 | API RP 40 |
| Toxicity, (ppm) | 86.5 | 12 | 85 | LC50, OECD 203 |

**Table 2.** Fatty Acid Composition Percentage of Jatropha oil and Palm Oil

| Name of the oil | % of oil in seed of Kernel | Myristic (Tetradecanoic) | Palmitic (Hexadecanoic) | Stearic (n-Octadecanoic) | Oleic | Linoleic |
|---|---|---|---|---|---|---|
| | | C14 | C16 | C18 | C18:1 | C18:2 |
| Palm Oil | 30-60 | 0.5-2.0 | 32.0-45.0 | 2.0-7.0 | 38.0-52.0 | 5.0-11.0 |
| Jatropha Curcas | 30-40 | 0.5-1.4 | 12.0-17.0 | 5.0-9.5 | 37-63 | 19-41 |

IOP Conf. Series: Earth and Environmental Science **164** (2018) 012006 doi:10.1088/1755-1315/164/1/012006

## 2. Methodology

### *2.1. Materials*

Diesel and Palm oil were purchased from a local supplier and Jatropha Oil was ordered from BATC Development BHD. 15 weight% hydrochloric acid, nonionic surfactants: Span80 and Tween80 were obtained from Benua Sains SDN BHD. A Linear alcohol based nonionic surfactant with an ethoxylation grade of 9 was obtained from a proprietary company. Steel samples were purchased from a local metal supplier.

### *2.2. Synthesis*

A Water in Oil type emulsified acid was made [11], having acid to oil ratio of 70:30, as this ratio gives ample mobility and retardation along with reducing the amount of oil cost, thus has been used in many research studies [12], [13], [14]. A surfactant blend was prepared using a highly hydrophilic surfactant (15 HLB) and a lipophilic surfactant (4.3 HLB) in order to create an emulsifier which provides a strong bond between the acid and oil phases [15], the surfactant mixture is mixed with the oil phase for 10 minutes. Hydrochloric acid is introduced into the oil and surfactant mixture which is being continuously stirred, the acid is sprayed in order to obtain fine droplet size [16]. After adding the acid, emulsion mixture is stirred at a speed of 1000 RPM for additional 20 minutes to obtain a stable Water in Oil emulsion.

### *2.3. Electrical Conductivity*

The emulsions were tested for near zero electrical conductivity to ensure a Water in Oil emulsion [17], [18], [19]. In case of conductivity the emulsion was mixed for another 30 minutes and was tested again after which the emulsion exhibited no indication of conduction [20].

### *2.4. Corrosion Testing*

Corrosion tests were using the weight loss analysis for Jatropha, Diesel, Palm oil based emulsified acids and 15wt. % HCl conducted at temperatures of 25°C and 70°C. Weight loss analysis is a gravimetric method to determine corrosion rate or metal loss. A coupon of metal sample for which the weight loss was to be determined was immersed in the corrosive sample fluid and was removed after a considerable amount of time. The coupon was then cleaned of all corrosion products and reweighed [21], [22]. Metal loss was determined using the following equation:

$$\text{Metal Loss} = \frac{W\,(g) \times K}{D\,(g/cm^3) \times A\,(cm^2)} \quad (1)$$

Where,

W= Weight Loss, K = K- Factor (constant=8.76 x $10^4$), D= Alloy Density, A= Exposed Area

## 3. Results and Discussion

### *3.1. Weight loss Analysis at 25°C*

The three emulsified acid samples and 15wt% HCl were measured for weight loss after a period of 24 hours. Figure 1 shows the lowest amount of metal loss was from the metal immersed in Jatropha curcas oil based emulsified acid, followed by metal taken out from Diesel oil based emulsion. These two emulsions had relatively close corrosion inhibition performance possibly because Diesel oil and Jatropha oil have long carbon chain lengths, which provide a highly viscous emulsion, keeping the acid reaction retarded with the metal surface to a certain extent more in comparison to the metal loss of Palm oil based emulsified acid which was higher than the other two emulsified acids. In order to evaluate the effect of acid in absence of a retarding medium the 15wt. % HCl solution was used as a reference which yielded a great amount of metal loss.

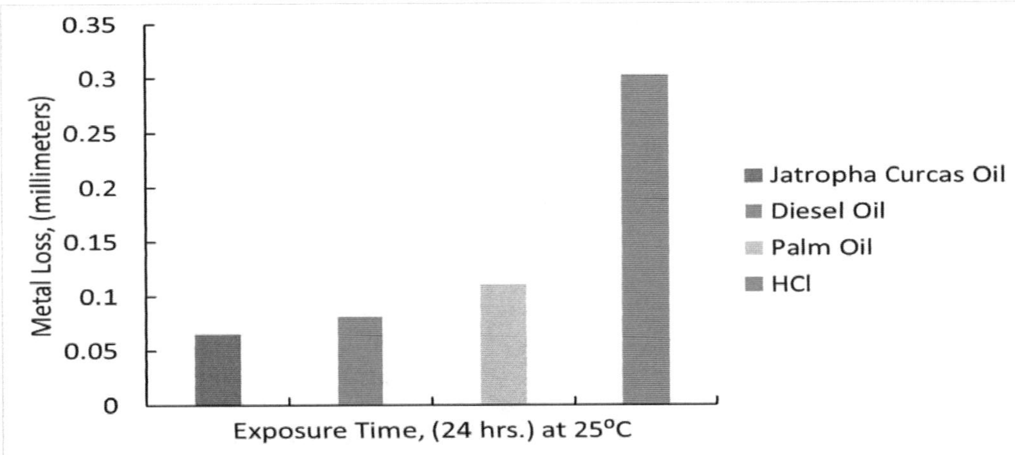


**Figure 1:** Weight loss Analysis at 25°C

*3.2. Weight loss Analysis at 70°C*

Following the weight loss analysis at temperature of 25°C, the weight loss test was conducted at a temperature of 70°C for the three emulsified acid samples and 15wt. % HCl solution. In this case the corrosion inhibition provided by Diesel oil emulsified acid was the highest followed closely by that of Jatropha oil based emulsion. When comparing the three emulsified acids Palm oil showed the lowest ability to inhibit the metal from corroding. As for the 15wt. % HCl solution, it completely dissolved the metal sample hence it was not mentioned in figure 2. At higher temperatures, acid reactivity increases, this is because the reactant particles get more energized causing them to move more quickly, resulting in an increment in particle collision rate [23]. Triglyceride oil based emulsions of Jatropha and Palm oil due to having a high fatty acid content, have a greater total acid value than that of Diesel oil based emulsified acid. Petrodiesel based rarely has olefin content and if present is normally less than 0.5mg of KOH/g, while Jatropha oil has total acid value of 32.8 mg of KOH/g [24]and around 6mg of KOH/g for Palm oil [25], [26] because of which the corrosion rate is found to be more. Jatropha oil based emulsion to its high viscosity compensates for its high acid value by giving a corrosion rate close to that of Diesel oil based emulsified acid.

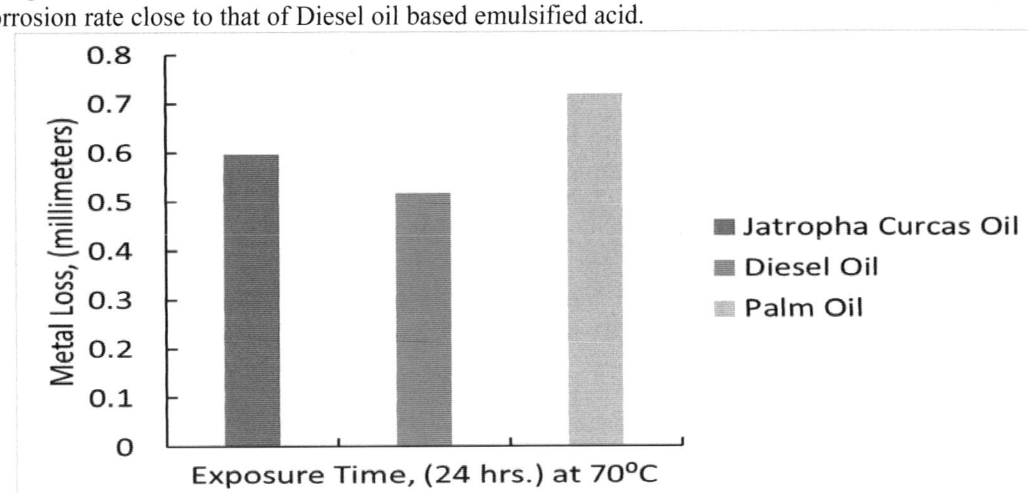


**Figure 2:** Weight loss Analysis at 70°C

## 4. Conclusion

As the environmental concerns arise the need for a green acidizing fluid, it is necessary that such fluid fulfils rest of the requirements for being a stable medium for acidizing wells. Corrosion inhibition is an important parameter to consider during an acidizing operation, it is crucial that the stimulating fluid causes minimum corrosion to metallic equipment. Based on the results obtained, it is safe to say that Jatropha curcas oil has the potential to replace Diesel oil as the oil phase in emulsified acids, giving corrosion inhibition performance relatively close to that of Diesel oil based emulsion courtesy of having large carbon chain and high fatty percentage in comparison to Palm oil based emulsion which due to being a medium chain triglyceride is unable to provide the same inhibition capability.

## References

[1] Williams, B.B., Gidley, J.L. S 1979 Acid Types and the Chemistry of Their Reactions *Acidizing Fundamentals* (Society of Petroleum Engineers of AIME) pp 10–7

[2] Navarrete R C, Holms B A, McConnell S B and Linton D E 2000 Laboratory, theoretical, and field studies of emulsified acid treatments in high-temperature carbonate formations *SPE Prod. Facil.* **15** 96–106

[3] Nierode D E and Kruk K F 1973 An Evaluation of Acid Fluid Loss Additives, Retarded Acids, and Acidized Fracture Conductivity *SPE 4549-MS, J. Pet. Technol.* 1–12

[4] Congress U S 2005 *Energy Policy Act of 2005, Subtitle C, Section 322*

[5] Bishop J and Bohlen S 2014 EPA letter to California State Water Resources Control Board and California Department of Conservation - 12/22/2014

[6] Adesina F, Anthony A, Gbadegesin A, Eseoghene O and Oyakhire A 2012 Environmental Impact Evaluation of a Safe Drilling Mud *SPE Middle East Health, Safety, Security, and Environment Conference and Exhibition, SPE-152865-MS* (Abu Dhabi, UAE) pp 1–9

[7] Fadairo A, Orodu D and Falode O 2013 Investigating the Carrying Capacity and the Effect of Drilling Cutting on Rheological Properties of Jatropha Oil Based Mud *SPE Nigeria Annual International Conference and Exhibition* pp 1–8

[8] Soffian N M, Jeirani Z, Hisham B, Jan M, Ali B S, Sareh M K and Rashidi M 2015 Jatropha Based Microemulsion Efficiency Screening Study for Enhanced Oil Recovery *Aust. J. Basic Appl. Sci.* **9** 118–25

[9] M.S.W. Nordiyana, Munawar Khalil, Badrul Mohamed Jan, Brahim Si Ali C W T 2016 Formation and Phase Behavior of Winsor Type III Jatropha curcas -Based Microemulsion Systems *J. Surfactants Deterg.* **19** 701–12

[10] Mohadyaldinn M E, Azad N and Kalam A 2017 Evaluation of Jatropha Curcas Oil as Corrosion Inhibitor of CO 2 Corrosion in Petroleum Production Environment *J. Appl. Environ. Biol. Sci.* **7** 28–34

[11] Sidaoui Z, Sultan A S and Qiu X 2016 Viscoelastic Properties of Novel Emulsified Acid Using Waste Oil : Effect of Emulsifier Concentration , Mixing Speed and Temperature *SPE Kingdom of Saudi Arabia Annual Technical Symposium and Exhibition, SPE-182845-MS* (Dammam, Saudi Arabia: Society of Petroleum Engineers) pp 1–11

[12] Al-Anazi H a., Nasr-El-Din H a. and Mohamed S K 1998 Stimulation of Tight Carbonate Reservoirs Using Acid-in-Diesel Emulsions: Field Application *SPE International Symposium on Formation Damage Control, SPE-39418-MS* (Society of Petroleum Engineers) pp 9–17

[13] Bazin B and Abdulahad G 1999 Experimental Investigation of Some Properties of Emulsified Acid Systems for Stimulation of Carbonate Formations *SPE Middle East Oil Show, SPE-53237-MS* (Bahrain) pp 1–10

[14] Navarrete R C, Holms B A, McConnell S B and Linton D E 1998 Emulsified Acid Enhances Well Production in High-Temperature Carbonate Formations *SPE European Petroleum Conference, SPE-0699-0047-JPT* (Hague, Netherland: Society of Petroleum Engineers) pp 391–406

[15] Sarma D K, Agarwal P, Rao E and Kumar P 2007 Development of a Deep-Penetrating Emulsified Acid and Its Application in a Carbonate Reservoir *15th SPE Middle East Oil &*

*Gas Show and Conference, SPE -05502-MS* (Bahrain: Society of Petroleum Engineers) pp 1–7

[16] Al-Mutairi S H, Nasr-El-Din H A and Hill A D 2009 Droplet Size Analysis of Emulsified Acid *SPE Saudi Arabia Section Technical Symposium, SPE- 126155-MS* pp 1–13

[17] Xiong C, Zhou F, Liu Y, Yang X, Liu X, Shi Y, Tan Y and Petrochina R 2010 Application and Study of Acid Technique Using Novel Selective Emulsified Acid System *CPS/SPE International Oil & Gas Conference and Exhibition, SPE-131216-MS* (Beijing, China: Society of Petroleum Engineers) pp 4–9

[18] Zakaria A S and Nasr-El-Din H A 2015 Application of Novel Polymer Assisted Emulsified Acid System Improves the Efficiency of Carbonate Acidizing *SPE International Symposium on Oil Field Chemistry, SPE-173711-MS* (Woodlands, USA: Society of Petroleum Engineers) pp 1–23

[19] Cairns A J, Al-muntasheri G A, Sayed M, Services A and Aramco C 2016 Targeting Enhanced Production through Deep Carbonate Stimulation : Stabilized Acid Emulsions *SPE International Conference & Exhibition on Formation Damage Control, SPE-178967-MS* (Lafayette, Louisiana, USA: Society of Petroleum Engineers) pp 1–24

[20] Sayed M A, Zakaria A ., Nasr-El-Din H A, Holt S and Al-Malki H 2012 Core Flood Study of a New Emulsified Acid with Reservoir Cores *SPE Int. Prod. Oper. Conf. Exhib. SPE-157310-MS* 1–17

[21] Anon Introduction to Corrosion Coupons and Weight Loss Analysis

[22] Samson A O 2015 *Assessment of Corrosion Rate of Dissimilar Welded Metals : Gravimetric Analysis*

[23] Jim Clark 2002 The Effect of Temperature on Reaction Rates *chemguide*

[24] Shambhu V B, Bhattacharya T K, Nayak L K and Das S 2013 Studies on Characterization of Raw Jatropha Oil and its Biodiesels with Relevance of Diesel *Int. J. Emerg. Technol. Adv. Eng.* **9001** 48–54

[25] Salhin A, Ali M and Abdurrhman A M 2013 Determination of Free Fatty Acids in Palm Oil Samples by Non-Aqueous Flow Injection Using Colorimetric Reagent *Chem. Mater. Eng.* **1** 96–103

[26] Anon 2014 *Determination of acid number of palm oil*

IOP Conf. Series: Earth and Environmental Science **164** (2018) 012007 doi:10.1088/1755-1315/164/1/012007

# Kinetic study on oil palm biomass conversion to levulinic acid via indium trichloride-ionic liquids

**Y W Tiong[1], C L Yap[1*], S Gan[2], W S P Yap[1]**

[1]Faculty of Science, University of Nottingham Malaysia Campus, Jalan Broga, 43500 Semenyih, Selangor, Malaysia
[2]Department of Chemical and Environmental Engineering, Faculty of Engineering, University of Nottingham Malaysia Campus, Jalan Broga, 43500 Semenyih, Selangor, Malaysia

Email: *Chiew-Lin.Yap@nottingham.edu.my

**Abstract.** This study investigates the kinetics of depolymerisation reaction of oil palm mesocarp fiber biomass conversion to levulinic acid via Bronsted-Lewis acidic ionic liquids catalyst. An eco-friendly catalyst was developed by coupling Lewis acid indium trichloride with a noncorrosive acidic mono-alkylated ionic liquids, 1-methylimidazolium hydrogen sulfate, for the production of levulinic acid in a greener perspective. The kinetic experiments were conducted at mild temperatures ranged from 135 to 175°C at varying time of 1.5 to 5h. A kinetic model based on the pseudo first-order reaction was established to describe the behaviour of catalytic LA production. The activation energy of the depolymerisation reaction was calculated to be 61.8 kJ $mol^{-1}$, which is comparatively lower than the previous reported values via various catalytic means of LA production from various biomass feedstock.

## 1. Introduction

Lignocellulosic biomass, such as agricultural waste, has become a particularly appropriate bioenergy feedstock due to its abundance, low cost and environmental benefits. As Malaysia has as a rich oil palm industry, a tremendous amount of oil palm biomass residues is generated each year. For instance, a palm fruit could generate up to 11% of mesocarp fiber (OPMF) biomass when the oil has been extracted [1]. This cellulosic biomass can be depolymerised into various chemicals, such as sugar, furfural, 5-hydroxymethylfurfural (5-HMF) and levulinic acid (LA) [2]. Among these, LA has been identified as a versatile intermediate chemical with dual functional groups, i.e. acidic carboxyl group (COOH) and ketone carbonyl group (C=O), which provide a wide range of functionality and reactivity for the conversion into various high end chemicals [3]. Specifically, LA can be upgraded into a variety of biofuel molecules such as 2-methylfuran,γ-valerolactone, valeric esters and levulinate esters [4]. Nonetheless, the economic ways of producing LA from biomass feedstock is yet to advance to a larger scale and hence is of high interest to researchers.

Conventionally, the conversion of LA from biomass can be enhanced by the addition of mineral acids or heterogenous acid catalysts. However, heterogenous acid catalysts are naturally unstable due to its rapid deactivation and mineral acids encounter corrosion issues [5]. These limitations drive the researchers to seek for alternative catalysts such as ionic liquids (ILs). ILs is a new generation of reaction media which is green and could serve in various applications, particularly in the processing of energy conversion materials by primarily disrupting the hydrogen bonds between the molecules [6].

Their negligible vapour pressure, non-flammable, reduce energy consumption in processing and recyclable, form the basis of being classified as "green" dual solvent-catalysts [7].

The present work focuses on the kinetic study of LA production from OPMF biomass conversion via ILs. A Bronsted-Lewis acidic ILs (BLAILs), which made of an eco-friendly Lewis acid indium trichloride ($InCl_3$) coupled with a noncorrosive acidic mono-alkylated ILs, 1-methylimidazolium hydrogen sulfate ([HMIM][$HSO_4$]), was used to catalyse the depolymerisation reaction of OPMF to LA. A kinetic model was established to depict the experimental data, alongside with the determination of the rate constant and activation energy of the reaction.

## 2. Materials and Methods

### *2.1. Materials*
OPMF sample was collected at Seri Ulu Langat Palm Oil Mill, Dengkil, Selangor. The ILs, [HMIM][$HSO_4$], was purchased from Sigma-Aldrich, USA, with a purity of ⩾95%. A standard analytical grade of LA (99%) and indium (III) chloride tetrahydrate, $InCl_3 \cdot 4H_2O$ (97%) were also purchased from Sigma-Aldrich, USA. Analytical grade ethyl acetate (RCI Labscan) was supplied by EOS Scientific Limited (Selangor, Malaysia).

### *2.2. Characterisation of OPMF composition*
The compositions of OPMF were determined by thermal gravimetric analysis (TGA) using a Mettler Toledo TGA/DSC1, followed by cellulose and hemicellulose determination analysis according to the method reported by Teramoto et al. [8]. For TGA analysis, the biomass samples were heated in a platinum cell with a scanning rate of 10°C/ min, from 30 to 700°C, under a nitrogen flow rate of 20 ml $min^{-1}$ [9].

### *2.3. Kinetic experimental study: Depolymerisation reaction*
Depolymerisation reactions were carried out at different temperatures (135 to 175°C), at a fixed ILs-to-biomass ratio of 6.6:1 (w/w) and water content of 22.7% (w/w), at varying reaction time (1.5 to 5h), in a pressure vessel (Chemglass, USA). Upon completion, the reaction mixture was isolated and analysed.

### *2.4. Products isolation*
The reaction mixture was extracted three times with 1:1 (v/v) ethyl acetate and transferred into a centrifuge tube. The solution was centrifuged at 2,000 rpm for 5 min. This resulted the formation of a two-layered solution, i.e., the upper layer of organic phase and the bottom layer of aqueous phase. The organic liquid layer was withdrawn from the upper layer and evaporated to dryness at 50°C for 30 min using a rotary evaporator (Heidolph, Germany). The content was reconstituted to 2 ml of ethyl acetate, and further diluted with ethyl acetate whenever necessary. The sample was then proceed for GC-MS analysis.

### *2.5. Products analysis*
LA content was determined by using GC (Perkin Elmer Clarus 680) equipped with a MS detector and fused silica capillary column (HP-FFAP, 30 m x 0.32 mm x 0.25 μm). The autosampler injection volume was 1.0 μl in the split mode ratio of 50:1. The injection port was maintained at 240°C, while the operating temperature for oven was initially set at 90°C for 3 minutes, then gradually increased to 230°C at a rate of 10°C/min, and held at the final temperature for 10 minutes. A solvent delay of 2 minutes was applied. Helium was used as carrier gas at a constant flow of 1.0 ml $min^{-1}$. LA content was identified with GC-MS library system (NIST MS search library) and further confirmed with injection of standard solution. The quantitative analysis of LA was conducted with a 6-point calibration curve ($R^2 > 0.99$), by using an internal standard of butyric acid (>99+%, Sigma-Aldrich, USA). The standard deviations (σ) of triplicate data obtained for all yields were ranged from 0.9-4.2%

with average at ±3%. The actual yield, theoretical yield and efficiency of LA were calculated by Equations (1)-(3):

$$\text{LA yield (\%)} = \frac{\text{Amount of LA}}{\text{Weight of initial biomass loading}} \text{ x } 100\% \quad (1)$$

$$\text{Theoretical LA yield (\%)} = \frac{\text{Cellulose loading in biomass x } 0.71}{\text{Initial biomass loading}} \text{ x } 100\% \quad (2)$$

$$\text{where } \frac{\text{Molecular weight of LA}}{\text{Molecular weight of cellulose}} = \frac{116}{162} = 0.71$$

$$\text{LA efficiency (\%)} = \frac{\text{LA yield}}{\text{Theoretical LA yield}} \text{ x } 100\% \quad (3)$$

where efficiency refers to the efficiency of biomass conversion to LA calculated based on cellulose content.

## 3. Results and discussions

### *3.1. Characterisation of OPMF compositions*

The thermal degradation of TGA gave three distinct stages of weight losses for both samples. The weight losses are due to the evaporation of residual water, volatilization of holocellulose and degradation of lignin. The TGA analysis, followed by cellulose and hemicellulose characterisation studies revealed that OPMF contains 4.31% moisture, 16.29% lignin, 23.12% ash, 29.96% cellulose and 26.32% hemicellulose [9].

### *3.2. Kinetic study*

The conversion of OPMF biomass to LA is a multi-step process which involves various intermediate products. The cellulose fraction (glucan) is depolymerised to glucose, which further dehydrated to 5-HMF and finally rehydrate into LA [10]. In this study, a simplified model was proposed based on sole LA efficiency, which exclude the role of intermediate products. A pseudo-first order irreversible reaction was established by assuming glucan was directly depolymerised into LA. The model is proposed as follows:

$$\text{Glucan (G)} \rightarrow \text{Levulinic acid (LA)} \quad (4)$$

Figure 1(a) illustrates the results of LA efficiency at different reaction temperature and time. The 5-HMF concentrations measured were low for all conditions studied, presumably the compound had been converted to LA. Some trace amount of by-products were not computed, as the possible side reactions were considered negligible. The LA efficiency was increased by increasing temperature and prolonging reaction time. This results were in the agreement with Ramli and Amin [11].

As shown in Figure 1(b), the line graph between –ln (1-X) and time at five different temperatures support the hypothesis of pseudo-first order reaction. Table 1 summaries the rate constant (*k*) and the corresponding correlation coefficients ($R^2$) of linear fitting for the temperatures studied. The activation energy (*Ea*) and exponential factor (*A*) were obtained by applying the Arrhenius equation as follows.

$$\ln k = \ln A - \frac{\mathbf{Ea}}{\mathbf{RT}} \quad (5)$$

where *k* is the rate constant, represents the number of effective collisions among the reactants, *A* is the exponential factor, represents the total number of collisions, R is the universal gas constant and *Ea* is the required activation energy.

Figure 1(c) shows strong linearity ($R^2$ = 0.99) between ln *k* and 1/T for the range of temperature studied. The *Ea* determined from the slope of the line was 61.8 kJ $mol^{-1}$, while *A* calculated from the intercept was 5.9 $x10^6$ $min^{-1}$. As such, the Arrhenius equation for this ILs catalytic system for the production of LA from OPMF biomass conversion could be written as:

$\ln k = 15.59 - 7.43 / T$

where $k$ is the rate constant ($min^{-1}$) and T is the temperature (K).

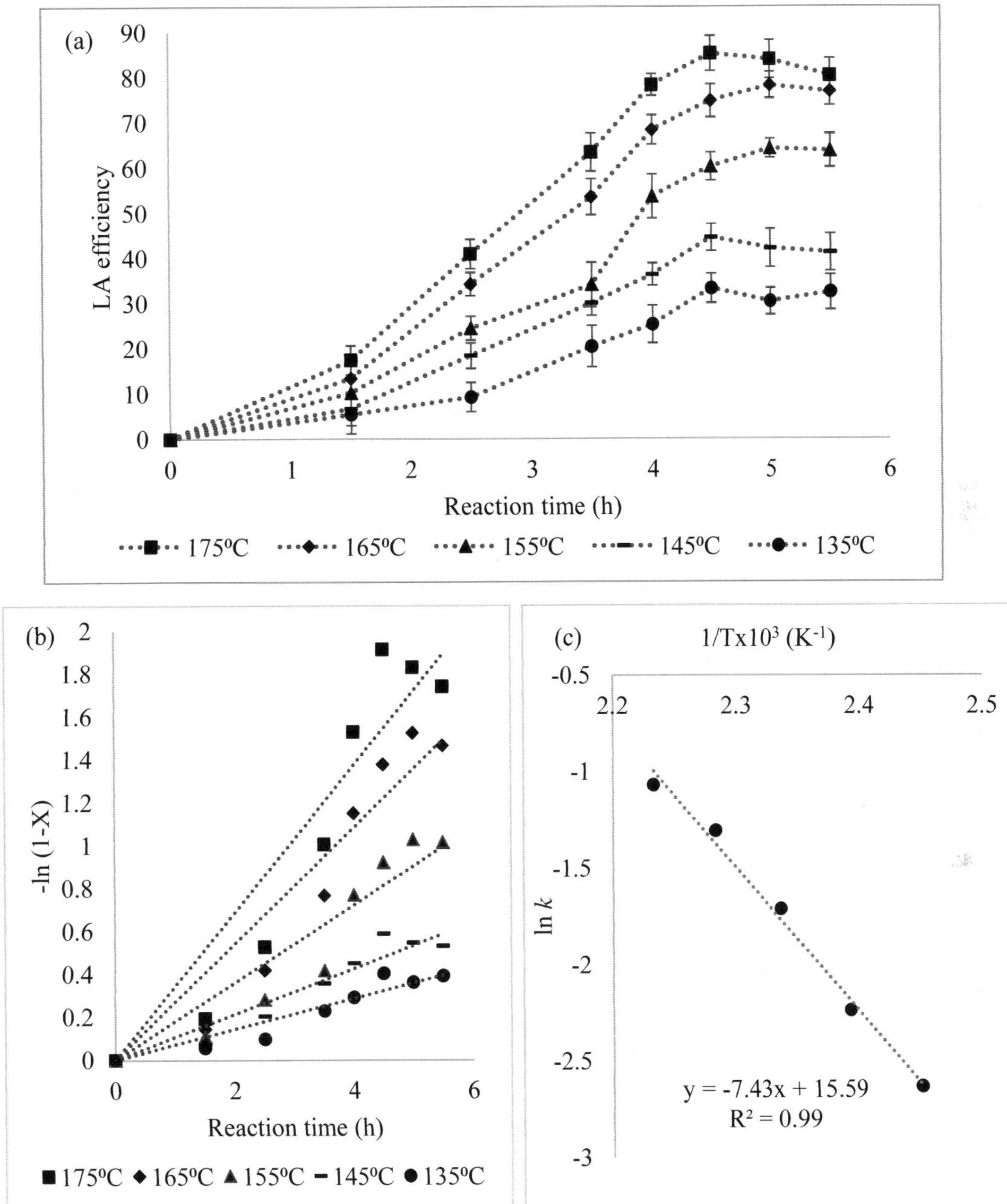


**Figure 1.** (a) LA efficiency from OPMF biomass conversion via ILs, (b) -ln (1-X) versus reaction time at different reaction temperatures, (c) Arrhenius plot of ln $k$ versus 1/T x $10^3$.

IOP Conf. Series: Earth and Environmental Science **164** (2018) 012007 doi:10.1088/1755-1315/164/1/012007

**Table 1.** Rate constant (*k*) and correlation coefficients ($R^2$) for the LA efficiency at different temperatures.

| Temperature (°C) | Rate constant, *k* ($min^{-1}$) | Correlation coefficient, $R^2$ |
|---|---|---|
| 135 | 0.0720 | 0.9072 |
| 145 | 0.1068 | 0.9222 |
| 155 | 0.1806 | 0.8915 |
| 165 | 0.2711 | 0.9131 |
| 175 | 0.3436 | 0.8924 |

The *Ea* and *A* obtained in the present study were lower than those reported for catalytic conversions of various biomass to LA via metal salts ($FeCl_3$) [10,12] and mineral acids ($H_2SO_4$) [13-15]. This suggested $InCl_3$-[HMIM][$HSO_4$] catalytic system has the immense potential to be used in a large scale sustainable LA production from renewable biomass feedstock.

## 4. Conclusions

OPMF biomass was converted to LA via $InCl_3$-[HMIM][$HSO_4$] for a greener perspective. The kinetic study suggested that the $InCl_3$-[HMIM][$HSO_4$] catalysed depolymerisation was a pseudo first-order reaction with a relatively lower *Ea* of 61.8 kJ $mol^{-1}$. A kinetic model was proposed to predict the LA efficiency for the conversions.

## 5. References

[1] Nordin N I A A, Ariffin H, Andou Y, Hassan M A, Shirai Y, Nishida H, Yunus, W M Z W, Karuppuchamy S and Ibrahim N A 2013 *Molecules* **18** 9132
[2] Chang C, Xu G and Jiang X 2012 *Bioresource Technology* **121** 93
[3] Morone A, Apte M and Pandey R A 2015 *Renewable and Sustainable Energy Reviews* **51** 548
[4] Koivisto E, Ladommatos N and Gold M 2015 *Energy & Fuels* **29** 5875
[5] Mukherjee A, Dumont M J and Raghavan V 2015 *Biomass and Bioenergy* **72** 143
[6] Eshetu G G, Armand M, Ohno H, Scrosati B and Passerini S 2016 *Energy & Environmental Science* **9** 49
[7] Badgujar K C and Bhanage B M 2015 *Bioresource Technology* **178** 2
[8] Teramoto Y, Lee S H and Endo T 2009 *Bioresource Technology* **100** 4783
[9] Tiong Y W, Yap C L, Gan S, and Yap W S P 2017 *Chemical Engineering Transactions* **56** 1021
[10] Dussan K, Girisuta B, Haverty D, Leahy J J and Hayes M H B 2013 *Bioresource Technology* **149** 216
[11] Zheng X, Zhi X, Gu X, Li X, Zhang R and Lu X 2017 *Fuel* **187** 261
[12] Ramli N A S and Amin N A S 2016 *Chemical Engineering Journal* **283** 150
[13] Zhi Z, Li N, Qiao Y, Zheng X, Wang H and Lu X 2015 *Industrial Crop and Products* **76** 672
[14] Chang C, Ma X and Cen P 2009 *Chinese Journal of Chemical Engineering* **17** 835
[15] Girisuta B, Dussan K, Haverty D, Leahy J J and Hayes M H B 2013 *Chemical Engineering* Journal **217** 61

## Acknowledgments

This work was financially supported by Ministry of Higher Education (MOHE), Malaysia, under the Fundamental Research Grant Scheme (FRGS), Grant No. NFHS0001. The University of Nottingham Malaysia Campus is acknowledged for its support towards this project.

IOP Conf. Series: Earth and Environmental Science **164** (2018) 012008 doi:10.1088/1755-1315/164/1/012008

# Numerical simulations for biomass burning aerosol transport in northern Thailand

**Nurzahziani[1] and C. Surussavadee[2]**

[1]Interdisciplinary Graduate School of Earth System Science and Andaman Natural Disaster Management, Prince of Songkla University, Phuket Campus, Phuket, Thailand.
[2]Telecommunication Engineering Department, Faculty of Engineering, King Mongkut's Institute of Technology Ladkrabang, Bangkok, Thailand.

zahziani@gmail.com

**Abstract.** This paper presents the evaluation of a numerical biomass burning aerosol transport simulation system developed for northern Thailand. The simulation system is composed of the Weather Research and Forecasting model coupled with Chemistry (WRF-Chem) employing its best physics options, global gridded analyses from the NCEP FNL (Final) Operational Global Analysis data, global emission data from the REanalysis of the TROpospheric chemical composition over the past 40 years (RETRO) and the Emission Database for Global Atmospheric Research (EDGAR), and biomass burning emission data from the Brazilian Biomass Burning Emissions Model (3BEM). The simulated hourly averaged particulate matters with diameters less than 10 micron (PM10) at 5-km resolution covering the 3-month period of February – April of 2015 were evaluated using measurements from 13 ground stations distributed in northern Thailand. Results show that the numerical biomass burning aerosol transport simulation system can predict the times when hourly averaged PM10 dry mass concentrations are high and low well. Simulations are negatively biased for concentrations above 200 $\mu g/m^3$ and are slightly positively biased otherwise. The simulations have good utility for concentrations above 100 $\mu g/m^3$ and are useful for all concentration ranges.

## 1. Introduction

Air pollution caused by agricultural burning and forest fires is one of the most crucial environmental issues in the northern region of Thailand [1-3]. During the dry season, the smoke haze situation severely affects the air quality in many areas in northern Thailand, including the provinces of Chiang Mai, Chiang Rai, Mae Hong Son, Lampang, Lamphun, Phrae, Nan, and Phayao [4], resulting in adverse health consequences [1]. The level of particulates matter (PM) concentrations in these provinces have been automatically measured at the Air Quality Monitoring (AQM) stations administered by the Pollution Control Department (PCD) of Thailand [5]. During the smoke and haze period, the levels of particulate matters with diameters less than 10 micron (PM10) measured at provincial stations rise above the standard level, i.e, 120 $\mu g/m^3$, set by PCD mostly in February and March of each year [6].

To better understand the impacts of biomass burning aerosols in northern Thailand, an accurate numerical biomass burning aerosol transport simulation system is required. Such system can provide details about the amounts and coverages of biomass burning aerosols as a function of time. Such information is useful for mitigating the impacts. Although it is a very important problem with big public impacts in northern Thailand, previous studies concerning biomass burning aerosol transport simulations in northern Thailand do not exist.

The main goal of this paper is to develop and evaluate a numerical biomass burning aerosol transport simulation system for northern Thailand. [7] has found the best physics options of the Weather Research and Forecasting (WRF) model for high-resolution weather forecasting for Thailand and nearby regions and has shown that the weather forecasts agree well with satellite observations. [8] and [9] have employed the WRF model to simulate and forecast near surface wind speeds and directions at 65 and 90 m above ground for wind stations in northeastern Thailand, respectively, and has shown that simulations and forecasts agree well with ground measurements.

The WRF model coupled with Chemistry (WRF-Chem) [10-11] is employed in this study. The biomass burning emission data from the Brazilian Biomass Burning Emissions Model (3BEM) [12] are used. Simulated hourly averaged PM10s covering a period of 3 months in year 2015 were evaluated using measurements from 13 ground stations distributed in northern Thailand.

Section II describes the method employed in this study, including details about ground measurements and the numerical biomass burning aerosol transport simulation system used in this study. Section III shows the results. Section IV summarizes and concludes the paper.

## 2. Methodology

### *2.1. Ground Measurements*

Hourly-averaged dry mass concentrations of PM10 employed in this study were collected by 13 PCD ground stations in northern Thailand. Figure 1 shows the study area and locations of the 13 ground stations. The PM10 concentrations are generally high in February and March of each year, which is in the cold season and the air is dry. In addition, the biomass burning activities in the 2 months are very active. The biomass burning is an agricultural practice for preparing the soil for the coming rainy season and a new rice planting season [13].

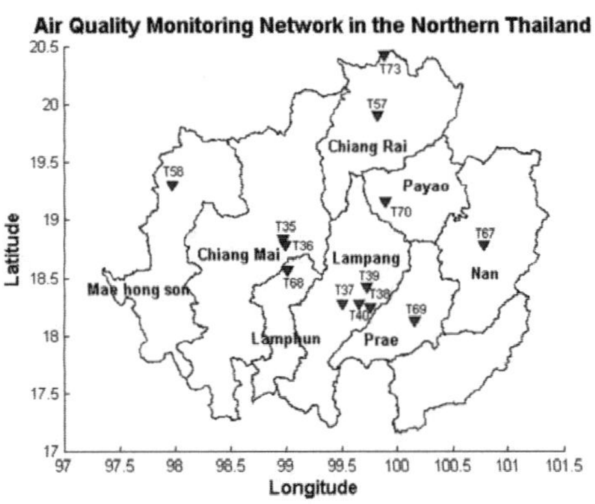


**Figure 1.** Locations of 13 ground stations in northern Thailand empoyed in this study.

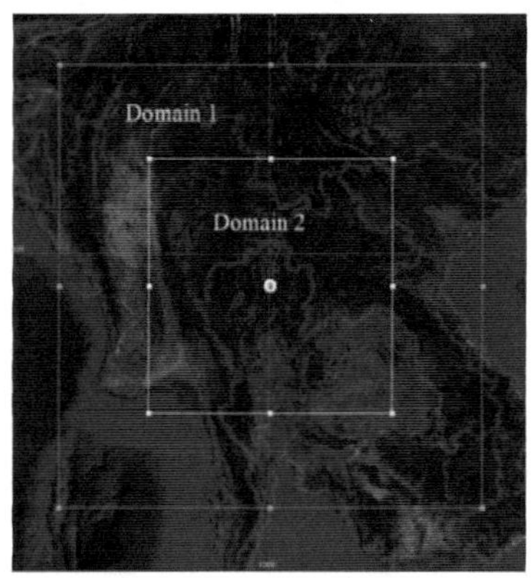


**Figure 2.** Two co-centered WRF domains employed in this study.

### *2.2. The Numerical Biomass Burning Aerosol Transport Simulation System*

To simulate the biomass burning aerosol transport in this study, WRF-Chem version 3.5.1 was employed. The WRF domain configurations are composed of 2 domains co-centered at 18.97°N and 99.4°E as shown in the Figure 2. The domains were selected such that domain 2 covers northern Thailand. The domains 1 and 2 have 99x99 and 120x120 grid points with 15 and 5 km resolutions, respectively. Both domains have 45 terrain-following levels extending to 50 mb. WRF physics options employed in this study follow those successfully used in [7-9]. The 1-degree gridded NCEP Final

Analysis every 6 hours were used as initial and boundary conditions for the WRF model. The global emission data from the REanalysis of the TROpospheric chemical composition over the past 40 years (RETRO) and the Emission Database for Global Atmospheric Research (EDGAR) [12], and the daily biomass burning emission data from 3BEM were used in this study.

The biomass burning aerosol transport for the study area was simulated continuously starting from January 30th and to April 30th of 2015 to ensure that chemical variables in the model integrate through the period. The 3BEM emission data were fed into the simulations daily. Simulations for the first 2 days were treated as a spin-up period and were not used. WRF-Chem simulated PM10 at 5-km resolution were used in this study. To compute PM10 at the ground stations, horizontally linear interpolation followed by vertically linear interpolation between neighboring levels were used.

## 3. Results

Table 1 shows mean errors (E[Simulations – Observations]; MEs), root mean square errors (RMSEs), and correlation coefficients (CCs) of simulated hourly averaged PM10 dry mass concentrations ($\mu g/m^3$) for each ground station during February–April, 2015. MEs for most stations except stations T35, T36, and T40, are less than 3 $\mu g/m^3$. The station with maximum RMSE is T58. Correlations between simulations and observations are quite good for most stations, where CCs are ranged from 0.4 to 0.69.

Since values of hourly averaged PM10 dry mass concentrations cover a wide range, Table 2 shows MEs and RMSEs of simulated hourly averaged PM10 dry mass concentrations for all 13 ground stations during February–April, 2015 computed separately for different concentration ranges divided using observations. RMSEs shown in boldface are those below the minimum of the concentration range listed in the first column, which indicate good utility. Simulated PM10 dry mass concentrations are useful for all ranges and have good utility for concentrations above 100 $\mu g/m^3$. Simulations are slightly positively biased for concentrations below 200 $\mu g/m^3$ and are negatively biased otherwise.

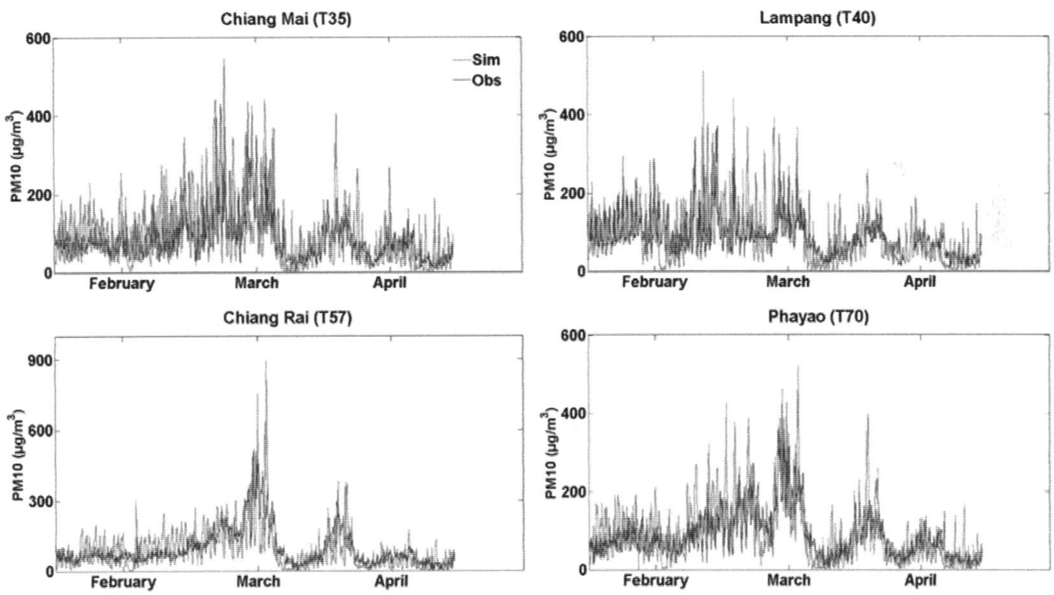


**Figure 3.** Comparisons of simulated (red) and observed (blue) hourly averaged PM10 dry mass concentrations ($\mu g/m^3$) for 4 ground stations during February–April, 2015.

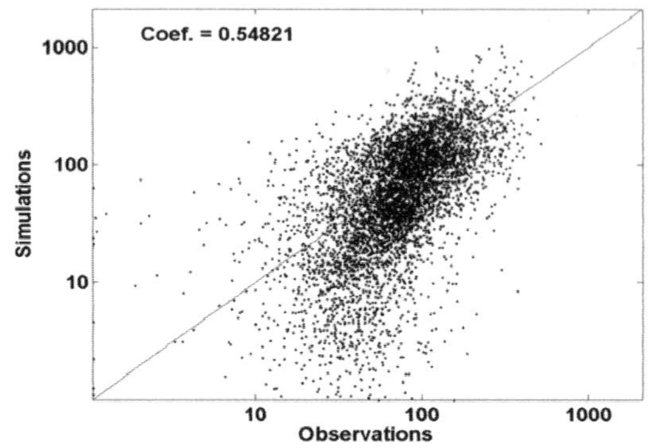


**Figure 4.** Scatterplots between simulated and observed hourly averaged PM10 dry mass concentrations (μg/m$^3$) for all 13 ground stations during February–April, 2015.

Figure 3 compares simulated and observed hourly averaged PM10 dry mass concentrations for the 4 ground stations for the full 3-month period of February-April 2015. PM10 concentrations from the end of February to the middle of March at most stations are relatively high and exceed the value of 50 μg/m$^3$, which is a threshold for polluted air quality according to the World Health Organization [14]. Simulated PM10 concentrations can predict lows and highs of observed PM10 concentrations quite well for all stations for the full 3 months.

Figure 4 shows scatterplots between simulated and observed hourly averaged PM10 dry mass concentrations for 13 ground stations for the full 3-month period of February-April 2015. Simulations agree well with observations. The overall correlation coefficient between simulations and observations is 0.55. The scatterplot is consistent with results shown in Table 2, that is, simulations are biased lower than observations for observed concentrations below 200 μg/m$^3$, and are biased higher otherwise.

**Table 1.** MEs (E[Simulations – Observations]), RMSEs, and CCs of Simulated Hourly Averaged PM10 Dry Mass Concentrations (μg/m$^3$) for Each Ground Station during February–April, 2015

| Stations | ME | RMSE | CC |
|---|---|---|---|
| T35 | 5.97 | 78.76 | 0.41 |
| T36 | -5.05 | 67.25 | 0.50 |
| T37 | 2.46 | 76.93 | 0.44 |
| T38 | 2.91 | 50.91 | 0.60 |
| T39 | 1.01 | 63.72 | 0.63 |
| T40 | -3.16 | 63.09 | 0.52 |
| T57 | -2.70 | 74.19 | 0.64 |
| T58 | 2.45 | 130.20 | 0.58 |
| T67 | 2.52 | 91.86 | 0.43 |
| T68 | 2.90 | 69.26 | 0.40 |
| T69 | -2.20 | 53.79 | 0.53 |
| T70 | 2.68 | 59.59 | 0.69 |
| T73 | 0.83 | 92.97 | 0.64 |

**Table 2.** MEs (E[Simulations – Observations]) and RMSEs of Simulated Hourly Averaged PM10 Dry Mass Concentrations (μg/m$^3$) for all 13 Ground Stations during February–April, 2015 for Different Observed Concentration Ranges

| Range (μg/m$^3$) | ME | RMSE |
|---|---|---|
| 0-100 | 3.06 | 54.61 |
| 100-200 | 2.76 | **95.72** |
| 200-300 | -28.67 | **168.71** |
| 300-400 | -72.78 | **201.40** |
| >500 | -148.00 | **214.16** |

Boldface: RMSEs below the minimum listed in column 1, indicating good utility.

## 4. Summary and Conclusions

A numerical biomass burning aerosol transport simulation system developed to be used for northern Thailand is evaluated. The system is composed of the WRF-Chem model with its best physics options, global gridded analyses from the NCEP Final Analysis used for initial and boundary conditions, the RETRO and EDGAR global emission data, and the 3BEM biomass burning emission data. Simulated hourly averaged PM10 dry mass concentrations at 5-km resolution for the full 3-month period of February-April 2015 were evaluated using measurements from 13 ground stations distributed in northern Thailand. The numerical biomass burning aerosol transport simulation system can simulate the times of highs and lows of hourly averaged PM10 dry mass concentrations well. Simulations have small positive biases for concentrations below 200 μg/m$^3$ and have negative biases for concentrations above 200 μg/m$^3$. The numerical biomass burning aerosol transport simulation system can provide useful simulated hourly averaged PM10 dry mass concentrations for all ranges of observed concentrations. The simulations have good utility for concentrations above 100 μg/m$^3$.

## 5. Acknowledgment

The authors with to thank the Pollution Control Department of Thailand for providing PM10 measurements, NCAR for making NCEP FNL data publicly available, the National Institute for Space Research (INPE) for providing emission data, and the Interdisciplinary Graduate School of Earth System Science and Andaman Natural Disaster Management, Prince of Songkla University, Phuket Campus, Thailand, for funding support.

## References

[1] Wiwanitkit V 2008 PM10 in the atmosphere and incidence of respiratory illness in Chiangmai during the smoggy pollution *Stoch. Environ. Res. Risk. Assess.* **22** 437–440.

[2] Amnauylawjarurn T, Kreasuwun J, Towta S and Siriwitayakorn 2010 Dispersion of particulate matter (PM10) from forest fires in Chiang Mai province, Thailand *Chiang Mai J. Sci.* **37(1)** 39-47.

[3] Kanabkaew T 2013 Prediction of hourly particulate matter concentrations in Chiang Mai, Thailand using MODIS aerosol optical depth and ground-based meteorological data *EnvironmentAsia* **6(2)** 65-70.

[4] Sooktawee S and Humphries U W 2015 Visualization and interpretation of PM10 monitoring data related to causes of haze episodes in Northern Thailand *App. Envi. Res.* **37(2)** 33-48.

[5] Chantara S PM10 and its chemical composition: a case study in Chiang Mai, Thailand, air quality - monitoring and modeling *accessed February 18, 2017*, available at http://www.intechopen.com/books/air-quality-monitoring-and-modeling/pm10-and-its-chemical-composition-acase-study-in-chiang-mai-thailand.

[6] Sirimongkonlertkul N, Upayokhin P, and Phonekeo V 2013 Multi-temporal analysis of haze problem in Northern Thailand: a case study in Chiang Rai province *Kasetsart J. (Nat. Sci.)* **47** 768-780.

[7] Surussavadee C and Aonchart P 2013 Evaluation of WRF physics options for high-resolution weather forecasting in Tropics using satellite passive millimeter-wave observations *Proc. IEEE Intern. Geosc. Remote Sens. Symp.* 2013 (Melbourne, Australia) p 2262-2265.

[8] Surussavadee C 2017 Evaluation of WRF Near-Surface Wind Simulations in Tropics Employing Different Planetary Boundary Layer Schemes *Proc. of 8th Int. Renewable Energy Congress* (Amman, Jordan) p 1-4.

[9] Surussavadee C 2017 Evaluation of Tropical Near-Surface Wind Forecasts Using Ground Observations *Proc. of 8th Int. Renewable Energy Congress* (Amman, Jordan) p 1-4.

[10] Grell G A, Peckham S E, Schmitz R. McKeen S A, Frost G, Skamarock W C and Eder B 2005 Fully coupled "online" chemistry within the WRF model *Atmos. Environ.*, **39** pp. 6957–75.

[11] Fast J D, Gustafson Jr W I, Easter R C, Zaveri R A, Barnard J C, Chapman E G, Grell G A and Peckham S E 2006 Evolution of ozone, particulates, and aerosol direct radiative forcing in the vicinity of Houston using a fully coupled meteorology–chemistry–aerosol model *J. Geophys. Res.* **111** D21305.

[12] Freitas S R, Longo K M, Alonso M F, Pirre M, Marecal V, Grell G, Stockler R, Mello R F and Sanchez Gacita M 2011 PREP-CHEM-SRC 1.0: a preprocessor of trace gas and aerosol emission fields for regional and global atmospheric chemistry models *Geosci. Model Dev.* **4** 419433.

[13] Duc H N, Bang H Q and Quang N X 2016 Modelling and prediction of air pollutant transport during the 2014 biomass burning and forest fires in peninsular Southeast Asia *Environ. Monit. Assess.* **188** 188:106.

[14] World Health Organization (WHO) 2006 WHO Air quality guidelines for particulate matter, ozone, nitrogen dioxide and sulfur dioxide: Summary of risk assessment WHO/SDE/PHE/OEH/06.02.

IOP Conf. Series: Earth and Environmental Science **164** (2018) 012009 doi:10.1088/1755-1315/164/1/012009

# An investigation of the vertical diffuse efficacy of solar irradiance for Thailand's climate

**Sutiwat Umar and Pipat Chaiwiwatworakul**

The Joint Graduate School of Energy and Environment, CHE Center for Energy Technology and Environment, King Mongkut's University of Technology Thonburi, 126 Pracha-Uthit Road, Bangmod, Tungkru, Bangkok 10140 Thailand.

sutiwatomar@gmail.com

**Abstract.** This paper presents a study result of the luminous efficacy of diffuse solar irradiance on vertical plans using records from a daylight measurement station located in Bangkok, Thailand. Hourly mean values of the vertical diffuse efficacy are first introduced. Variations of the efficacy are also presented for different sky conditions. Our results indicate that the tropical diffuse efficacy was higher than those in high latitude regions. In this study, the authors proposed a method for the efficacy prediction. The proposed method was evaluated its performance against some selected existing models. The results show that the authors' method predicted well the tropical efficacy value and outperformed other evaluated models with their original coefficients.

## 1. Introduction

Daylighting is perceived to be a key measure that makes buildings towards environmental and energy sustainability. Appropriate daylighting technologies help reduce substantial electrical energy for both lighting and air-conditioning in buildings particularly those in tropical region where the sky is highly luminous and daytime is long throughout a year [1-2]. Different from practical electric lighting for indoor illumination, the daylighting is more complicate as amount of light from its sources varies in natural with geography and outdoor climate. Hence, daylight data presents its essence for establishing solutions in order to reap a full benefit from the available daylight in a location. Unfortunately, the daylight data is still scarce today, as the routine observation program at most meteorological station includes only solar irradiance. An alternative method to produce illuminance data is via the use of luminous efficacy. The efficacy can be thought simply of a multiplicative factor of the solar irradiance. Not a constant, the efficacy value varies with prevalent climate condition. A number of researches have been conducted in Europe and North America to establish the accurate efficacy models [3-5]. The main objective of this paper is to investigate the luminous efficacy of solar irradiances on vertical planes of the north, east, south, and west in Thailand. To fulfil the objective, the specific activities carried out in the study include: (i) characterization of the luminous efficacy of the vertical diffuse solar irradiances, (ii) modeling of the vertical diffuse efficacy and (iii) performance evaluation of the developed efficacy models against some selected existing models. At present, the scope of this efficacy study was limited only to the diffuse component of the solar irradiance.

## 2. Daylight and solar irradiance measurement

A daylight measurement station was established under a graduate program in Energy Division of the Joint Graduate School of Energy and Environment (JGSEE), King Mongkut's University of Technology, Thonburi (KMUTT), Thailand. The station is located on a roof deck of a seven-story building in Bang Khun Tien campus of the university. The campus is situated in southern suburban area of Bangkok (latitude 13.58°N and longitude 100.44°E). The station has measured solar irradiance and daylight illuminance of global, diffuse horizontal and beam normal components; including total irradiance and illuminance on vertical planes facing the four cardinal orientations (N, E, S, and W). All the mentioned data have been recorded every five minutes.

## 3. Characteristics of the tropical vertical efficacy

The hourly data aggregated from the five-minute illuminance and irradiance records in year 2015-2016 from the station were used for the analysis. Values of diffuse efficacy ($K_d$) were derived by dividing the hourly values of diffuse illuminance (lux) with its corresponding diffuse irradiance (W/m$^2$). This section presents results from statistical analysis of the vertical diffuse efficacy data. Correlations between the diffuse efficacies and sky conditions will be presented next.

Hourly mean values of the vertical diffuse efficacy were analysed for four cardinal orientations. However, in this paper, only that of the south surface was presented as shown in Table 1. It can be observed that the diffuse efficacy on the south varied from about 107.1 lm/W to 172.5 lm/W. The efficacy values were relatively high during noon. The efficacy was also high during November-February when the sky was rather clear compared to other period. The yearly average of diffuse hourly mean value was approximately 138.6 lm/W.

**Table 1.** Hourly mean values of the *south* diffuse efficacy (lm/W) by calendar months

| Time | Jan | Feb | Mar | Apr | May | Jun | Jul | Aug | Sep | Oct | Nov | Dec |
|---|---|---|---|---|---|---|---|---|---|---|---|---|
| 6-7 | | | 126.40 | 150.89 | 129.86 | 125.62 | 130.27 | 127.41 | 136.01 | 134.85 | 125.10 | 147.06 |
| 7-8 | 129.78 | 126.22 | 135.53 | 167.16 | 130.03 | 129.05 | 130.66 | 134.50 | 144.07 | 153.24 | 140.07 | 136.86 |
| 8-9 | 144.50 | 157.78 | 165.33 | 148.67 | 128.52 | 128.15 | 129.56 | 133.57 | 151.12 | 161.24 | 146.46 | 138.28 |
| 9-10 | 150.34 | 161.10 | 172.52 | 144.58 | 128.11 | 127.53 | 129.48 | 136.14 | 154.12 | 154.51 | 142.08 | 137.71 |
| 10-11 | 149.75 | 157.40 | 164.44 | 146.04 | 126.82 | 125.89 | 128.45 | 131.01 | 142.75 | 143.34 | 145.55 | 150.87 |
| 11-12 | 152.31 | 156.70 | 160.82 | 152.66 | 127.37 | 125.76 | 128.46 | 130.71 | 148.42 | 143.93 | 165.48 | 167.12 |
| 12-13 | 148.75 | 152.40 | 161.00 | 155.52 | 131.50 | 126.82 | 134.66 | 132.25 | 143.20 | 139.96 | 158.47 | 153.67 |
| 13-14 | 147.11 | 147.19 | 152.98 | 144.14 | 126.51 | 126.45 | 130.77 | 127.55 | 145.66 | 138.87 | 148.13 | 143.62 |
| 14-15 | 142.74 | 143.82 | 150.66 | 137.58 | 127.23 | 126.19 | 130.30 | 127.93 | 141.42 | 135.85 | 141.98 | 136.58 |
| 15-16 | 139.69 | 139.42 | 145.21 | 135.54 | 126.73 | 126.65 | 129.58 | 131.17 | 140.22 | 137.42 | 140.04 | 132.62 |
| 16-17 | 132.83 | 129.50 | 134.91 | 155.57 | 127.84 | 127.36 | 129.93 | 133.09 | 135.94 | 137.67 | 134.54 | 124.14 |
| 17-18 | 115.12 | 108.02 | 120.78 | 155.19 | 127.29 | 126.95 | 131.54 | 133.73 | 128.81 | 132.84 | 120.44 | 107.12 |
| 18-19 | | | 133.43 | 158.72 | 122.41 | 122.92 | 129.74 | 130.86 | 123.88 | | | |

## 4. Efficacy Modelling and Evaluation

In order to further the study, a method was developed to predict the tropical vertical efficacy for different sky conditions. Some existing models introduced by other researchers were chosen to evaluate against our tropical data. In this section, the authors also proposed a simple method for the vertical diffuse efficacy prediction. The authors' method was compared their prediction performance with the existing models.

Table 2 summarizes the existing efficacy models to be evaluated in this study. The first model DE1 was introduced by Muneer, Gul and Kubie [6] with a polynomial form. Ruiz, Soler and Robledo developed the three models from their studies denoted as DE2-DE4 [7, 8]. Among the three models,

DE3 and DE4 included solar altitude as their parameter input. The original coefficients of the models derived from the efficacy records in their study locations were included in Table 2.

In a step of the study, models DE1-DE4 were regressed with our tropical efficacy data to obtain the optimal local coefficient values. Table 2 also gives the local coefficient values of each model. As shown, models DE1-DE3 employed a single set of coefficients for the efficacy prediction, while model DE4 employed separate coefficient set each for a particular orientation. From comment of the earlier studies [9] that the coefficient separation offered a better prediction, the regression of the local coefficients was thus made for each of the orientations (north, east, south and west).

**Table 2.** The vertical diffuse efficacy models with their original coefficients and the local coefficient values regressed from the tropical data.

| Model | Functional form | Coefficient | North | East | South | West |
|---|---|---|---|---|---|---|
| DE1 | $K_d = a + b \cdot k_t + c \cdot k_t^2$ | Original | a = 130.2 b = 39.83 c = 49.98 | | | |
| | | Local | a = 143.6<br>b = -19.88<br>c = 35.86 | a = 138<br>b = 0.5676<br>c = 34.55 | a = 140.1<br>b = -82.98<br>c = 146.1 | a = 142.3<br>b = -80.38<br>c = 145.3 |
| DE2 | $K_d = a + b \cdot k_d + c \cdot k_d^2$ | Original | a = 160.61 b = 47.05 c = 196.94 | | | |
| | | Local | a = 160.3<br>b = -38.29<br>c = 15.39 | a = 166.8<br>b = -35.56<br>c = 5.068 | a = 184.9<br>b = -125.4<br>c = 69.58 | a = 197.2<br>b = -152.5<br>c = 86.89 |
| DE3 | $K_d = a \cdot (\sin\alpha)^b \cdot k_d^c$ | Original | a = 86.97 b = -0.143 c = 0.218 | | | |
| | | Local | a = 133.3<br>b = -0.036<br>c = -0.0837 | a = 134.7<br>b = -0.0348<br>c = -0.1087 | a = 124.9<br>b = -0.0053<br>c = -0.1604 | a = 123.4<br>b = -0.0291<br>c = -0.1927 |
| DE4 | $K_d = a \cdot (\sin\alpha)^b \cdot \Delta^c$ | Original | a = 96.04<br>b = -0.08<br>c = -0.039 | a = 112.2<br>b = -0.061<br>c = -0.009 | a = 108<br>b = 0.088<br>c = -0.171 | a = 112.63<br>b = -0.041<br>c = -0.027 |
| | | Local | a = 125.8<br>b = 0.0025<br>c = -0.0941 | a = 136.3<br>b = 0.0084<br>c = -0.0575 | a =121.4<br>b = 0.0631<br>c = -0.1206 | a = 116.9<br>b = 0.0569<br>c = -0.1611 |

The prediction performance was first evaluated by comparing the calculated efficacy values of the models using the original coefficients with that using the local coefficients through scattered plots. Fig. 1 shows exemplarily that the local model DE1 gave higher predicted efficacy values than the original model. The difference became larger when the sky was clearer (higher value of $k_t$). In order to evaluate the model prediction performance, the two statistical parameters of Mean Bias Difference (MBD) and Root Mean Square Difference (RMSD) were employed. The parameter can be defined as:

$$MBD = \sum_{i=1}^{N} (C_i - M_i) / N \tag{1}$$

$$RMSD = \left[ \sum_{i=1}^{N} (C_i - M_i)^2 / N \right]^{1/2} \tag{2}$$

where $C_i$ is the calculated value, $M_i$ is observed values and $N$ is the number of data. By their definition, RMSD gives weight to points far away from the mean. RMSD gives only positive values. MBD can give positive or negative values, corresponding to that positive value implies an overestimation of the model.

From Table 3, the original model DE1 underestimated the tropical diffuse efficacy by 14.0-22.8 lm/W while the local model slightly overestimated by 0.28-0.69 lm/W. The local model gave RMSD value of 13.36-17.63 lm/W, lower than the original model. Table 3 also shows that the original model DE2 possessed large positive MBD and RMSD values, indicating it overestimated the efficacy values and performed worse compared with the original model DE1. The comparison also exhibits that the local model DE2 performed slightly better than model DE1 for all four orientations. This might imply that $k_d$ is better for diffuse efficacy parameterization than $k_t$.

For the original model DE3, Table 3 shown that the model underestimated the tropical diffuse efficacy values. It performed worse than DE1 but better than DE2. For the local model DE3, its prediction performance was comparable with the local models DE1 and DE2; although the model included solar altitude as additional input parameter. However, the model DE3 offered the smallest discrepancy between the predicted values and the observed values. For the last model DE4, it had the functional form identical to that of model DE3, but brightness index (Δ) was used as the weather parametrization input. By observing the results in Table 3, models DE4 did not perform any significant prediction improvement.

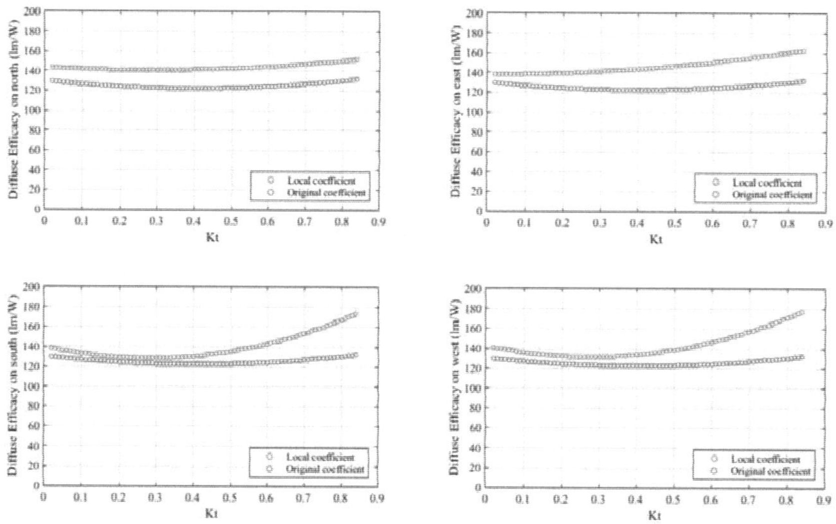


**Figure 1.** Comparison of diffuse efficacy from the original and local model DE1

**Table 3.** Evaluation of the vertical diffuse efficacy models

| Model | North | | East | | South | | West | |
|---|---|---|---|---|---|---|---|---|
| | Mean | 143.4 | Mean | 147.0 | Mean | 138.2 | Mean | 141.5 |
| | Dev | 15.5 | Dev | 21.9 | Dev | 21.1 | Dev | 22.2 |
| | MBD | RMSD | MBD | RMSD | MBD | RMSD | MBD | RMSD |
| Original coefficients | | | | | | | | |
| DE1 | -19.18 | 15.31 | -22.79 | 21.66 | -14.00 | 20.35 | -17.28 | 21.50 |
| DE2 | 75.97 | 53.95 | 72.37 | 58.13 | 81.15 | 59.97 | 77.87 | 61.49 |
| DE3 | -34.28 | 15.96 | -37.89 | 21.46 | -29.10 | 19.79 | -32.38 | 19.34 |
| DE4 | -36.43 | 15.76 | -28.73 | 22.39 | -8.15 | 19.67 | -21.37 | 22.44 |
| Local coefficients | | | | | | | | |
| DE1 | 0.44 | 13.36 | 0.28 | 20.09 | 0.69 | 15.76 | 0.59 | 17.63 |
| DE2 | 0.40 | 12.88 | 0.18 | 19.78 | 0.65 | 15.34 | 0.21 | 16.66 |
| DE3 | 0.30 | 13.00 | 0.17 | 19.69 | 0.60 | 15.48 | 0.17 | 16.54 |
| DE4 | 0.35 | 12.87 | 0.55 | 20.30 | 0.79 | 16.87 | 0.51 | 18.40 |
| DE5 | 0.26 | 13.17 | 0.28 | 20.62 | 0.43 | 15.85 | 0.05 | 16.27 |

As already mentioned, the authors aimed to introduce their own prediction methods. To be in line with our past daylight studies, Perez's clearness and sky brightness indices were to be used to parameterize insolation condition for the efficacy prediction. Table 4 summarizes the mean values of the vertical diffuse efficacy under eight Perez's clearness bins.

It can be observed that the efficacy value tended to increase with Perez's clearness. Under overcast condition (Clearness bin #1), the mean values of the diffuse efficacy ranged from 128.5 lm/W for south to 137.9 lm/W for east. The highest efficacy value would occur for clear sky condition with the mean values of 157.7 lm/W for north to 170.9 lm/W for west. The proposed prediction method was the direct use of the mean values of the diffuse efficacy in Table 4. In Table 3, this proposed method was denoted as the local model DE5. The table shows that the model could predict the diffuse efficacy values as comparable as model DE1-4. From the literatures, some studies introduced the use of a constant value to convert solar irradiance to its corresponding illuminance. However, Table 3 indicated that the author's method could improve the overall prediction performance by 5.8-26.7% (Reduced RMSD values) compared to the use of a constant efficacy value.

**Table 4.** The mean values of diffuse luminous efficacy on north, east, south and west surfaces

| Clearness bin no. | North | East | South | West |
|---|---|---|---|---|
| 1.Overcast | 137.37 | 137.11 | 128.51 | 130.46 |
| 2 | 138.15 | 140.79 | 128.71 | 130.52 |
| 3 | 140.28 | 145.48 | 132.12 | 132.75 |
| 4 | 144.40 | 149.75 | 137.95 | 140.86 |
| 5 | 148.32 | 155.65 | 146.13 | 147.64 |
| 6 | 152.92 | 157.55 | 152.99 | 161.94 |
| 7 | 157.68 | 161.46 | 161.71 | 170.88 |
| 8. Clear | 151.01 | 156.98 | 170.57 | 184.67 |

## 5. Conclusion

This paper characterized the vertical diffuse efficacy in Thailand's tropical climate. The statistical analysis showed that the mean values of the vertical diffuse efficacy were within a range of 107.1 lm/W to 172.5 lm/W. The evaluation of some existing efficacy models from higher latitude regions indicated that the tropical diffuse efficacy was comparatively high, and thus the daylight was more abundant. The characterization also showed that the diffuse efficacy value increased with clear sky condition. In the efficacy modeling, the existing models were regressed with our tropical efficacy records to obtain the optimal local coefficients. A prediction method was also introduced by the authors. The author's method determined the vertical efficacy values according to the eight Perez's clearness bins. The evaluation results showed that the author's methods and the existing model with local coefficients performed much better than the original efficacy models.

## 6. References

[1] Chirarattananon S, Chaiwiwatworakul P and Pattanasethanon S 2002 *Renewable Energy* Daylight availability and models for global and diffuse horizontal illuminance and irradiance for Bangkok. Vol **26**.pp. 69-89

[2] Chaiwiwatworaku P, Chirarattananon S 2013 *Renewable Energy* Luminous efficacies of global and diffuse horizontal irradiances in a tropical region. Vol **53**.pp. 148-58

[3] Cucumo M, De Rosa A, Kaliakatsos D and Marinelli V 2008 *Energy Conversion and Management* Experimental data of globl and diffuse luminous efficacy on vertical surfaces at Arcavacata di Renda and comparison with calculation models. Vol **50**.pp. 166-73

[4] Cucumo M, De Rosa A, Kaliakatsos D and Marinelli V 2008 *Energy and Buildings* Simplified

correlations of global, direct and diffuse luminous efficacy on horizontal and vertical surfaces. Vol **40**.pp. 1991-2001

[5] Lam J C and Li D H W 2001 *Light Res Technology* Predicting vertical luminous efficacy using horizontal data. Vol33.pp. 25-42

[6] Muneer T, Gul MS and Kubie J. 2000 *J Sol Energy Eng Trans ASME* Models for estimating solar radiation and illuminance from meteorological parameters. Vol **122**.pp. 146–53

[7] Ruiz E, Soler A and Robledo L 2001 *J Sol Energy Eng Trans ASME* Assessments of Muneer's luminous efficacy models in Madrid and a proposal for new models based on his approach. Vol **123**.pp. 220–4.

[8] L Robledo and A Soler. 2003 *Energy Conversion & Management* Modeling the luminous efficacy of solar radiation on inclined surfaces for all sky conditions. Vol **44**.pp. 177-89

[9] DHW Li, TNT Lam, KL Cheung and HL Tang. 2008 *Renewable Energy* An analysis of luminous efficacies under the CIE standard skies. Vol **33**.pp. 2357-65

# CFD and experimental analysis of solar crop dryer with waste heat recovery system of exhaust gas from diesel engine

**R Meenakshi Reddy[1], E Siva Reddy[1], C Uma Maheswari [2]and K Krishna Reddy[3]**

[1]Department of Mechanical Engineering, G. Pulla Reddy Engineering College, Kurnool, Andhra Pradesh, India
[2]Department of Mechanical Engineering, Sri Venkateswara College of Engineering & Technology, Chittoor, Andhra Pradesh, India
[3]Department of Mechanical Engineering, Brindavan Institute of Technology and Science, Kurnool, Andhra Pradesh, India

[1]Corresponding author: rmreddy123@gmail.com

**Abstract**. Crop drying is the method used to preserve the seasonal food crops during the non-seasonal periods in agricultural sector. In olden days, solar energy was used to dehydrate and preserve the food. Nowadays many new advanced technologies have been introduced. But due to electricity unavailability in rural areas and to utilize the abundantly available solar energy and the waste heat recovery of exhaust gas from the diesel engines used in the agricultural fields for water pumping. Our present work focuses on the CFD and experimental analysis of the solar crop dryer attached with flat plate collector and attached with diesel engine exhaust pipe heat exchanger. A comparison study was made between the crop dryer with and without exhaust pipe heat exchanger and found that when the exhaust pipe was attached there was an improvement in the increase in temperature compared without exhaust pipe. The copper tubes of exhaust pipe were also added with the fins to improve its heat exchanging efficiency.

## 1. Introduction

Drying of agricultural product is a vital post-harvest operation since it needs high energy consumption. Right smart energy savings are often achieved by victimization alternative energy as energy supply for drying of agriculture product. Ancient sun drying has several disadvantages like insect infestation, catalyst reactions, being growth so drying by solar drier is a choice. Besides standard sun drying takes longer length to succeed in the required wet content. Solar dryers are employed in all elements of globe and intensive analysis has been dole out, aiming at the advance of those systems. From the literature, it's found that the majority of those solar drying systems use flat plate air collectors of assorted styles and solely some of them thought-about a distinct style of collectors, one in every of them being natural and made convection solar tunnel driers. The success of the technology is measured by the numbers of dryers put in throughout the past few years. Few tries are created to boost the technology. The amount of such varieties of dryers put in within the province with regard to amount of paddy to be dried is negligible, solely regarding 0.55% of annual production [1]. The observation within the gift study implies that the temperature of cupboard solar drier is beyond that of open sun drying. Cupboard solar drying is quicker than open sun drying. The performance of solar drying system was extremely captivated with radiation and close temperature [2]. Only a few researchers have done experimental study of hybrid solar dryers, and compared to different dryers, their numerous style modifications and sweetening techniques. The Hybrid dryers are value effective, and are straight forward to fabricate and use [3]. This shows that drying rate is up in drier integrated with fin sort collector followed by V-type collector and flat plate collector. [4]. A large-scale greenhouse solar drier with LPG burner has been developed and its performance for drying osmotically dehydrated tomato are investigated [5]. A comprehensive review of the varied style details of construction and operational principles of the big variety of much realized styles of alternative energy drying system were given and additionally various styles of solar dryers for drying food materials has been mentioned as well as non- concentrating and concentrating sort [6]. Solar drying system is employed to dry food and save the consumption of the precious standard fuels. Studied that solar dryers not solely save energy compared to different industrial dryers however additionally save plenty of your time, improve quality of the merchandise and create the merchandise a lot of economical and additionally cut back the negative environmental impacts like pollution, ozonosphere depletion [7]. Solar collectors with exhausted tubes [EVT] are a special sort, providing vital benefits compared to flat-plate collectors, like higher efficiencies. EVT collectors, flat-plate ones alike, are classified into 2 broad classes, water and air-collectors, supported their heat transfer fluid. A two-pass solar drier was found to be technically and economically appropriate for drying of bamboo shoots beneath the specific conditions in central Vietnam [8]. The experiments given have shown that a batch of one thousand kilo of paddy are often simply dried to a target megacycle of Bastille Day inside the primary day throughout the time of year, achieving a top quality in terms of head rice yield. However, a second day of drying is usually necessary throughout the time of year [9]. The review state that Passive solar dryers are terribly economical compared with active solar dryers. In keeping with the review of numerous analysis articles the passive sort solar dryers are a lot of appropriate for drying various agricultural products [10]. The target of this study was to dry 2 commercially vital tropical African fish species African sharp tooth cat fish and Nile tilapia victimization completely different improved affordable solar driers. This is often with the aims of evaluating the performance, efficiency, and effectiveness of those improved affordable solar driers in terms of wet loss from the fish samples and testing the organoleptic characteristics of the dried product. [11]. As compared to sun dried product and that they are applicable for production of quality dried fruits, vegetables, spices, herbs and healthful plants, and fish. The potency of solar drying system is full of the properties of drying materials e.g. wet content, size, form and pure mathematics additionally as close conditions, that embrace radiation and temperature, ratio, speed and gas pressure of close air. [12]. The forced convection system has higher air flow through the reflector and drying chamber thus, it's higher drying capabilities compared to the natural convection system.[13]. The solar cupboard drying of Bitter Gourd was found to be quickest drying technique as compared to sun drying resulted in very best quality. Solar drying allows vegetables export worthy processed foods with long time period meeting the healthful standards of the mercantilism countries. Thus solar drying is incredibly viable and valuable one [14]. Here project is targeting decreasing loses in husbandry. Associate experimental study of but this drier fare compared to different dryers [15]. The objective of this paper is to do the CFD and experimental analysis to select the appropriate specification and mass flow rate to get efficient solar crop drier when diesel engine exhaust gas heat is also used for heating in addition to the solar energy as to increase the efficiency of the solar crop drier.

## 2. Numerical Analysis

The solid models of solar crop drier with exhaust gas pipe are created in SOLIDWORKS and imported to ANSYS –workbench for performing CFD analysis. Figure 1 shows the geometry model of air domain and exhaust gas domain created for analysis and Figure 2 shows the assembly of both the domains.

### *2.1 Modeling*

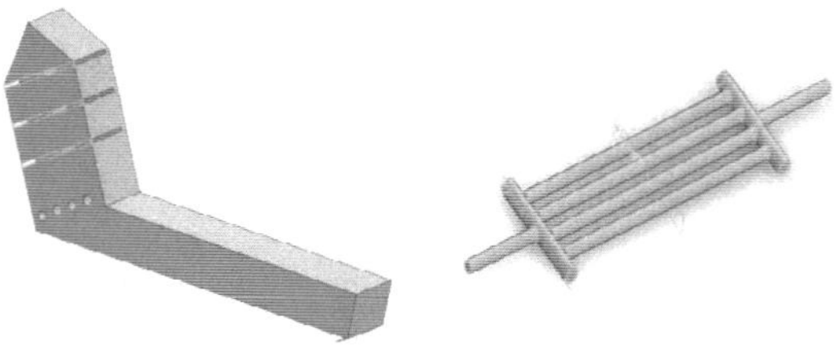

**Figure1.** Part Model

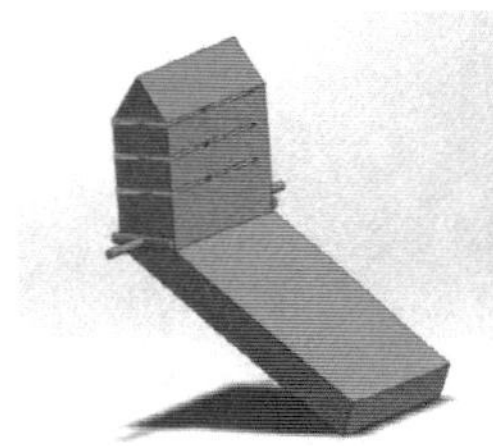

**Figure2.** Assembly

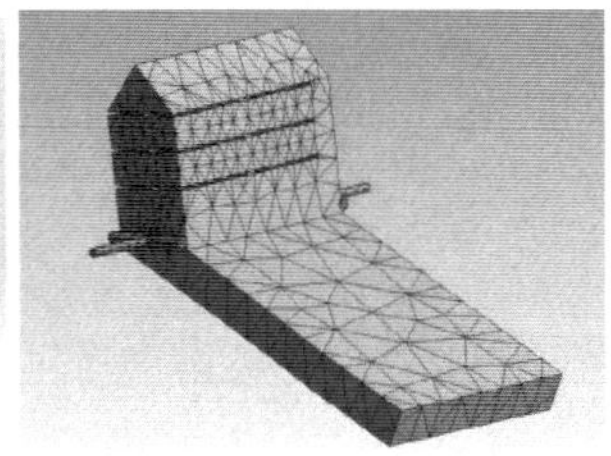

**Figure 3.** Meshed Model

*2.2 Meshing*

The solid model was meshed in ICEM-CFD using tetrahedral volume mesh. Figure 3 shows the meshed model of the assembled domains.

*2.3 Boundary Conditions*

Figure 4 shows the boundary condition of the meshed model which is imported to CFX-Pre. Air inlet is applied with air mass flow rate of 2 kg/s at temperature of 30 $^0$C and out let at average static pressure of 1 bar. Exhaust gas inlet is applied with 2 kg/sec mass flow rate at inlet temperature of about 50 $^0$C and outlet with static pressure of 1 bar. Heat flux was also applied on the glass top. Mass and momentum is selected with free slip wall. CFD analysis was performed to compute the air flow temperature by taking the inlet mass flow rate of about 2 kg/s.

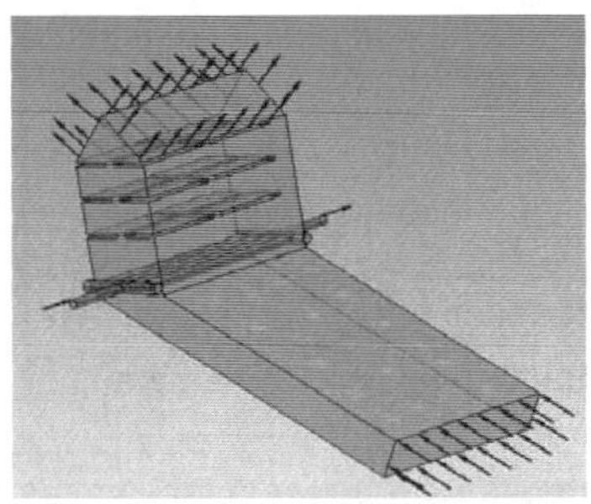

**Figure 4.** Boundary conditions

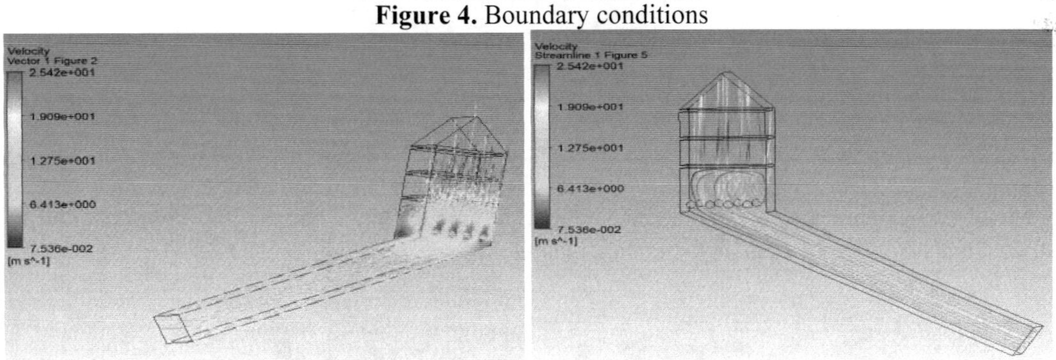


**Figure 5.** Velocity vector and streamline

*2.4 CFD results*

Figure 5 shows the velocity vector and streamline flow of air. Figure 6 represents the temperature distribution in solar drier without exhaust gas heat exchanger. The figure shows that the temperature at the food plate collector is very less. Figure 7 shows the temperature distribution when exhaust gas heat exchanger was inserted and it proves that there is a valuable increase in the temperature flow at the food plates.

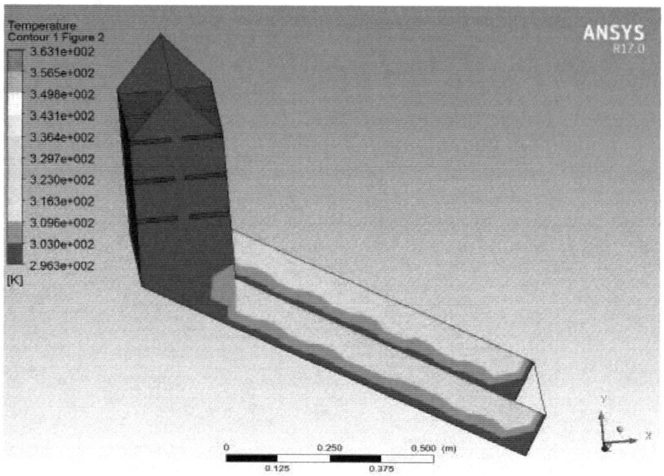


Figure 6. Temperature distribution

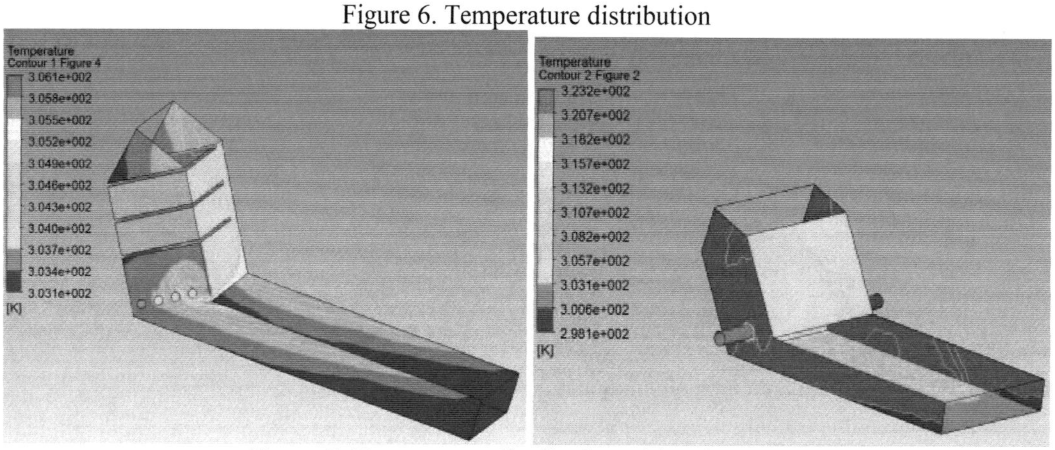


Figure 7. Temperature distribution with exhaust gas

## 3. Experimental Analysis:

The experimental model of solar crop drier was constructed with wooded chambers having slots to insert three food plate trays. Exhaust gas pipes were inserted at the bottom of the trays. The flat plate collector was fixed at $30^0$ to the ground and facing towards the south. The test was conducted at full summer during April and June to have more efficiency of the drier. Figure 8 shows the constructed experimental model of solar crop drier with the exhaust gas heat exchanger. The temperatures are air inlet and at each food plate trays are measured using thermocouples for every hour. The mass flow rate of air was varied from 0.5, 1, 1.5 and 2 kg/s. The slices of cucumber were placed on the trays and moisture content was checked by weighing them.

Figure 8. Experimental set up of solar crop drier with exhaust gas heat exchanger

## 4. Results and discussions

### *4.1. Natural Convection*

Table 1 gives the data of the removal of moisture by weighing the slices of the cucumber. It shows that as the solar radiation increases the moisture removal increases that can be assured by the weight loss in the crop. Figure 9 show the temperature distribution in three food plate trays during the natural convection.

**Table 1.** Loss of Weight

| Time (hours) | 10:00-11:00 | 11:00-12:00 | 12:00-13:00 |
|---|---|---|---|
| Weight Loss (Grams) | 169 | 150 | 110 |

### *4.2. Exhaust gas heat exchanger*

Figure 10 show the temperature distribution in three food plate trays during the exhaust gas flow heat exchanger was used. It was found that copper finned tube is more efficient than copper tubes without fins that exchanges heat from the exhaust gas of the diesel engine. Hence it was continued with all three crops with finned copper tube. Table 2 represents the moisture removed in solar crop drier in natural and exhaust gas drying.

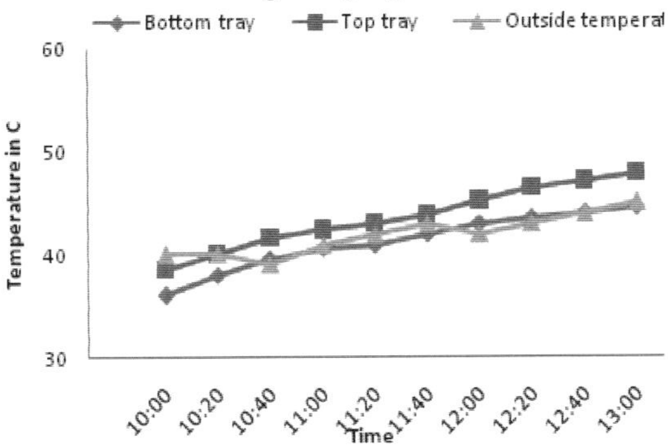


**Figure 9.** Temp distribution in each tray

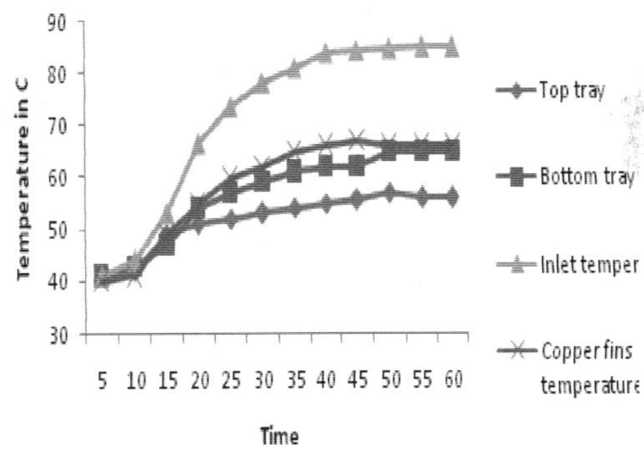


**Figure 10.** Temp distribution in finned copper tube

**Table 2.** Weight loss comparision between Natural and Exhaust gas Drying

| Time (Minutes) | Weight loss in Natural drying (Grams) | Weight loss in Exhaust Drying (Grams) |
|---|---|---|
| 20 | 192 | 186 |
| 40 | 183 | 168 |
| 60 | 169 | 146 |

## 5. Conclusion

The CFD and experimental analysis of the solar crop dryer attached with flat plate collector and attached with diesel engine exhaust pipe heat exchanger was done. A comparison study was made between the crop dryer with and without exhaust pipe heat exchanger and found that when the exhaust pipe was attached there was an improvement in the increase in temperature compared without exhaust pipe. The waste heat from the diesel engine was found more effective when it was used with the attachment of the solar crop drier. The copper tubes of exhaust pipe were also added with the fins to improve its heat exchanging efficiency.

## References

[1] Bhandari R and Gaese H 2008 Evaluation of Box Type Paddy Dryers in South Sumatra , Indonesia *Agric. Eng.* **X** 1–16

[2] Navale S R, Harpale V M and Mohite K C 2015 Comparative study of open sun and cabinet solar drying for fenugreek leaves *Int. J. Renewable Energy Technology Research* **4** 1–9
[3] Jawahar S and Siva Reddy E 2016 Experimental Investigation of Hybrid Food Dryer Using Solar and Exhaust Gas *Imperial Journal of Interdisciplinary Research* 496–9
[4] Patil S U and Patil V H 2015 Comparative Analysis of Cabinet Type Solar Dryer with Different Collector Geometries *Int. J. Modern Trends in Engineering and Research* **2** 324–34
[5] Janjai S 2012 A greenhouse type solar dryer for small-scale dried food industries: Development and dissemination *Int. J. Energy Environ.* **3** 383–98
[6] Selvaraj M and Sadagopan P 2017 A Review of Solar Energy Drying Technology with Air Based Solar Collectors *Advances in Natural and Applied Sciences* **11** 472–78
[7] Vijay P, Yadav V and Arya S An Overview of Solar Drying Technology *Journal of Renewable Energy and Resources* **2** 1–8
[8] Banout J and Ehl P 2010 Using a double-pass solar drier for drying of bamboo shoots. *J. Agric. Rural Dev. Trop. Subtrop.* **111** 119–27
[9] Salvatierra-Rojas A, Nagle M, Gummert M, de Bruin T and Müller J 2017 Development of an inflatable solar dryer for improved postharvest handling of paddy rice in humid climates *Int. J. Agric. Biol. Eng.* **10** 269–82
[10] Balasuadhakar A 2016 A Review on Passive Solar Dryers for Agricultural Products *Int. J. for Innovative Research in Science & Technology* **3** 64–70
[11] Mustapha M K, Ajibola T B, Salako A F and Ademola S K 2014 Solar drying and organoleptic characteristics of two tropical African fish species using improved low-cost solar driers *Food Sci. Nutr.* **2** 244–50
[12] Pradeepkumar H V and Ravikumar G M P 2017 A Review on Solar Drying System and its Applications *Int. J. for Scientific Research & Development* **5** 662–6
[13] Babagana G, Silas K and Mustafa B G 2012 Design and construction of forced/natural convection solar vegetable dryer with heat storage *ARPN J. Eng. Appl. Sci.* **7** 1213–7
[14] Hii C L, Jangam S V, Ong S P and Mujumdar A S 2012 Solar Drying: Fundamentals,Applications and Innovations *ARPN J. Eng. Appl. Sci.* **7** 149
[15] Sudharshan V, Siva Reddy E and Meenakshi Reddy R 2005 Experimental investigation of food dryer using solar and exhaust gas with finned copper tubes *Int. J. Scientific Research and Management Studies* **3** 144–51

IOP Conf. Series: Earth and Environmental Science **164** (2018) 012011 doi:10.1088/1755-1315/164/1/012011

# Multiple Distribution Generation Location in Reconfigured Radial Distribution System

## Distributed generation in Distribution System

**Ashwani Kumar, SMIEEE[1, a], Ravi Teja Bhimarasetti[2, b]**

Department of Electrical Engineering, NIT Kurukshetra, Kurukshetra, India

E-mail: [a]ashwani.k.sharma@nitkkr.ac.in; [b]bsvrteja@gmail.com

**Abstract.** Distribution system reconfiguration is one of the major approaches to reduce the losses in the distribution system. With the integration of Distributed Generation (DG) in distribution system, there can be further improvement in voltage profile and further loss reduction in the reconfigured system. The main contribution of this paper is: (i). Proposing novel and modified novel power loss sensitivity methods for finding optimal locations for placement of multiple DGs, (ii) obtaining the sizes of distributed sources for reduced losses and better voltage profile, and (iii) Study of variation of multiple DG sizes taking the effect of reconfiguration of the distribution system, (iv) Study of the impact of DG sizes taking into consideration the power factor. The proposed method has been implemented and tested on IEEE 33 bus distribution system.

## 1. Introduction

Distributed Generation placement, load management, Network Reconfiguration and so on. The combination of Distribution system Reconfiguration and optimal placement of DG units is the effective way to reduce the losses in distribution network [1]. The network reconfiguration problem in a distribution system is to find a best configuration of radial network that gives minimum power loss while the imposed operating constraints are satisfied, which are voltage profile of the system, current capacity of the feeder, and radial structure of the distribution system[2].

A branch exchange-type heuristic algorithm has been suggested by S. Civanlar, *et al* [3], where a simple formula has been derived to determine how a branch exchange affects the losses. W. C. Wu, *et al.* [4] proposed an effective approach based on the particle swarm optimization with enhanced integer coding to determine the switch operation schemes for feeder reconfiguration. LW de Oliveira, *et al.* [5] proposed an optimal distribution network reconfiguration combined with optimal capacitor allocation to minimization of energy loss on radial distribution systems. In Ref. [6], a heuristic harmony search algorithm was used to solve reconfiguration problem in presence of DGs. The genetic algorithm developed using the edge window decoder technique for network representation and building up spanning trees, as well as efficient genetic operators in order to explore the search space for power distribution system reconfiguration with minimal losses [7]. Distribution feeder reconfiguration considering distribution generation based on particle swarm optimization presented in [8].The optimal location of DGs, analytical methods for DG allocation, allocation based on losses reduction, heuristic methods, and a novel approach was presented in [9-14]. V.V.S.N.Mury, *et al* [15] proposed modified novel method to calculate the optimal location and DG sizes based on both real and reactive power losses at lagging power factor.

IOP Conf. Series: Earth and Environmental Science **164** (2018) 012011 doi:10.1088/1755-1315/164/1/012011

This paper proposed a method for finding optimal location for placing multiple DGs and their sizes to reduce the losses and to improve the voltage profiles. This paper is the extension of the previous work which was reported for single DG placement, however, the reconfiguration of the distribution system has not been considered [15]. Also the effect of multiple DGs allocation after reconfiguration of distribution has been studied in this paper. The reconfiguration of distribution system has been implemented based on the methodology reported in [18]. The data of IEEE 33 bus RDS is taken from [16-17]. The base MVA as100 and base KV as 12.66 are taken for both the systems.

## 2. Network Reconfiguration

The Reconfiguration of 33 bus distribution systems have been considered to obtain the DG sizes and their impact on reduction of losses and improving the voltage profile. In the first stage, all candidate switches are closed to form Mesh system. After performing power flow, switches are ranked based on increasing current magnitude. The top ranked switch is considered to be opened first by keeping all other switches closed. Power flow is performed with the new feasible configuration and various constraints are checked. The two switches in the neighborhood of the opened switch are stored in a list to use them in second stage. This procedure is repeated till the radial system is obtained. In the second stage, a series of branch exchange operation will be performed by using the neighborhood switch list to obtain the optimal configuration. For 33 bus RDS [17], the real and reactive power losses are 202.66kW and 135.13kVAr with s33, s34, s35, s36 and s37 switches opened. After obtaining optimal configuration, the real and reactive power losses are 139.55kW and 102.3kVAr with s7, s9, s14, s32 and s37 switches opened. The minimum voltage is increased from 0.9131p.u to 0.9378p.u.

## 3. Methodology

### *3.1 Optimal location and sizing of DGs using Modified Novel power loss sensitivity Method*

Modified Novel power loss sensitivity method [15] was proposed for single DG allocation at unity and lagging power factor. In this paper, the method is extended for multiple DGs allocation and their optimal sizes are determined at unity and lagging power factors.

*3.1.1 Single DG allocation.* Loss reduction by a single DG allocation. Let us consider the following: $TLP$ = total active power loss, $TLQ$ =total reactive power loss. $I_i$ =branch current, $R_i$ =branch resistance, $I_{ai}$ =active component of branch current, $I_{ri}$ =reactive component of branch current,$TLP_a$=loss associated with active component of branch current,$TLP_r$= loss associated with reactive component of branch current

$$TLP = \sum_{i=1}^{br} I_i^2 R_i \qquad (1)$$

$$TLP = \sum_{i=1}^{br} I_a^2 R_i + \sum_{i=1}^{br} I_r^2 R_i \qquad (2)$$

$$TLP = TLP_a + TLP_r \qquad (3)$$

*a) DG at unity power factor placed at bus 'k'.*

$I_{adgk}$=active component of current supplied by DG at node 'k',The optimal DG current for maximum loss reduction is given by

$$I_{adgk} = -\frac{\sum_{i=1}^{k} I_{ai} R_i}{\sum_{i=1}^{k} R_i} \qquad (4)$$

$V_{dgk}$= voltage magnitude of DG at node 'k'. Optimal size of DG at unity power factor is given as

$$P_{dgk} = I_{adgk} * V_{dgk} \qquad (5)$$

*b) DG at lagging power factor placed at bus 'k'.*

$I_{adgk}$=active component of the current to be supplied by DG for maximum loss reduction at bus 'k'

$I_{rdgk}$= reactive component of the current to be supplied by DG for maximum loss reduction at bus 'k'

$$I_{adgkp} = -\frac{\sum_{i=1}^{k} I_{ai} R_i}{\sum_{i=1}^{k} R_i} \quad (6) \qquad I_{rdgkp} = -\frac{\sum_{i=1}^{k} I_{ri} R_i}{\sum_{i=1}^{k} R_i} \quad (7)$$

$$I_{adgkq} = -\frac{\sum_{i=1}^{k} I_{ai} X_i}{\sum_{i=1}^{k} X_i} \quad (8) \qquad I_{rdgkq} = -\frac{\sum_{i=1}^{k} I_{ri} X_i}{\sum_{i=1}^{k} X_i} \quad (9)$$

$$I_{adgk} = \sqrt{{I_{adgkp}}^2 + {I_{rdgkp}}^2} \quad (10)$$

$$I_{rdgk} = \sqrt{{I_{adgkq}}^2 + {I_{rdgkq}}^2} \quad (11)$$

$P_{dgk}$=optimal real power supplied by DG at power factorcos $\emptyset$ at bus 'k',$Q_{dgk}$=optimal reactive power supplied by DG at power factor cos $\emptyset$at bus 'k'

$$P_{dgk} = I_{adgk} * V_{dgk} * \cos\emptyset \quad (12)$$

$$Q_{dgk} = I_{rdgk} * V_{dgk} * \sin\emptyset \quad (13)$$

The optimal size of DG can be calculated as

$$S_{dgk} = \sqrt{P_{dgk}^2 + Q_{dgk}^2} \quad (14)$$

Using eqns. (5), (12) and (13), optimal DG sizes can be obtained for unity and lagging power factor.

*3.1.2 Multiple DGallocation.* The concept of loss reduction by a single DG allocation can be extended for multiple DGs allocation. Let us consider the following:

kdg= number of buses compensated by DG

$I_{dg}$ =kdg-dimensional vector consisting of DG currents

$\propto_j$= Set of branches from the source bus to the $j^{th}$ DG bus (j=1, 2… k). In Figure1, if three DGs (k=3) are placed at buses 6, 15 and 30, the branch set

$\alpha_1$= [1 2 3 4 5],$\alpha_2$= [1 2 3 4 5 6 7 8 9 10 11 12 13 14]

$\alpha_3$= [1 2 3 4 5 25 26 27 28 29]

*a) DGs at unity power factor*

The optimal DG currents for the maximum loss reduction is given by

$$[G][I_{adg}] = [B] \quad (15)$$

Where G is a kdg x kdg square matrix and B is a kdg-dimensional vector. The elements of G and B are given by

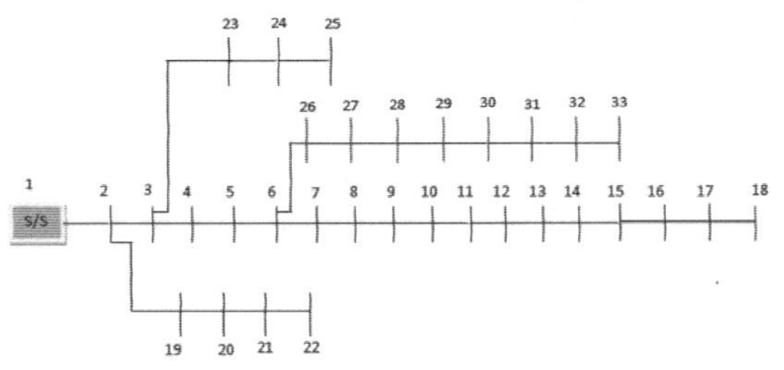


**Figure1.** Single line diagram of 33-bus system.

$$G_{jj} = \sum_{i\in\alpha_j} R_i \ (16), G_{jm} = \sum_{i\in(\alpha_j\cap\alpha_m)} R_i \quad (17)$$

$$B_j = \sum_{i\in\alpha_j} I_{ai} R_i \quad (18)$$

Only the branch resistances and real currents in the original system are required to find the elements of G and B. The DG currents at unity pf for the maximum loss reduction can be obtained from eqn. (15)

$$[I_{adg}] = [G]^{-1}[B] \quad (19)$$

Once the DG currents are known, the optimal DG sizes can be written as

$$[P_{dgk}] = [V_{dg}][I_{adg}] \quad (20)$$

Here $V_{dg}$is the voltage magnitude vector of DG buses.

*b) DGs at lagging power factor*

The optimal DG currents at lagging power factor for the loss reduction is given by

$$[G][I_{pdg}] = [P] \ (21), [H][I_{qdg}] = [Q] \quad (22)$$

Where G and H are kdg x kdg square matrix and B, C, D,E,P and Q are kdg-dimensional vector. The elements of G and H are given by

$H_{jj} = \sum_{i\in\alpha_j} X_i$ (23), $H_{jm} = \sum_{i\in(\alpha_j\cap\alpha_m)} X_i$ (24)

$C_j = \sum_{i\in\alpha_j} I_{ri} R_i$ (25), $D_j = \sum_{i\in\alpha_j} I_{ai} X_i$ (26)

$E_j = \sum_{i\in\alpha_j} I_{ri} X_i$ (27), $P = \sqrt{B^2 + C^2}$ (28)

$Q = \sqrt{D^2 + E^2}$ (29), $I_{pdgk} = [G]^{-1}[P]$ (30)

$I_{qdgk} = [H]^{-1}[Q]$ (31)

By calculating DG currents at lagging pf using eqns.(30-31), real and reactive powers supplied by DG can be calculated as:

$$P_{dg} = I_{pdg} * V_{dg} * \cos\emptyset \quad (32)$$

$$Q_{dg} = I_{qdg} * V_{dg} * \sin\emptyset \quad (33)$$

The optimal DG sizes can be written as

$$S_{dg} = \sqrt{P_{dg}^2 + Q_{dg}^2} \quad (34)$$

## 4. Algorithm for DG Placement

The computational steps involved in finding the optimal DG size and location to minimize the loss in distribution system are summarized as:

1) Run the base case load flow and obtain the base case power losses. Find the DG sizes at each node using eqn. (5) at unity pf and eqns. (12-13) at lagging pf.
2) Compensate each node (except the source node)with the DG sizes obtained in step 1 then select the node which gives maximum loss reduction, as a first DG location.
3) Update load data after placing the DG. Find the DG sizes at each node then identify the next compensating node at which maximum loss reduction is obtained.
4) Compensate all selected DG locations by using corresponding DG sizes found from eqn. (20) at unity power factor and eqns. (32-33) at lagging pf.
5) Repeat step 3-4 to determine next potential location for DG placement until it is found that there is no significant loss reduction that can be achieved by further DG placement.

Flow chart of proposed method for multiple DG allocation is shown in Figure 2.

## 5. Results and Discussions

Before reconfiguration of 33 bus distribution system, the base case losses are 202.665kw and 135.133kVAr. Using Novel method, DG with optimal size will be placed at each and every node and then the bus at which total power loss is minimum will be considered as optimal location. It is observed from the Figs. 3 and 4, the optimal location for placement of single DG is bus 6. After placing the DG of size 2487.44 kW at bus 6, the real and reactive power losses are reduced by 48.649% and 44.684% from its base case respectively. The minimum voltage has been increased from 0.91309p.u to 0.9498p.u. By assuming single DG is connected at bus 6 of base system and then repeat the same procedure for the next optimal DG location and it is found that the bus 15 is next optimal location. Simultaneously placing two DG units at buses 6 and 15 with the sizes of 1902.4556kW and 564.926kW, the real and reactive power losses are reduced by 55.66 % and 53.427% from its base case respectively. The minimum voltage has been increased to 0.95319p.u. Similarly by assuming two DG units are connected at buses 6 and 15 of base system, the same procedure has been repeated for finding next optimal location for placement of DG and it is found that bus 25 is next optimal location. By placing three DG units at buses 6, 15 and 25 with the sizes of 1696.669kW, 564.9261kW and 772.1976kW respectively, the real and reactive power losses are reduced by 60.95% and 58.86% from its base case respectively. The minimum voltage has been increased to 0.95331p.u. The voltage profile obtained with and without DG units has been shown in Figure 5.

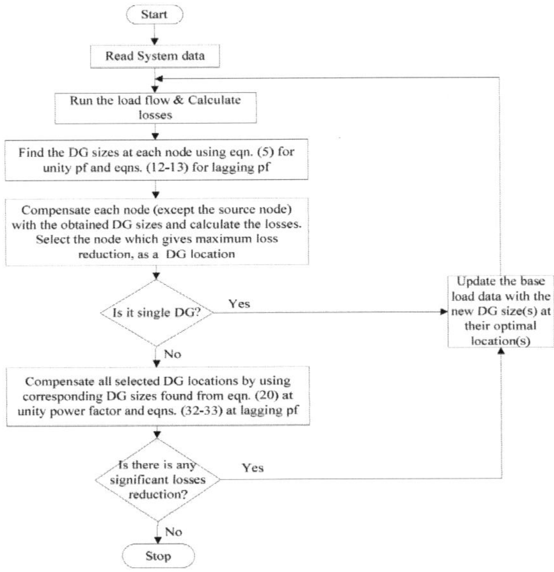


**Figure 2.**Flow chart of proposed method for multiple DG allocation.

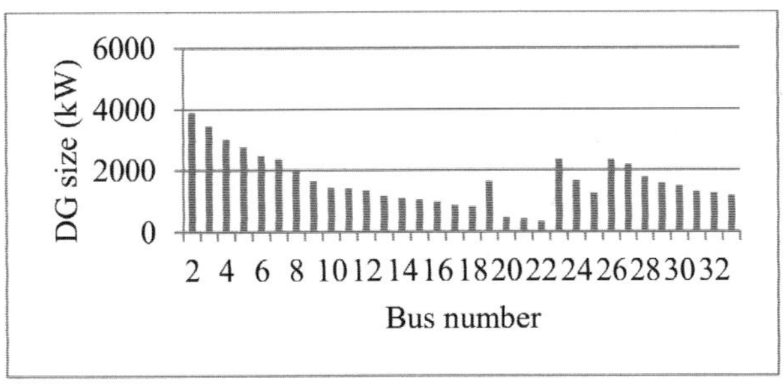


**Figure 3.**Optimal DG sizes for 33 bus system at UPF – before reconfiguration.

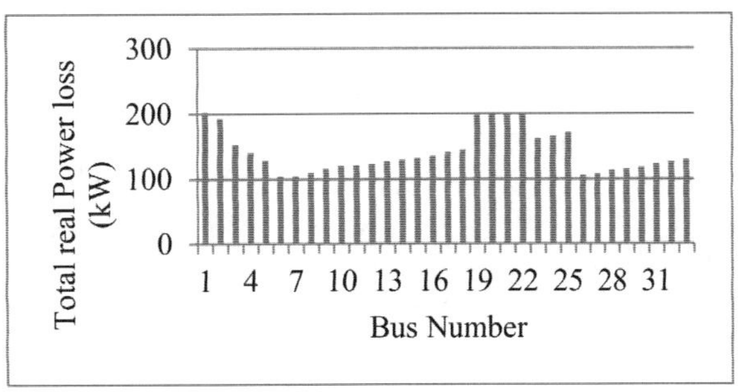


**Figure 4.**Variation of Total real power loss w.r.t DG size at UPF – before reconfiguration.

However, after reconfiguration the base case losses are 139.549kW and 102.304kVAr. The optimal location of for placement of single DG is bus 30 and its size is 1065.797 kW Figure 6-7. The real and reactive power losses are reduced by 29.707% and 25.39% from its base case. The minimum voltage has been increased from 0.93782p.u to 0.9478p.u. With the placement of two DG units of sizes 1048.965kW and 934.547kW at locations 30 and 8 buses, the real and reactive power losses are reduced by 47.88% and 48.096% from its base case and the minimum voltage is increased to 0.973237p.u. With placement of three DG units of sizes of sizes 930.76kW, 914.3285kW and

IOP Conf. Series: Earth and Environmental Science **164** (2018) 012011 doi:10.1088/1755-1315/164/1/012011

1052.458kW at buses 30, 8 and 24 respectively, the losses are reduced by 57.79% and 56.74% from its base case and the minimum voltage has been increased to 0.973467p.u.

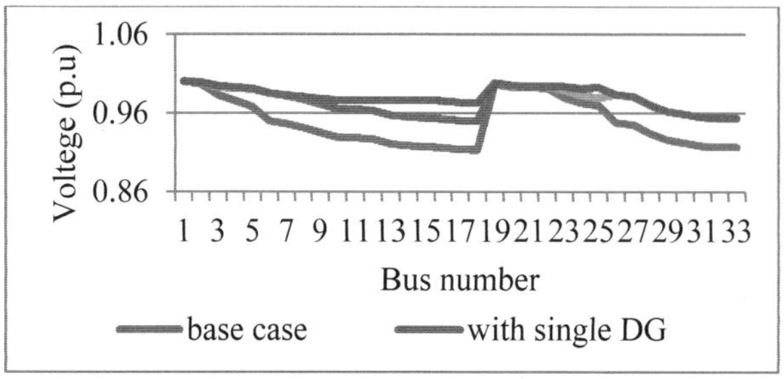


**Figure 5.**Voltage profile of 33 bus RDS without and with DGs at UPF– before reconfiguration.

The voltage profile obtained with and without DG units has been shown in Figure 8. Thus, before reconfiguration by placing three DG units of total size 3033.7927kW, the losses are reduced to 79.1376kW and 55.5914kVAr. However after reconfiguration by placing three DG units of total size 2897.7094kW, the real and reactive power losses are reduced to 58.899kW and 44. 255kVAr. It is observed that the minimization of power loss is more with the DG placement after network reconfiguration .

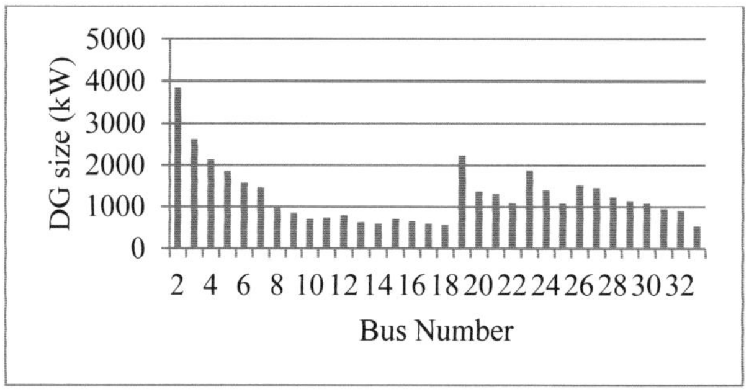


**Figure 6.**Optimal DG sizes for 33 bus system at UPF – after reconfiguration.

The results are also obtained for DG allocation at 0.9 lagging power factor and are shown in Table 2. Before reconfiguration of 33 bus distribution system, the base case losses are 202.665 kW and 135.133kVAr. Using modified Novel method, DG with optimal size will be placed at each and every node and then the bus at which total power loss is minimum will be considered as optimal location. The optimal location for placement of single DG is bus 6. After placing the DG of size 3000.99kVA at bus 6, the real and reactive power losses are reduced by 68.03% and 62.63% from its base case respectively. The minimum voltage has been increased from 0.91309p.u to 0.9649p.u. By assuming single DG is connected at bus 6 of base system and then repeat the same procedure for the next optimal DG location and it is found that the bus 31 is next optimal location.

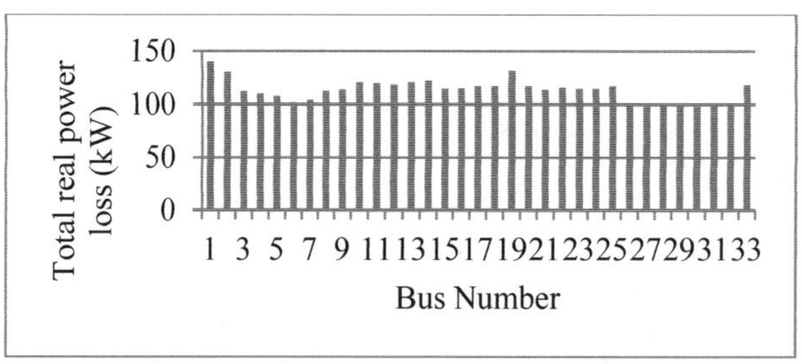


**Figure 7.**Variation of Total real power loss w.r.t DG size at UPF – after reconfiguration.

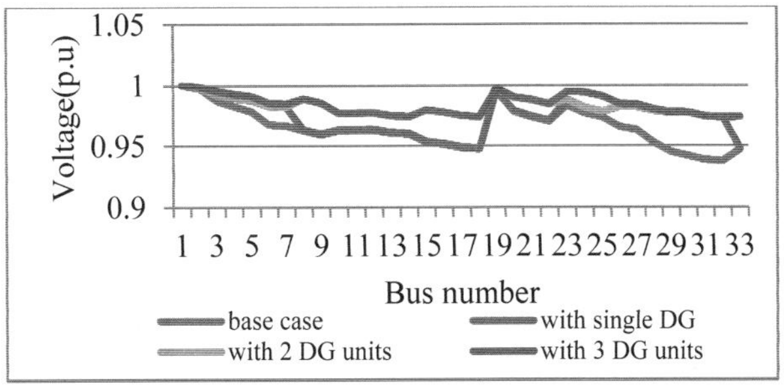


**Figure 8.** Voltage profile of 33 bus RDS without and with DGs at UPF– after reconfiguration.

Simultaneously placing two DG units at buses 6 and 31 with the sizes of 2005.405kVA and 962.188kVA, the real and reactive power losses are reduced by 78.645% and 74.62% from its base case respectively. The minimum voltage has been increased to 0.96479p.u. Similarly by assuming two DGs are connected at buses 6 and 31 of base system, the same procedure has been repeated for finding next optimal location for placement of DG and it is found that bus 25 is next optimal location. By placing three DG units at buses 6, 31 and 25 with the sizes of 1787.983kVA, 962.1886kVA and 848.4035kVA respectively, the real and reactive power losses are reduced by 85.05% and 81.245% from its base case respectively. The minimum voltage has been increased to 0.96494p.u.. However, after reconfiguration the base case losses are 139.549kW and 102.304kVAr. The optimal location of for placement of single DG is bus 30 and its size is 1452.52kVA. The real and reactive power losses are reduced by 49.367% and 42.676% from its base case. The minimum voltage has been increased from 0.93782p.u to 0.9482p.u. With the placement of two DG units of sizes 1434.93kVA and 1027.382kVA at locations 30 and 8, the real and reactive power losses are reduced by 71.303% and 70.08% from its base case and the minimum voltage is increased to 0.97985p.u. With placement of three DG units of sizes of sizes 1312.71kVA, 1007.149kVA and 1144.936kVA at buses 30, 8 and 24, the losses are reduced by 83.391% and 80.758% from its base case and the minimum voltage has been increased to 0.980202p.u. Thus before reconfiguration by placing three DG units of total size 3598.575kVA, the losses are reduced to 30.2917kW and 25.34372kVAr. However after reconfiguration by placing three DG units of total size 3464.7947kVA, the real and reactive power losses are reduced to 23.1778kW and 19.68457kVAr. It is observed that the minimization of power loss is more with the DG placement after network reconfiguration and is given in the Table 2.

**Table 1.** Results for 33 bus system without and with DG at UPF

| | Before Reconfiguration upf | | | | | | | After Reconfiguration upf | | | | | | |
|---|---|---|---|---|---|---|---|---|---|---|---|---|---|---|
| | Without DG | 1 DG | 2 DG | | 3 DG | | | Without DG | 1 DG | 2 DG | | 3 DG | | |
| DG loc. | - | 6 | 6 | 15 | 6 | 15 | 25 | - | 30 | 30 | 8 | 30 | 8 | 24 |
| DG size in kW | - | 2487.441 | 1902.4556 | 564.92613 | 1696.669 | 564.9261 | 772.1976 | - | 1065.797 | 1048.965 | 934.547 | 930.7651 | 914.2856 | 1052.4587 |
| Tpl (kW) | 202.665 | 104.07 | 89.856845 | | 79.13762 | | | 139.549 | 98.092 | 72.72954 | | 58.89953 | | |
| Tql (kW) | 135.133 | 74.749 | 62.93498 | | 55.59139 | | | 102.304 | 76.321 | 53.09944 | | 44.25469 | | |
| Min V (p.u) | 0.91309 | 0.9498 | 0.953199 | | 0.953315 | | | 0.93782 | 0.9478 | 0.973237 | | 0.973467 | | |
| bus | 18 | 18 | 33 | | 33 | | | 32 | 33 | 32 | | 32 | | |

**Table 2.** Results for 33 bus system without and with DG units at 0.9 PF (lag)

| | Before Reconfiguration 0.9 LPF | | | | | | | After Reconfiguration 0.9 LPF | | | | | | |
|---|---|---|---|---|---|---|---|---|---|---|---|---|---|---|
| | Without DG | 1 DG | 2 DG | | 3 DG | | | Without DG | 1 DG | 2 DG | | 3 DG | | |
| DG location | - | 6 | 6 | 31 | 6 | 31 | 25 | - | 30 | 30 | 8 | 30 | 8 | 24 |
| DG size in kVA | - | 3000.999 | 2005.4059 | 962.1887 | 1787.983 | 962.1886 | 848.4035 | - | 1452.5225 | 1434.9341 | 1027.3826 | 1312.7097 | 1007.1490 | 1144.936 |
| Tpl (kW) | 202.665 | 64.792 | 43.2792 | | 30.29177 | | | 139.549 | 70.658 | 40.0465 | | 23.17783 | | |
| Tql (kW) | 135.133 | 50.502 | 34.29204 | | 25.34372 | | | 102.304 | 58.645 | 30.60828 | | 19.68457 | | |
| Min V (p.u) | 0.91309 | 0.9649 | 0.964797 | | 0.96494 | | | 0.93782 | 0.9482 | 0.979855 | | 0.980202 | | |
| bus | 18 | 18 | 18 | | 18 | | | 32 | 33 | 14 | | 14 | | |

## 6. References

[1] B. Esmailnezhad and H. Shayeghi "Simultaneous distribution network reconfiguration and DG allocation for loss reduction by invasive weed optimization algorithm", IEEE SGC, pp. 166-172, 2013.

[2] R. SrinivasaRao, S.V.L.Narasimham, M.RamalingaRaju, and A. SrinivasaRao, "Optimal Network Reconfiguration of large-scale distribution system using harmony search algorithm", IEEE Trans. on Power Syst., vol. 26, pp. 1080-1088, 2011.

[3] S. Civanlar, J.J.Grainger, H,Yin and S.S.H.Lee, "Distribution feeder reconfiguration for loss reduction", IEEE Trans. on Power Del., vol. 3, no.3, pp.1217-1223, 1988.

[4] W.C. Wu and M.S. Tasi., "Application of enhanced integer coded particle swarm optimization for distribution system feeder reconfiguration", IEEE Trans. on Power Syst., vol. 26, no.3, pp.1591–1599, 2011.

[5] L.W. de Oliveira, S.CarneiroJr, E. J. de Oliveria, J. L. R. Pereira, I. C. Silva Jr and J. S. Costa., "Optimal reconfiguration and capacitor allocation in radial distribution systems for energy losses minimization", Elect Power Ener. Syst., vol. 32, pp.840-848, 2010.

[6] R. S. Rao, K. Ravindra, K. Satish and S.V.L. Narasimham., "Power loss minimization in distribution system using network reconfiguration in the presence of distributed generation", IEEE Trans. on Power Syst., vol. 28, No. 1, pp. 317-325, 2013.

[7] J. Torres, J.L.Guardado, F. R. Davalos, S. Maximov and E. Melgoza., "A genetic algorithm based on the edge window decoder technique to optimize power distribution systems reconfiguration", Elect. Power Ener. Syst., vol. 45, pp. 28–34, 2013.

[8] J. Olamaei, T. Niknam and G. Gharehpetian., "Application of particle swarm optimization for distribution feeder reconfiguration considering distributed generators", Applied Mathe. and Compu., vol. 201, Issues 1-2, pp. 575-586, 2008.

[9] C. Wang, M. H. Nehrir., "Analytical approaches for optimal placement of distributed generation sources in power systems", IEEE Trans. on Power Syst., vol. 19, no. 4, pp. 2068-2076, 2004.

[10] N. Acharya, P. Mahant and N. Mitulananthan., "An analytical approach for DG allocation in primary distribution network", Elect. Power and Ener. Syst., vol. 28, Issue 10, pp. 669-678, 2006.

[11] R.K. Singh and S.K. Goswami, "Optimal allocation of distribution generations based on nodal pricing for profit, loss reduction, and voltage improvement including voltage rise issue", Elect. Power and Ener. Syst., vol. 32, pp. 637-644, 2010.

[12] L.D. Arya, AtulKoshti, and S.C. Choube "Distributed generation planning using differential evolution accounting voltage stability consideration" Elect. Power and Ener. Syst. Vol. 42, pp.

196–207, 2012.D. Singh, D. Singh, K.S.Verma., "Multiobjective optimization for DG planning with load models", IEEE Trans. on Power Syst., vol. 24, no.1, pp. 427-436, 2009.

[13] M.M. Aman, G.B.Jasmon, A.H.A.Bakar and H.Mokhlis, "A new approach for optimum DG placement and sizing based on voltage stability maximization and minimization of power losses", Energy Conv. and Manag.,vol. 70, pp.202–210, 2013.

[14] K.Nagaraju, S.Sivanagaraju, T. Ramana, S. Satyanarayana, P.V. Ramana, "A Novel method for optimal distributed generator placement in radial distribution system", *Journal on Distr. Gener. and Alter. Ener.*,vol. 26, no.1, pp.7-19, 2011.

[15] V.V.S.N.Murthy and Ashwani Kumar, "Comparison of Optimal DG Allocation Methods in Radial Distribution systems based on Sensitivity Approaches", Elect. Power and Ener. Syst., vol. 53, pp. 450–467, 2013.

[16] M.E. Baran and F.F. Wu, "Network reconfiguration in distribution systems for loss reduction and load balancing" IEEE Trans. on Power Del., vol. 4, no. 2, pp. 1401-1407, 1989.

[17] S. Sivanagaraju, N. Visali, V. Sankar, and T. Ramana, "Enhancing Voltage Stability of Radial Distribution Systems by Network Reconfiguration", Elect. Power Comp. and Syst., vol. 33, no. 5, pp. 539-550, 2005.

[18] G.K.ViswanadhaRaju and P. R. Bijwe, "Efficient reconfiguration of balanced and unbalanced distribution systems for loss minimization", IET Gener., Trans. and Dist., vol. 2, Issue 1, pp. 7-12, 2008.

IOP Conf. Series: Earth and Environmental Science **164** (2018) 012012 doi:10.1088/1755-1315/164/1/012012

# Electricity Consumption and Associated Risks in European Countries - Convergence Analysis

**M Szajt[1], R Kucęba[2], M Zawada[1] and J Kowalik[1]**

[1] Technical University Of Czestochowa, Faculty of Management, Department of Econometrics and Statistics, 39b Al. Armii Krajowej 42-200 Częstochowa, Poland
[2] Technical University Of Czestochowa, Faculty of Management, Institute of Management Information Systems, 39b Al. Armii Krajowej 42-200 Częstochowa, Poland

marszajt@poczta.onet.pl

**Abstract.** Demand for electricity increases with economic development. At the same time, it is expected to provide the cleanest energy. This paper attempts to analyze the convergence of electricity consumption in European countries, reduce transmission losses and reduce the pollution associated with it - CO2. In the presented research European states were taken into account. All countries and individual European Union and non-EU unions were analyzed. The study used descriptive statistics and B-Convergence analysis. The aim of the study was to compare changes in these variables in the EU Member States and non-affiliated countries. The results indicate in general terms the convergence of both energy consumption and CO2 emissions reduction. Specific studies indicate that non-EU countries have no convergence in terms of energy consumption and reduction in transmission losses. The only significant convergence of CO2 emissions concerns. This may be due to financial constraints that inhibit the efficiency of the tested energy systems.

## 1. Introduction

Permanent economic development, which is observable and increasingly common [1] on the one hand triggers other actions, on the other hand is determined by certain limitations and emerging opportunities. Among the opportunities mentioned above are the innovations accompanying humanity from the dawn of history, improving the work and improving the quality of life. It should be noted that in the long term, improvement of quality of life, increase of wealth of societies, prolongation of life and increase of population are observed. There are also opportunities for cultural and social change in different groups, but they are secondary to innovation. Development constraints are usually resource-based. In the production process, capital, labor, land and scientific and technological progress are usually mentioned as resources. Contemporary labour is not only about people but also about their knowledge, skills, and experience. Obviously, in parallel, it is important to remember that performing work involves the use of existing solutions, technology and equipment. In turn, due to the commitment of adequate amount of capital and use of technical progress, they are continually modified and improved in the area of efficiency and / or optimization of the profit effect for entrepreneur. At the same time, wealth of societies is growing due to the increase in wages (in the ideal model) reflecting the payment for competences. Wealth also grows in less quantifiable dimensions, due to the ease of work and daily living resulting from the use of modern equipment, which also results in increased free time, consumption of which generates further demand shifts and development especially in the service sector. As a result of specialization, we have at every stage emergence of new industries and services that in many cases bring people into their day-to-day operations. With this development, importance of using new devices is growing. These devices are powered by energy they consume. So it seems natural that increase in consumption of electricity - especially in highly developed countries is observed. These include among others European countries. Given their high level of integration and relatively similar development, it can be argued that the level of electricity consumption in these countries should be similar. This paper attempts to analyze the convergence of electricity consumption in European countries, reduction of transmission losses and reduction of pollution associated with it - CO2. The data used in the analysis come from the World Bank [2].

## 2. Basic aspects of electricity management

The reduction of electricity resources is related to several issues. The first is access to electricity. The main sources today are: hard coal and lignite, gas, nuclear reactors, wind turbines and water turbines, photovoltaics. Especially in the last types of activity is seen the hope of engaging prosumer [3].

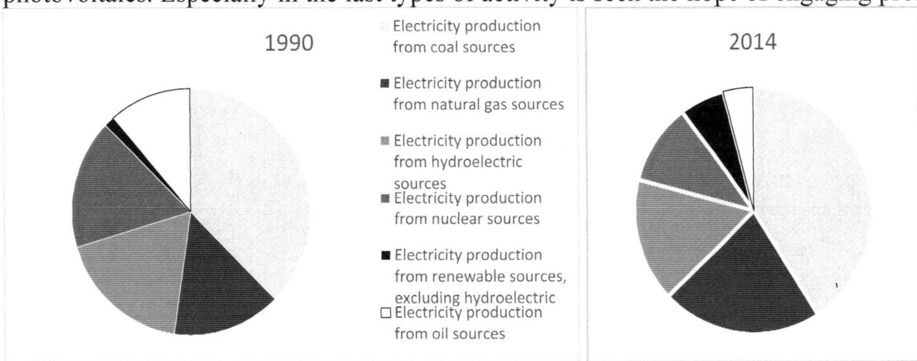


**Figure 1.** Electricity sources in Europe in 1990 and 2014 year.

In the case of lignite energy, one of the main constraints is the distance from power generating plants to sources of extraction - due to limited transport capacity. In the case of hard coal - increasing costs of getting out of ever deeper decks. In both cases we also face the problem of mining damage and environmental degradation. To make matters worse, the production of electricity from fossil fuels is associated with the combustion process, and thus the emission of many harmful substances into the atmosphere, among which are the following: sulphur dioxide SO2, nitrogen dioxide NO2, carbon oxide CO, benzene C6H6, ozone O3, dust PM10, lead Pb in PM10, arsenic As in PM10, cadmium Cd in PM10, nickel Ni in PM10, benzo(a)pyrene B(a)P in dust PM10, dust PM2,5[4]. In Europe, under the many regulations [5] [6] [7], the largest polluters pay higher energy production costs due to contaminated production. This results in a reversal of the carbon-based energy policy of the [8] in some countries. However, globally, the role of coal-based energy is growing steadily and its current share in total electricity production is as high as 40.7%. The second source is natural gas with a significant increase in share (by 46.5% between 1990 and 2014), which now accounts for 21.6% of electricity production. Another place is hydro energy - here we are dealing almost with supply saturation. Possible new investments and more efficient technologies are not able to keep pace with rising demand. In the case of nuclear energy, we are seeing a marked decrease in overall production from 16.9 in 1990 to 10.6% in 2014. Dangers of nuclear power failure can lead to increased concerns about this source of energy and virtually halting the development of nuclear power in Europe. The global economy is constantly looking for solutions that will help optimize the use of existing resources and more cost-effective technology. However, the steady increase in the use of electronics, the increase in the standard of living of societies and the population determines the increase in the consumption of electricity.

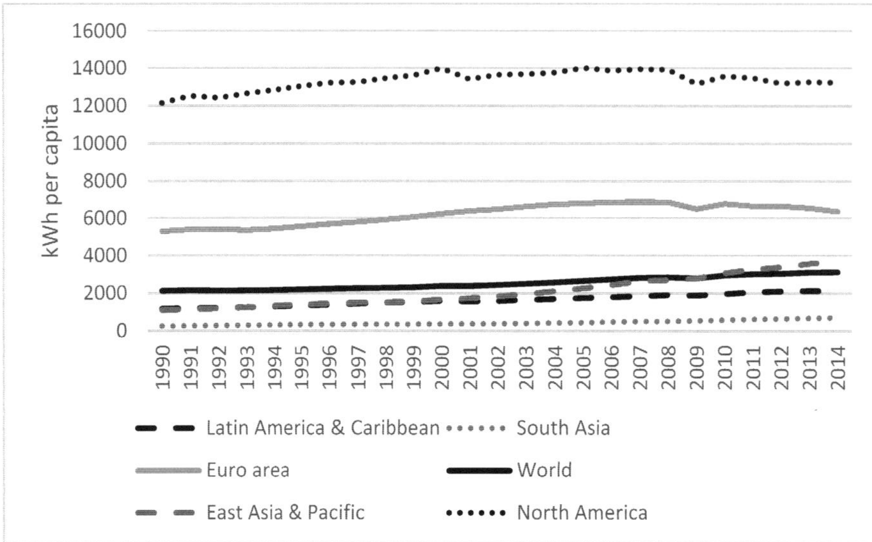


**Figure 2.** Change of electric power consumption (kWh per capita) in the years 2000-2010 in various parts of the world.

Electricity consumption varies in different parts of the world, but has a common feature (in the long run). Its consumption is constantly increasing. What is important, not only aggregated consumption but also per capita. The largest increase - 5.2% annually was observed in East Asia and Pacific. On average it is 0.4% globally and 0.8% in Europe. Unconsciously, using information technology, using the mass media, extending the working day and activity time, using entertainment, we increase the consumption of electricity. Importantly, current forecasts for the availability of existing liquid fuels resources commonly used to drive transport are not optimistic. Also, motor vehicles tend to use electricity. Therefore, attempts to anticipate the nearest demand for electricity carried out by various government agencies and private are justified.

## 3. Convergence analysis

By observing the European economy, it is easy to see some differences between their levels [9]. One would assume that economic leaders are treated as some sort of pattern for others. The weaker economies (most of which functioned at the end of the twentieth century in the Eastern Bloc) still stand out from the leaders, although the possibility of very rapid growth is evident - especially in the early stages of market economy implementation. In practice, we see three groups of states in Europe: economically developed (most of the "old" members of the EU and associated countries in the European Economic Area), developing continually - mainly new members of the European Union and countries where the road to development is open. In this respect, further analyzes were carried out taking into account the division of states.

**Table 1.** Convergence of electric power consumption (kWh per capita) in EU countries and broken down into groups of countries.

| country grup / measure | 44 countries | EU 28 | 15 „old" EU + 3 EOG | EU „new" 13 | others |
|---|---|---|---|---|---|
| **θ** | -3,232 | -0,579 | -0,760 | -1,497 | -5,042 |
| **β** | 0,012 | 0,002 | 0,004 | 0,012 | 0,048 |
| **p** | 0,002 | 0,567 | 0,458 | 0,163 | 0,000 |

With regard to consumption of electricity per capita, we see a clear convergence across Europe ($\beta$ = 0.012, p = 0.002). However, when analyzing irrelevant convergence in the European Union countries with significant and very clear in other countries ($\beta$ = 0.048, p <0.001), it must be acknowledged that it is a result of catching up (in economic sense) of the Union countries on the part of other European countries. In 2014, the average electricity consumption per capita was 8397kWH in the European Union and EEA, but this was overstated by Sweden, Finland, Norway and Iceland for an average consumption of 26,391 kWh. Average for other EU Member States is 5731 kkWh per person. Member States (old) are characterized by consumption at 6785 kWh with an average of 4597 for new members. Other countries are characterized by consumption at an average of 3849 kWh per person. Clearly higher consumption in northern states is due to the length of the day and the need for lighting and average daily temperatures, which also determine the consumption of electricity by both residents and industry. It should be noted here that especially the states located in the Scandinavian Peninsula are characterized by a high level of economic development, which at the same time results in an increased consumption of electricity from the economy. Consumption also affects the habits and wealth of the people, climate, geographical location, etc.

Convergence in the field of electricity management is more clearly seen in the factors that the energy producer directly affects. Such factors include electric power transmission and distribution losses.

**Table 2.** Convergence of electric power transmission and distribution losses in EU countries and broken down into groups of countries.

| country grup / measure | 44 countries | EU 28 | 15 „old" EU + 3 EOG | EU „new" 13 | others |
|---|---|---|---|---|---|
| **θ** | -2,268 | -2,165 | -2,209 | -1,129 | -2,155 |
| **β** | 0,016 | 0,024 | 0,027 | 0,025 | 0,029 |
| **p** | 0,029 | 0,039 | 0,042 | 0,283 | 0,054 |

In the case of electric power transmission and distribution losses, we see a clear convergence on almost all levels. The strongest is seen in the most developed countries ($\beta$ = 0.027, p = 0.042). The convergence level recorded in the "new" 13 and non-member states was not considered statistically significant. Closer analysis shows, however, an ambiguous situation.

In the "old" EU Member States, the annual growth rate of the electric power transmission and distribution losses below 0.5% is observable among five countries, the highest observed in Netherlands +0.39%. In other countries we observe the electric power transmission and distribution losses decline, including three significant ones: Denmark -1.8, Luxembourg -3.4, Iceland -4.7. In the "new" 13 we observe quite large decreases in electric power transmission and distribution losses - in ten countries – of which 7 are higher than 1% annually, the strongest in Slovakia -4.6%. However, three countries break down here: Estonia - growth of 0.1%, Romania - growth of 0.7% and Lithuania - growth of as much as 6.0% a year. Hence the lack of convergence in this group. The situation in the group of non-member countries is even more diverse. We observe either the electric power transmission and distribution losses - 7 countries with the strongest decrease in Montenegro -4.7% and countries with significant growth - 6 countries with the strongest growth in Moldova + 4.5%. This is largely due to the way the system is managed. These countries mostly use old transmission lines and older technologies. In the case of those that have already implemented changes in the system, we are seeing an improvement, while the rest of the older infrastructure is generating more and more losses. Older infrastructure is also often accompanied by higher emissions of harmful substances.

**Table 3.** Convergence of $CO_2$ emissions (kg per 2011 PPP $ of GDP) in EU countries and broken down into groups of countries.

| measure \ country grup | 44 countries | EU 28 | 15 „old" EU + 3 EOG | EU „new" 13 | others |
|---|---|---|---|---|---|
| **θ** | -5.156 | -4.468 | -1.652 | -1.854 | -4.382 |
| **β** | 0.021 | 0.021 | 0.013 | 0.019 | 0.041 |
| **p** | 0.000 | 0.000 | 0.118 | 0.091 | 0.002 |

In terms of CO2 emissions, there is a clear convergence in the European countries ($\beta = 0.021$, $p < 0.001$), in the European Union ($\beta = 0.0217$, $p < 0.001$) and non-associated countries ($\beta = 0.041$; $p = 0.002$). It is the result of general care to reduce the amount of emitted pollutants, mainly due to regulations. EU directives impose penalties on Member States if certain levels of emissions are exceeded and international standards (not just EU ones) are aimed at limiting global pollution. Of the 44 countries surveyed, only Bosnia and Herzegovina saw an increase in emissions in the years 1990-2014 by 0.3% per year. In all others we observe a decrease in emissions, and in 32 countries higher than 1%, of which 16 is more than 3% annually. Current CO2 emissions (kg per 2011 PPP $ of GDP) are 0,200 on average for the European Union, 0,165 for the old 15 + EEA countries and 0,250 for the new 13. The level of pollutant emissions for the other states is significantly higher and in 2014 was on average 0.391 (kg per 2011 PPP $ of GDP). In this aspect, we are dealing with a very unfavorable situation. By assuming higher electricity consumption with the development of these countries, we can also expect higher emissions. So nowadays, the authorities of these countries should endeavor to introduce appropriate environmental standards and emission reduction technologies, including CO2. However, it should be noted that such activities are undertaken. Average emissions in 2000 for the old Member States were 0.232, for the new 13 0.397 and for the remaining 0.637. It is therefore clear that emissions from all groups of countries are reduced. This is all the more important in the case of the old 13 states, a decrease in CO2 emissions of 29% was accompanied by an increase in energy consumption by 12%, for new 12 countries, a 37% decrease in new CO2 emissions was accompanied by increase of 24% of energy consumption, and for non-associated countries a decrease in CO2 by 39% was accompanied by an increase in consumption of electricity by 38%. This is a very clear indication of the policy pursued in almost all countries aimed at reducing the nuisance of electricity production processes.

## 4. Conclusions

When analyzing data on electricity consumption in European countries, apart from the noticeable increase in consumption, environmental care is also recognized. Despite the ever-increasing consumption, the fastest rate in the countries undergoing structural changes, the level of emissions of harmful substances - especially CO2 - is constantly decreasing. Convergence is generally observed in terms of electricity consumption, CO2 emissions (in a positive sense) and ECDTL minimization. However, in detail, it turns out to be the largest, permanently unresolved problem, especially for less developed countries. This is likely not only due to the shortage of funds, but also to the problems of repair or replacement of existing infrastructure during its use.

## 5. References

[1] Piłatowska M Włodarczyk A Zawada M 2015 CO2 Emissions, Energy Consumption and Economic Growth in the EU Countries: Evidence from Threshold Cointegration Analysis (2015 12th International Conference on the European Energy Market (EEM), Lisbon) pp 1-5

[2] https://data.worldbank.org

[3] Kucęba R Bylok F Pabian A Zawada M 2014 Prosumer Energy Dimension in the Conditions of Sustainable Micro-region Development in the EU (ICSSAM, International Conference on Social Science and Management. ISEPSS. International Symposium on Education Psychology and Social Sciences May 2014 Kyoto Japan Conference Proceedings) pp 1040–1051

[4] Zawada M and Szajt M 2016 Analysis and Evaluation of Air Pollution Level in Czestochowa in the Years 2012-2015 (Acta Oeconomica Cassoviensia Vol 9) pp 14-28

[5] The Directive of the European Parliament and the Council 2004/107/WE of 15 December 2004 on arsenic, cadmium, mercury, nickel and polycyclic aromatic hydrocarbons in the air (O. J. EU L 23 of 26.01.2005) p 3
[6] The Directive of the European Parliament and the Council 2008/50/WE of 21 May 2008 on air quality and cleaner air for Europe (O. J. EU L 152 of 11.06.2008) p 1
[7] The Council Implementing Decision 2011/850/WE of 12 December 2011 setting the rules for the use of directives 2004/107/WE & 2008/50/WE of the European Parliament and the Council with regard to a system of information and reports exchange on the air quality (O. J. L 335 of 17.12.2011) pp 86 - 106
[8] http://www.bbc.com/news/uk-england-york-north-yorkshire-35803048
[9] Zawada M 2007 Economic Development and Electric Energy Consumption in the European Union Countries-Copmarative Analysis (Acta Universitatis Lodziensis Folia Oeconomica nr 212) pp 259-272

IOP Conf. Series: Earth and Environmental Science **164** (2018) 012013 doi:10.1088/1755-1315/164/1/012013

# Performance Analysis of Multi-Photovoltaic (PV)-Grid Tied Plant in Malaysia

**Laith M Halabi** [1, a], **Saad Mekhilef**[1, b]

[1] Power Electronics and Renewable Energy Research Laboratory (PEARL), Department of Electrical Engineering, Faculty of Engineering, University of Malaya, 50603 Kuala Lumpur, Malaysia.

Emails: [a]l.halabi@outlook.com; [b]saad@um.edu.my

**Abstract.** This paper evaluates and analyzes the performance of three different types of photovoltaic (PV) arrays connected to utility-grid without any storage system in Malaysia. The three types modules include Poly-Crystalline, Mono-Crystalline and Thin-Film modules, each works separately and are connected to three DC/AC inverters where it monitored and controlled over an IP network system. The system installed and monitored for one-year period and this is the first work describing such systems of different PV types at the same plant in the proposed location. The system continuously collects and stores all data sets of DC/AC current, voltage, frequency and power for a 5-minutes period. Besides, the weather parameters that include ambient temperature, module temperature, solar irradiation and wind speed. The performance of this plant is evaluated by analyzing the collected data sets and then applying performance indices such as final energy output, array yield, final yield, reference yield, system efficiency and performance ratio. The results show that the PV performance is depending directly on the solar irradiation as well as the surrounding environmental conditions. Also, the results show a very-good efficiency of the proposed modules and demonstrate the potential of using such systems in any area with the similar conditions.

## 1. Introduction

Fossil fuels are the main causes of environmental and global warming. Besides, it is suffering from leakage and transportation problems which make the need for an alternative source of energy a significant issue in the meantime [1, 2]. In addition, oil prices are exposed to change due to internal and regional conflicts [3, 4]. Thus, many countries are now moving toward using renewable sources in energy generation sector. Malaysia is a vital example of using renewable energy (RE) sources as a main source of energy [5]. Whereas, the global accumulative installed capacity increased from 1.4 GW in 2000 to around 102.2 GW in 2012 [6]. PV technology is expected to play a significant role in future developments in developed and developing countries[7]. As more attention is being focused on their viability from technical and economical assessment [8]. In the recent years, many works were established to study the performance of PV plant around the world in both small and large scale systems [9-12]. Another study established in Malaysia to study the performance of two decentralized hybrid PV/diesel generators/batteries stations. The results demonstrate the significant role of analyzing the operational performance of hybrid renewable energy systems in order to avoid the disadvantages and get all proposed advantages associated with such systems [13-15]. The literature shows the important of studying the performance of the PV systems. This study is the first study describes and analyzes the performance of three different PV arrays with a total of 6.725 $kW_p$, installed on the roof of Electronics And Renewable Energy Laboratory (PEARL) at University of Malaya (UM), Kuala Lumpur, Malaysia. Poly-Crystalline, Mono-Crystalline and Thin-Film modules are connected to the utility grid without storage system. IEC 61724 standard's [16] parameters are using in evaluating the performance of this plant including; final energy output, performance ratio, array, final, reference yields and system efficiency.

## 2. Method and system description

Gird tide PV system is describing in this paper, is installed on the roof of PEARL Laboratory which located on (3.118 N) latitude, (101.654 E) longitude and the wind speed sensor located at (62) m high above sea level. The PV system consists of three types; Poly-Crystalline PV modules (MITSUBISHI brand), Mono-Crystalline PV models (SHELL brand) and Thin-Film (SHARP brand). All modules are unshaded facing the south at an azimuth of 180° and a fixed angle of 10° where all are fixed with an open back so it allows the air to flow and cool the modules. The system is using three DC/AC inverters, one for each type of PV modules. The three inverters types are SMA string inverters, two rated of 1600W and one of 2500W at rated 230V and 50Hz. Figure 1 simplified block diagram shows the whole system combination. The performance of the studied system is carrying out using performance parameters included in IEC 61724 standard [16] such as final energy output, array, final and reference yields, system efficiency, and performance ratio.

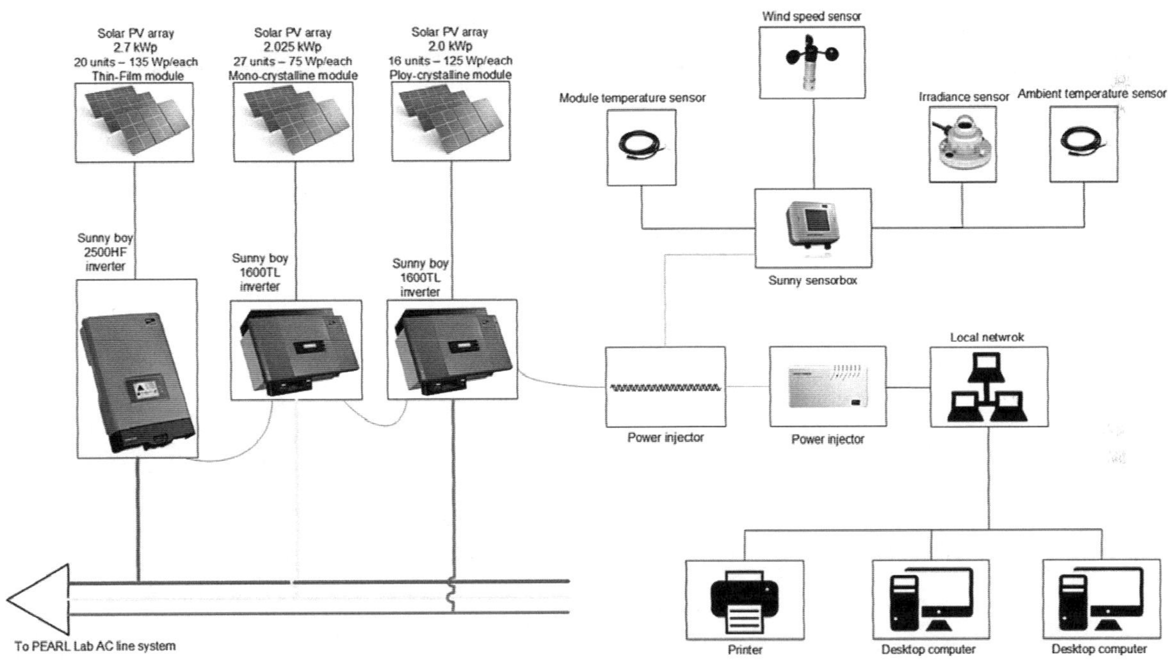


**Figure 1.** A simplified block diagram of PEARL Laboratory's grid-tied PV system at University of Malaya, Malaysia.

## 3. Results and discussion

The daily average irradiation, ambient temperature, modules temperature and wind speed are varying in the range of (4.28-3.26) kWh/m$^2$/day, (30.23-32.96) °C, (35.77-39.64) °C and (0.75-1.33) m/s respectively. All PV modules are connected to the three inverters (one for each type), both modules and inverters were explained in the previous section (2). However, the performance of these devices is analyzed in this section. 16 poly-crystalline modules form one string connected in series, 27 mono-crystalline modules are connected in series to form one string, and 20 Thin-Film modules are connected to form 4 strings in parallel each contains 5 modules in series. One-year data sets are used to describe the performance where the monthly variation is shown in Figure 2 based on daily average energy.

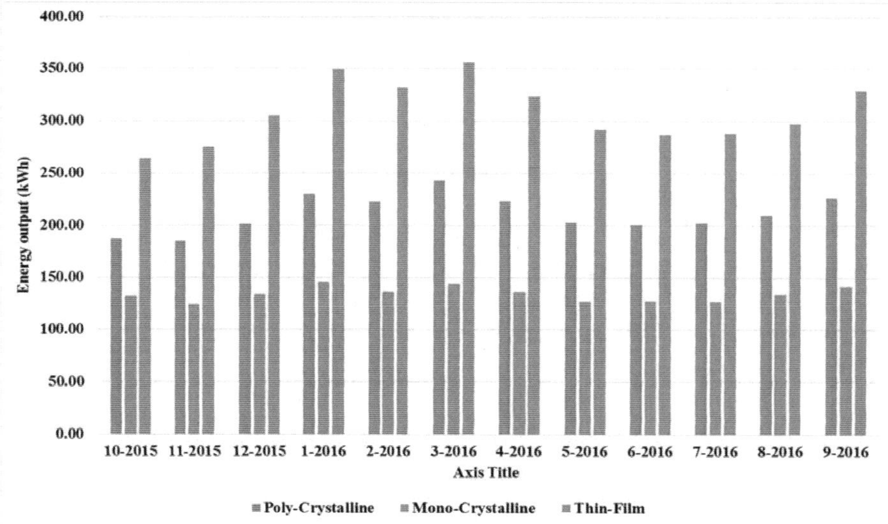


**Figure 2.** Monthly average generated energy variation of all PV modules.

Poly- crystalline modules output energy is ranged from 185.33 kWh in November to 242.75 kWh in March. Whereas, the mono-crystalline output is ranged from 124.78 kWh in November to 146.04 kWh in January. However, the Thin-Film output is ranged from 264.27 kWh in October to 356.35 kWh in March. The output varies in accord to the solar irradiation based on daily average variation. The daily average array yield and system yield based on monthly variation are shown in Figure 3 (a) and (b) respectively. Where, the total generated annual energy is 7849.58 kWh.

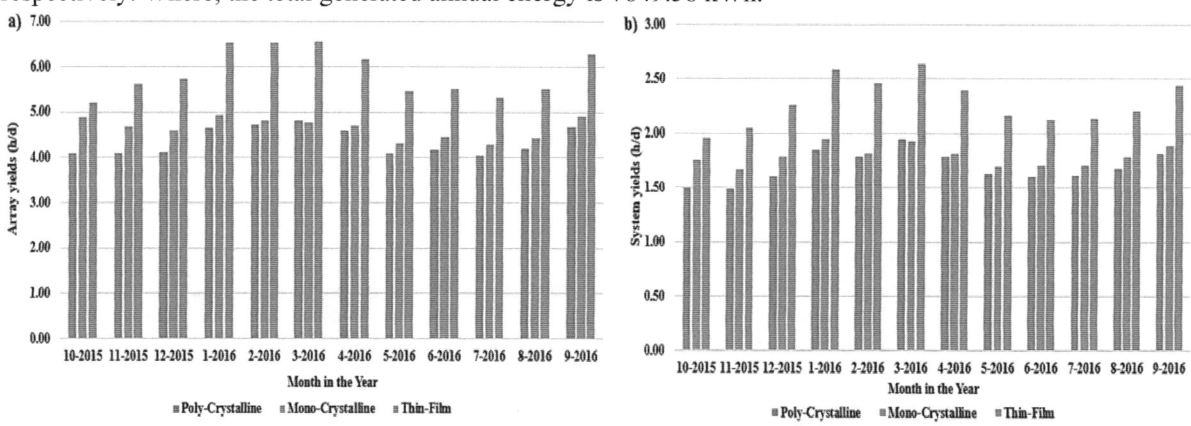


**Figure 3.** (a) Array and (b) System yields.

The array and system yields of polycrystalline vary from a minimum of 4.82 h/d to a maximum of 4.04 h/d. Whereas, vary in a range of 4.29 h/d and 4.92 h/d of monocrystalline and 5.20 h/d and 6.56 h/d of Thin-Film modules. It's clear that the array and system yields are directly proportional to the daily average solar radiation were both are proposed to reference yield that is presented in Figure 4.

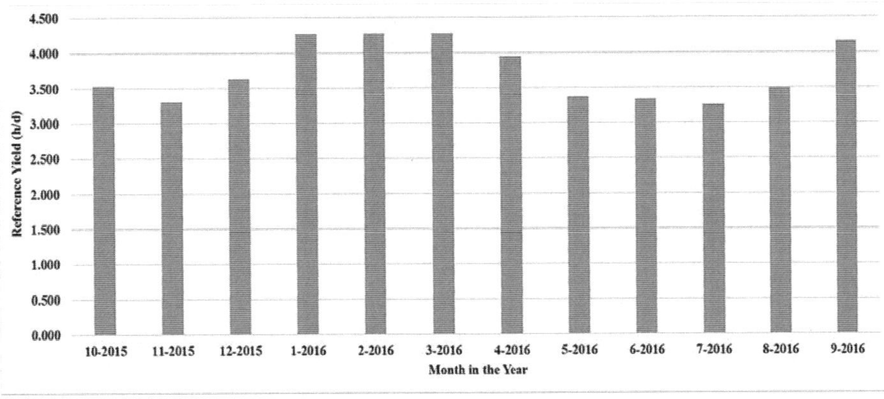


**Figure 4.** System reference yield.

During the monitored period there was an expected difference due to DC/AC conversion losses by the inverters where the performance ratio indicates the overall effect of losses on the rated output due to many reasons of inverters efficiency, mismatch, dust, snow, etc. and equipment failure [8]. The efficiency for all modules are varied in a range of (4-11) % with an average of ~ 7.5%, since the system was initially run and the system did not show the maximum efficiency of the modules among this period. The inverters efficiency is varied between 85.36% to 97.15% based on monthly bases and the average value for one year of the inverter system of 93.93%. However, the DC performance ratio based on yearly variation is shown in Figure 5. It varies from 91.3% in 2015 to 88.2% in 2016 with a total average of 89.75% for the whole monitored period.

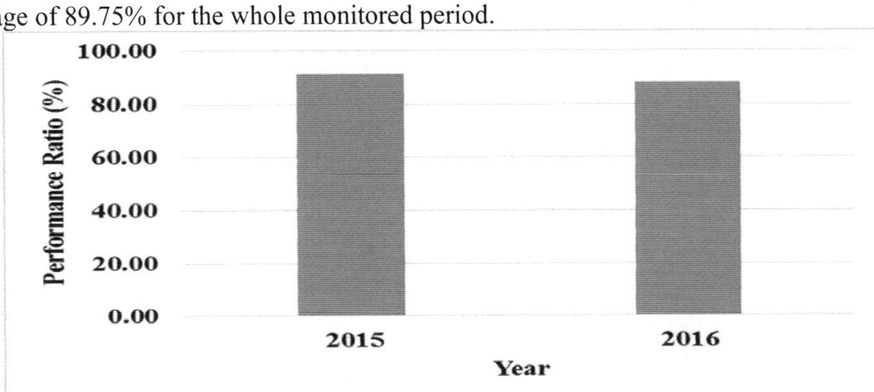


**Figure 5.** Yearly performance ratio (%).

## 4. Conclusion

A gird connected PV system installed on the roof of Power Electronics And Renewable Energy Laboratory (PEARL) at University of Malaya (UM), Kuala Lumpur, Malaysia is presenting in this paper. The installation and operational performance of a three PV array models including; is Poly-Crystalline (2.0 $kW_p$), Mono-Crystalline (2.025 $kW_p$) and Thin-Film (2.7 $kW_p$) are considered. A description of the complete photovoltaic power plant settings is also presented in this work which shows the PV arrays, controlling system and inverters unit. The analysis of this study showed that sunshine hours have the major effects on the PV modules during the year. However, clouds, haze, rains, etc. which slightly affect the final monthly output of the modules since Malaysia has no season variation where one hot, humid and rainy season is found all the year. The operational modes of the entire system while integrating with the grid have been described in details. Month-wise energy generated varied between (185.33 -242.75) kWh, (124.78 - 146.04) kWh and from (264.27 - 356.35) kWh for Poly-Crystalline, Mono-Crystalline and Thin-Film modules respectively, and the total annual generated energy output is 7849.58 kWh. The efficiency of the PV module varied between 4% and 11% and the inverter efficiency varied between 85.36% to 97.15% based on monthly bases and the average value for one year of the inverter system of 93.93%. Based on the monitored values from the system, the performance ratio (PR) was calculated for every month and the average annual PR is found to be 89.75%. The results show that the PV output energy is varied in accord with the solar irradiation and show almost the same performance among the year as there is no season variability in the studied location where one hot, humid and rainy season is found among the year. Thus, it is observing that the performance not only depending on the solar insolation PV array size, but also on the surrounding environment. This study indicates the high interest of such a system to be installed and used in this region.

## 5. References

[1] Phuangpornpitak, N. and S. Kumar, *PV hybrid systems for rural electrification in Thailand.* Renewable & Sustainable Energy Reviews, 2007. 11(7): pp. 1530-1543.

[2] Olatomiwa, L., et al., *Techno-economic analysis of hybrid PV–diesel–battery and PV–wind–diesel–battery power systems for mobile BTS: the way forward for rural development.* Energy Science & Engineering, 2015. 3(4): pp. 271-285.

[3] Adaramola, M.S. and E.E. Vågnes, *Preliminary assessment of a small-scale rooftop PV-grid tied in Norwegian climatic conditions.* Energy Conversion and Management, 2015. 90: pp. 458-465.

[4] Olatomiwa, L., S. Mekhilef, and O.S. Ohunakin, *Hybrid renewable power supply for rural health clinics (RHC) in six geo-political zones of Nigeria.* Sustainable Energy Technologies and Assessments, 2016. 13: pp. 1-12.

[5] Mekhilef, S., et al., *Solar energy in Malaysia: current state and prospects.* Renewable and Sustainable Energy Reviews, 2012. 16(1): pp. 386-396.

[6] Association, E.P.I. *Global Market Outlook for Photovoltaics 2013–2017.* 2013.

[7] Dreidy, M., H. Mokhlis, and S. Mekhilef, *Application of Meta-Heuristic Techniques for Optimal Load Shedding in Islanded Distribution Network with High Penetration of Solar PV Generation.* Energies, 2017. 10(2): pp. 150.

[8] Adaramola, M.S., *Techno-economic analysis of a 2.1 kW rooftop photovoltaic-grid-tied system based on actual performance.* Energy Conversion and Management, 2015. 101: pp. 85-93.

[9] Allouhi, A., et al., *Grid-connected PV systems installed on institutional buildings: Technology comparison, energy analysis and economic performance.* Energy and Buildings, 2016. 130: pp. 188-201.

[10] Emziane, M. and M. Al Ali, *Performance assessment of rooftop PV systems in Abu Dhabi.* Energy and Buildings, 2015. 108: pp. 101-105.

[11] Aarich, N., et al. *"Propre. Ma" project: roadmap & preliminary results for grid-connected PV yields maps in Morocco.* in *Renewable and Sustainable Energy Conference (IRSEC), 2014 International.* 2014. IEEE.

[12] Sidi, C.E.B.E., et al., *Performance analysis of the first large-scale (15MW p) grid-connected photovoltaic plant in Mauritania.* Energy Conversion and Management, 2016. 119: pp. 411-421.

[13] Halabi, L.M., et al., *Performance analysis of hybrid PV/diesel/battery system using HOMER: A case study Sabah, Malaysia.* Energy Conversion and Management, 2017. 144: pp. 322-339.

[14] U. K. Das, T. K. Soon, M. Seyedmahmoudian, S. Mekhilef, M. Y. I. Idris, W. V. Deventer, B. Horan and A. Stojcevski, *Forecasting of photovoltaic power generation and model optimization: A review*, Renewable and Sustainable Energy Reviews, 2018. vol. 81, pp. 912-928.

[15] M. Hossain, S. Mekhilef, and L. Olatomiwa, *Performance evaluation of a stand-alone PV-wind-diesel-battery hybrid system feasible for a large resort center in South China Sea, Malaysia*, Sustainable Cities and Society, 2017. vol. 28, pp. 358-366.

[16] IEC61724, *Photovoltaic system performance monitoring—guidelines for measurement, data exchange and analysis.* 1998, International Electrotechnical Commission.

IOP Conf. Series: Earth and Environmental Science **164** (2018) 012014 doi:10.1088/1755-1315/164/1/012014

# Control and energy management of grid-connected pv system with battery-supercapacitor hybrid energy storage

**A Guichi[1], A Talha[1], El Berkouk[2] and N Sabeur [3]**

[1]University of Sciences and Technology Haouari Boumediene, ALGERIA
[2]Laboratoire de Commande de Processus, ENP Alger, ALGERIA
[3]Power Electronics and Renewable Energy Research Laboratory (PEARL), Department of Electrical Engineering, University of Malaya, Kuala Lumpur, Malaysia

guichi.omar@gmail.com

**Abstract.** An optimized energy management technique to reduce the cost and increase the effectiveness of a photovoltaic (PV) system is proposed in this paper. It is based on the optimum sizing of the energy storage system, to meet the needs of the system in normal conditions, while in the critical cases, the AC residential load will be supplied by the grid itself. In addition, the surplus of the PV power will be injected to the grid. However, to extend the battery life and reduce its size, a supercapacitor has been combined with the batteries, to quickly respond to the short-time peak current demand, while the battery will deliver a smooth current. Furthermore, the proposed algorithm will supervise the battery state of charge (SOC) and satisfy the highest priority load demand under different situation. A simulation work is carried out to evaluate the control of the whole PV system and determine the performance of proposed algorithm. Finally, the obtained results show the effectiveness of the proposed management algorithm, including the control techniques of different components of the whole system.

## 1. Introduction

The climate change, the fossil fuel exhaustion and the shifting to the low-carbon economy are the driving forces behind the accelerating deployment and the development of renewable energy[1], In the recent years, photovoltaics (PV) have impressively invade the international energy markets at exponential growth rates [2], it has been shifted from being simple power source for those who do not have grid access to an alternative of conventional energy sources. However, the impressive expansion of the photovoltaic systems has requested an enhancement of the power quality to meet different loads requirements, such as voltage stability, supplying continuous and instantaneous power, fast transient response. Lowest cost way that permits the PV system to supply various loads and meet their needs, is to combine batteries and supercapacitor to create hybrid storage system.

By utilizing the battery-supercapacitor hybrid storage system, the PV system can effectively gain beneficial properties of both components, which leads to reducing the size of the battery banks and increase the battery lifetime. The battery banks are able to store high energy density and release it over a long time; on the contrary, the supercapacitor has high power density, it is capable to charge very fast and release large quantity of power in short period of time[3],[4].

In this work, the aforementioned hybrid storage system is integrated in a PV-grid connected system in order to be controlled and tested under different cases, which is shown in Figure 1. Connecting this system to the grid has two main benefits; injecting the surplus energy to the grid and using the grid as a backup system to supply the load in the critical cases, this will help to avoid the batteries oversized. Then, to manage and to optimize the photovoltaic power under different scenarios, a newly algorithm has been proposed, it takes into account, the PV extracted power, the battery state of charge (SOC), the quick response of the supercapacitor, the load needs, and the grid-injected power as shown in Figure 1.

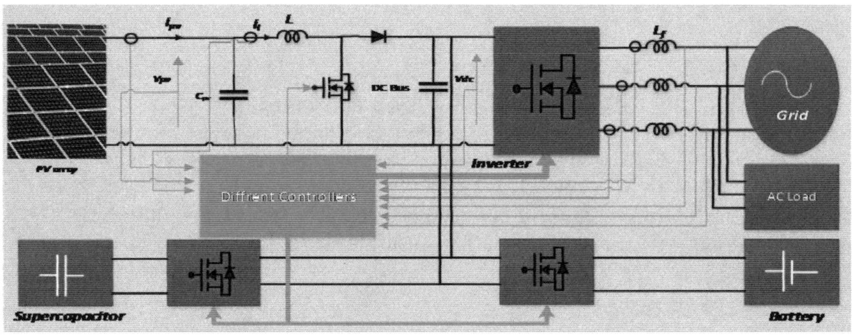


**Figure 1.** Block diagram of the proposed PV system

The simulation of the hole system is carried out in MATLAB Simpowersystems, the control of the different converter will be discussed in the next section.

## 2. PV array modeling

In this work, a commonly one diode, five parameters solar cell model has been used [5], which is shown in Figure 2. The output current is the difference between the photocurrent $I_{ph}$ and the diode current $I_D$ with the shunt resistance current $I_{sh}$ as presented in equation (1) .

$$I_{pv} = I_{ph} - I_0\left[exp\left(\frac{V_{pv}+R_sI}{V_{therm}\,m}\right) - 1\right] - \frac{V_{pv}+R_sI}{R_{sh}} \quad (1)$$

Where $I_0$ is the diode dark saturation current , $V_{therm}$ ($=KT/q$) is the thermal voltage of the cell, $K$ is Boltsmann constant ($1.3806503 \times 10^{-23}$ J/K), T (Kelvin) is the temperature of the p-n junction of the diode, $q$ is the electron charge; Rs and Rsh are the series and shunt resistance of the cell, respectively; and $m$ is the ideality factor of the diode.

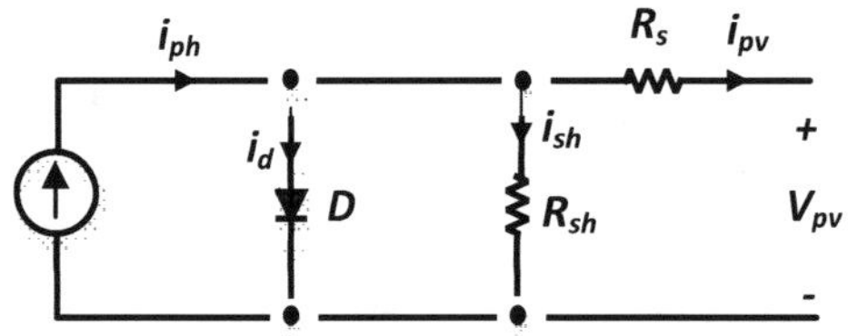


**Figure 2.** One diode photovoltaic cell model

## 3. PV system description

Figure 1 depicts the configuration of the hole system, the PV generator is connected to DC-DC boost converter, which is used to extract the PV array's maximum power, the battery and the supercapacitor are combined to create a hybrid storage buck-up system, the remaining energy is stored and recovered from the storage system through bidirectional buck-boost converter, it operates in buck mode when the battery starts saving power, while the boost mode is used for injecting power to the AC load or to the grid. Besides the DC-DC converters, the DC bus linked the PV to the AC load through a three phase

IOP Conf. Series: Earth and Environmental Science **164** (2018) 012014 doi:10.1088/1755-1315/164/1/012014

inverter and to the grid using the same inverter connected a inductor filter $L_f$, this later is used to filter out the ripples in the injected current.

## 4. Hybrid storage system configuration

The optimal active hybrid system topology was chosen to connect the battery and supercapacitor in one combination. It has the advantage of combining their advantages and solving the disadvantages of supercapacitor voltage variation and matching by placing DC-DC converter between the supercapacitor and the DC bus [6]. The power requested $P_{str\_ref}$ by the supervisory is met by supply base power from the battery $P_{bat}$ and the transient power from the supercapacitor $P_{sc}$ as shown in Figure 3.a, the references of the both powers are obtained from low pass filter as shown in Figure 3.b. while Figure 3.c depicts the output powers of the battery and the supercapacitor according to the given reference.

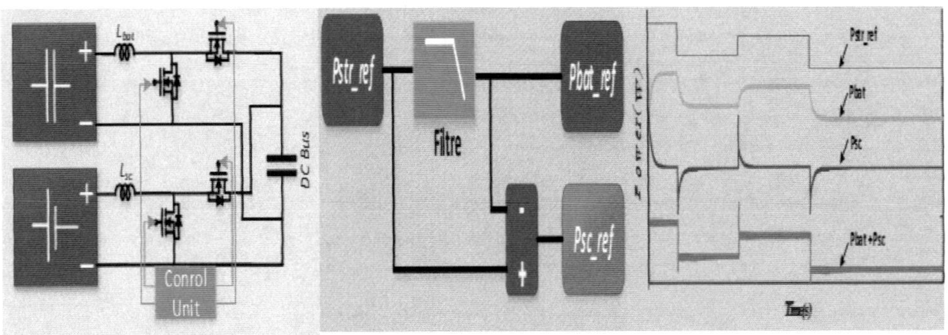


**Figure 3.**Topology and control of Storage system (a) parallel topology (b) power control (c)delivered power

## 5. Control techniques

To maintain the system in the maximum power point (MPP), Perturbation and Observation (P&O) algorithm has been used [7], in addintion, for keeping the DC bus voltage at a constant value and reaching a unity power factor, the voltage orinted control (VOC)technique has been applied to the three phase inverter[8], while the phase locked loop (PLL) technique was used to synchronize the inverter with the grid, finally, the battery and supercapacitor control diagram is shown in Figure 4.

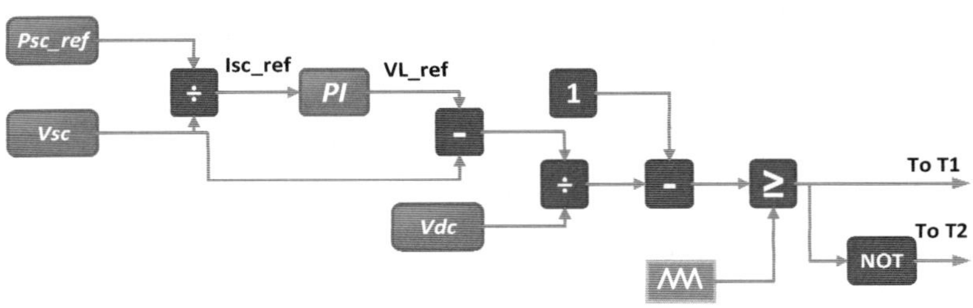


**Figure 4.** Supercapacitor control diagram

## 6. Results and discussion

In order to validate the performance of the proposed system, to check the behavior of all components that are working together, and to verify the management algorithm under all the possible cases, a typical day has been simulated, by adopting 1 hour to 1 second and simulating the system for 24 seconds. In addition, the battery capacity (Ah) was divided by 3600 Ah, so that it will be possible to simulate one cycle. Comparing Figure 5 of the solar irradiation and Figure 7 of the PV output power, it can be clearly seen that maximum power has been extracted over the day. For the other results and figures, it is useful to analyze them according to load variation and to the battery SOC.

IOP Conf. Series: Earth and Environmental Science **164** (2018) 012014 doi:10.1088/1755-1315/164/1/012014

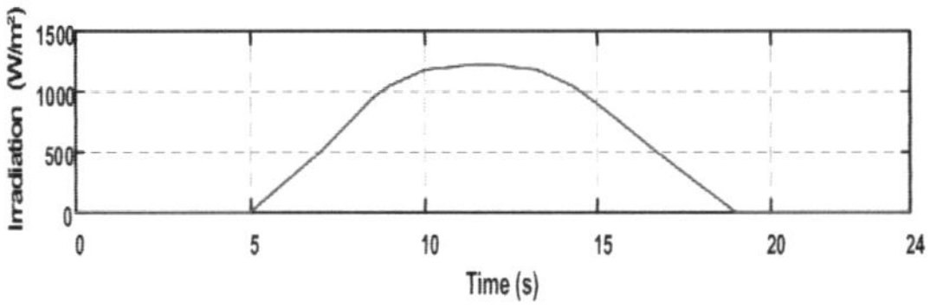


**Figure 5**. Solar irradiation

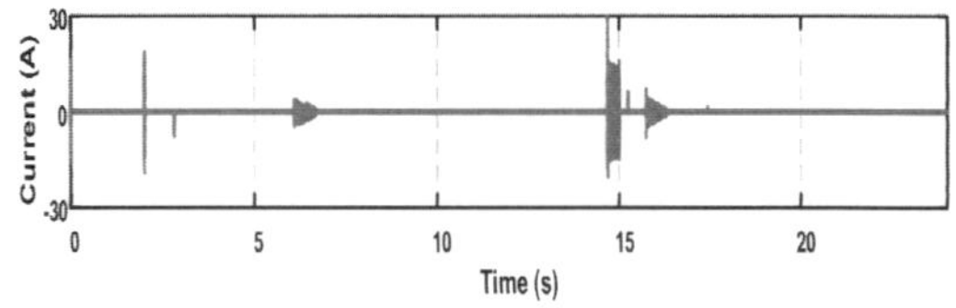


**Figure 6**. Grid current

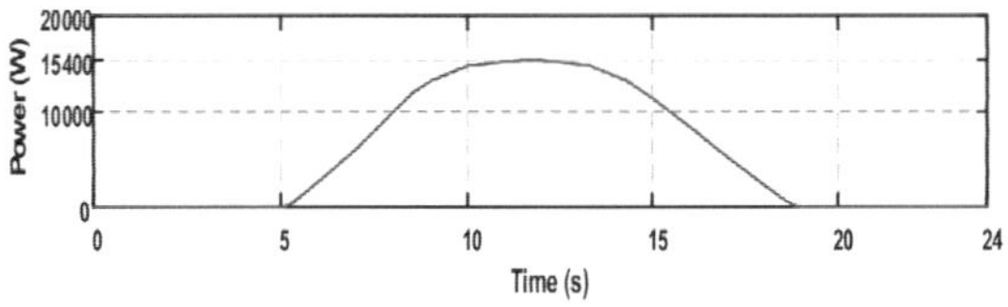


**Figure 7**. PV output power

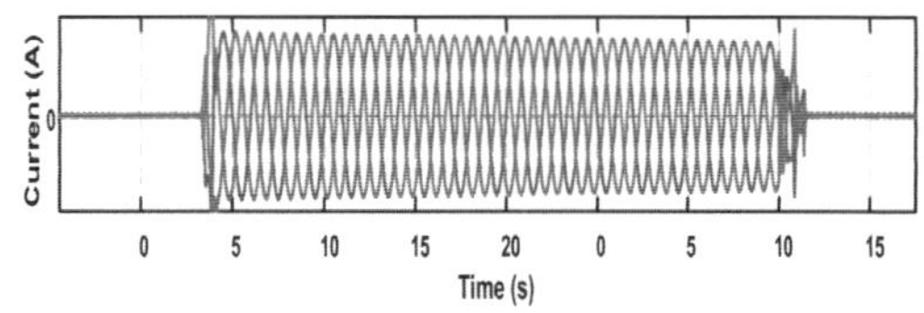


**Figure 8**.Zoom of the grid current

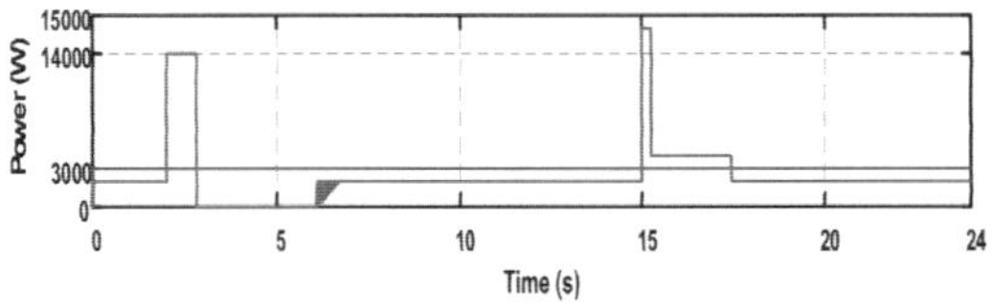


**Figure 9**.Loads consumed power

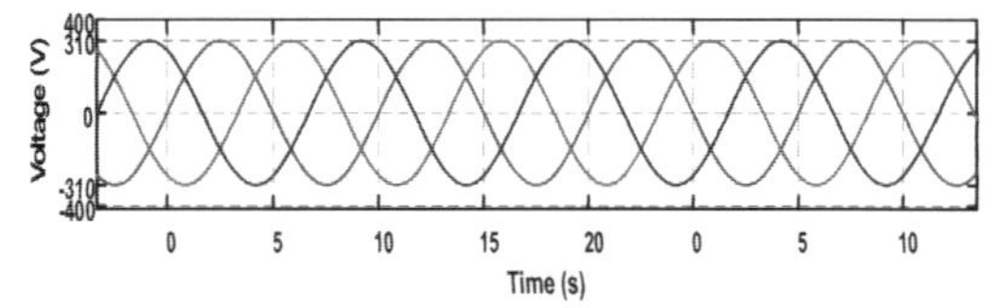


**Figure 10**. Grid voltage

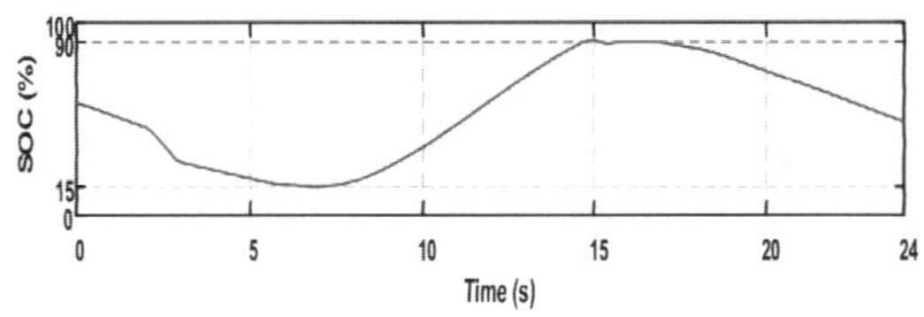


**Figure 11**.Battery SOC

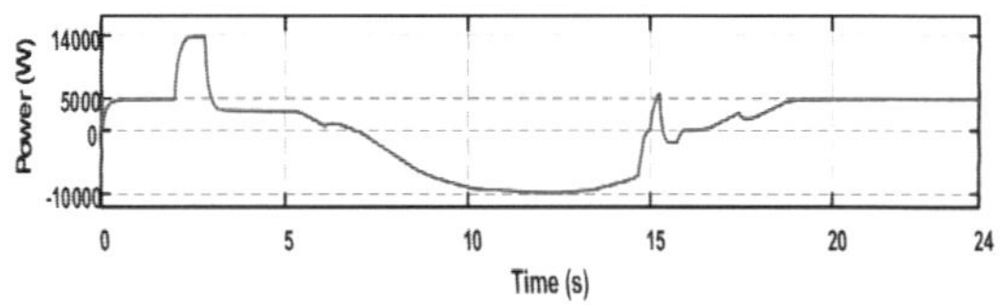


**Figure 12**. Battery power

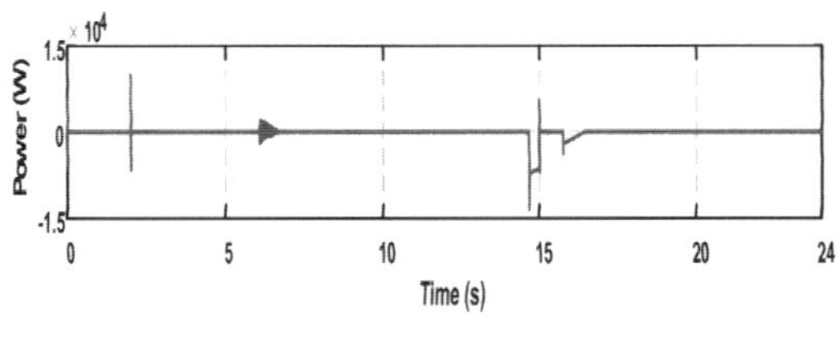


**Figure 13**. Delivered grid power

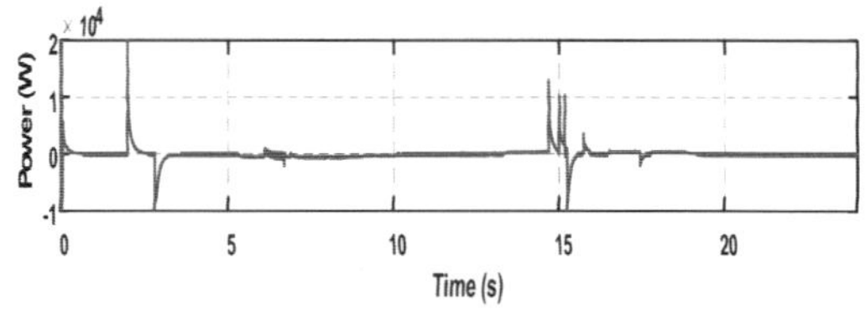


**Figure 14**. Supercapacitor power

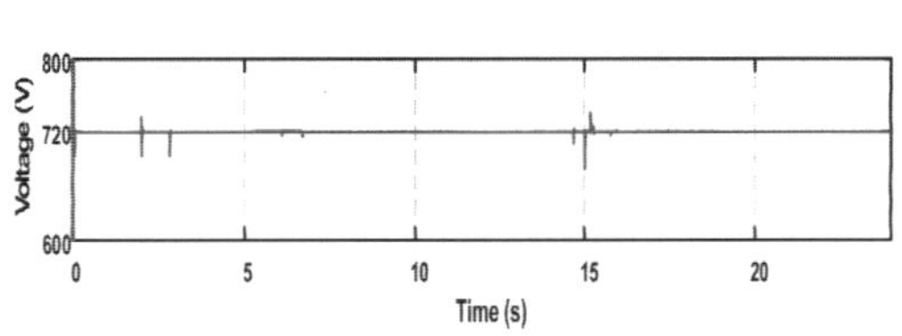


**Figure 15**. DC bus voltage

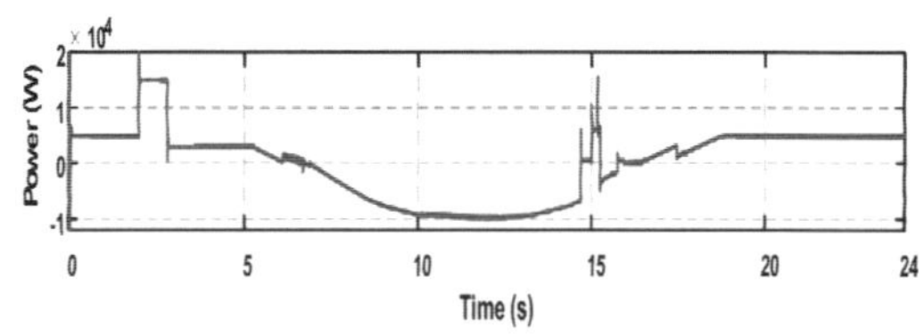


**Figure 16**. Hybrid system storage power

**From 1s to 5s:** In this period, it is assumed to be night time, both loads were supplied by the storage system. However, for protecting the battery against deep discharging, the load of lowest priority had switched off at *t=2.8s*, when the *SOC* reach 30%.

**From 5s to 15s:** The PV started generating power but the storage system is still needed to supply the highest priority load, at *t=6s*, the delivered power reached the needs of the load and the supervisory switched back on the second load. It is clear from Figure 11 that the battery started charging also; meanwhile, at *t=14s*, when the *SOC* reached 90%, the supervisory stopped the battery charging to be protected against the overcharging and the surplus energy delivered to the grid.

**From 16s to 24s:** It can be clearly seen from the three Figures 9,11 and 13 that during this interval, the battery reached again 90% after contributing with the PV to deliver a huge amount of energy to the loads, where a small power was injected to the grid, after that, the battery came back to support the PV.

Furthermore, Figure 15 shows how the used regulator kept the DC bus voltage constant despite the huge power demand for short time period. Meanwhile, Figure 12 and 14 evidently demonstrate the high complementarity between the battery and the supercapacitor. The current injected to the grid is shown in Figure 6 with its zoom in Figure 8, while the grid voltage is given in Figure 10.

## 7. Conclusion

The simulation results confirm the high efficiency of the proposed management algorithm, as well as the benefits of using hybrid energy storage system. The PV system has been simulated in the worst cases, where the load requested a peak current, even though, the hybrid storage system responded to this demand without destabilizing the system performance.

## References

[1] C. Bustos, D. Watts, and M. Ayala, "Financial risk reduction in photovoltaic projects through ocean- atmospheric oscillations modeling," *Renew. Sustain. Energy Rev.*, vol. 74, no. October 2015, pp. 548–568, 2017.

[2] S. Jacobsson and V. Lauber, "The politics and policy of energy system transformation — explaining the German diffusion of renewable energy technology," vol. 34, pp. 256–276, 2006.

[3] C. Kagiri and X. Xia, "Optimal control of a hybrid Battery / Supercapacitor storage for Neighborhood Electric Vehicles," *Energy Procedia*, vol. 105, pp. 2145–2150, 2017.

[4] T. Mesbahi, F. Khenfri, N. Rizoug, P. Bartholomeüs, and P. Le Moigne, "Combined Optimal Sizing and Control of Li - Ion Battery / Supercapacitor Embedded Power Supply Using Hybrid Particle Swarm - Nelder - Mead Algorithm," vol. 3029, no. c, 2016.

[5] M. G. Villalva, J. R. Gazoli, and E. R. Filho, "Comprehensive Approach to Modeling and Simulation of Photovoltaic Arrays," vol. 24, no. 5, pp. 1198–1208, 2009.

[6] A. Kuperman and I. Aharon, "Battery – ultracapacitor hybrids for pulsed current loads : A review," *Renew. Sustain. Energy Rev.*, vol. 15, no. 2, pp. 981–992, 2011.

[7] R. Kadri, J. P. Gaubert, and G. Champenois, "An Improved maximum power point tracking for photovoltaic grid-connected inverter based on voltage-oriented control," *IEEE Trans. Ind. Electron.*, vol. 58, no. 1, pp. 66–75, 2011.

[8] A. Bouharchouche, E. M. Berkouk, and T. Ghennam, "Modeling and Control of an Integrated Photovoltaic Generator," no. 1, pp. 16–17, 2013.

IOP Conf. Series: Earth and Environmental Science **164** (2018) 012015 doi:10.1088/1755-1315/164/1/012015

# A New Under-Frequency Load Shedding Scheme Based on Adaptive Neuro-Fuzzy Inference System and Evolutionary Programming Shedding Priority

**Mohammad Dridy [*,1], Hazlie Mokhlis[**,1], Saad Mekhilef [***,1]**

[1] Power Electronics and Renewable Energy Research Laboratory (PEARL), Department of Electrical Engineering, Faculty of Engineering, University of Malaya, 50603 Kuala Lumpur, Malaysia.

Authors' Emails: * mohamaddradi@gmail, **hazli@um.edu.my, ***saad@um.edu.my

**Abstract.**Frequency stability of islanded distribution system is a topic of interest, due to the significant penetration of distributed generation (DG). This paper proposes a new under frequency load shedding technique, which address the over-shedding issue caused by fixed priority. The proposed scheme consists of two units; first, the power imbalance estimation unit (PIEU), which use adaptive neuro-fuzzy inference system (ANFIS) to estimates the power imbalance. Second, the load shed unit (LSU), which is based on binary evolutionary programming (BEP) technique to shed the optimal loads. To validate the performance of the proposed UFLS scheme, different simulation studies have been conducted using PSCAD software. Moreover, the response of the proposed scheme is compared with another UFLS scheme having fixed priority loads. From the simulation results, the proposed scheme can shed the optimal loads, which is not achieved in the existing load shedding schemes.

## 1. Introduction

Recently, controlling the frequency within the permissible limits during islanding operation is the most important technical challenge being investigated worldwide. Hence, alternative techniques are required to enhance the reliability of today's complex distribution network. This can be mitigated by using accurate load shedding (UFLS) scheme, which balances the system. Accurate load shedding depends upon two main factors; accurate estimation of power imbalance and accurate amount of load to be shed. In the literature, several load shedding schemes have been proposed: conventional, adaptive, and intelligent. The conventional UFLS is the most applicable in this case [1-3]. In this technique, a pre-determined load values will be shed based on frequency value. Although this technique is very simple, it is suffering from over-shedding issue. For this reason, the adaptive UFLS technique comes with advantage of using swing equation to calculate the imbalance of power. Further efforts to accurately estimate the power imbalance were the application of computational intelligence based techniques such as artificial neural network (ANN) [4-6], fuzzy logic control [7-9], particle swarm optimization technique (PSO) [10, 11], and genetic algorithm (GA) [12, 13]. From reviewing different load shedding techniques, there is inaccurate load shedding issue since every technique is bounded by fixed priorities shedding load. Fixed load shedding priority is linked to the loads that are shed sequentially based on lookup table which consist of critical and non-critical loads. For this reason, an UFLS scheme called fixed and random load shed priority (FRLSP) is proposed to shed the optimal combination of loads [14]. However, study proposed in [14] takes quite a while, since it goes through all the possible combination of loads. In this paper, an optimal UFLS scheme is proposed to stabilize the frequency of islanded distribution network. This scheme consists of two main units: (a) power imbalance estimation unit (b) load shed unit. For the former, an adaptive neuro-fuzzy inference system (ANFIS) is used to estimates the power imbalance. While for the latter, a BEP technique is used to shed the optimal number of loads from fixed and variable priority loads [15].

## 2. Method and system description

This paper proposed a new UFLS strategy, which combine precision in estimating the imbalance power and the appropriate shedding loads. Proposed UFLS strategy is based on both frequency and *df/dt* values to estimates the power imbalance. The ANFIS uses these values as input and intelligently estimates the power imbalance during disturbances. After estimating the power imbalance, this value is sent to LSU for shedding loads according to load priority using BEP. Where, loads from 1-10 are selected to take random priority, and loads from 11-12 are selected to take fixed priority. Figure 1. illustrates the overview of load shedding scheme for an islanded distribution network, which is modeled using PSCAD software. This network consists of two mini-hydro generators, each rated 2MVA, and four photovoltaic units each rated, 0.5 MW.

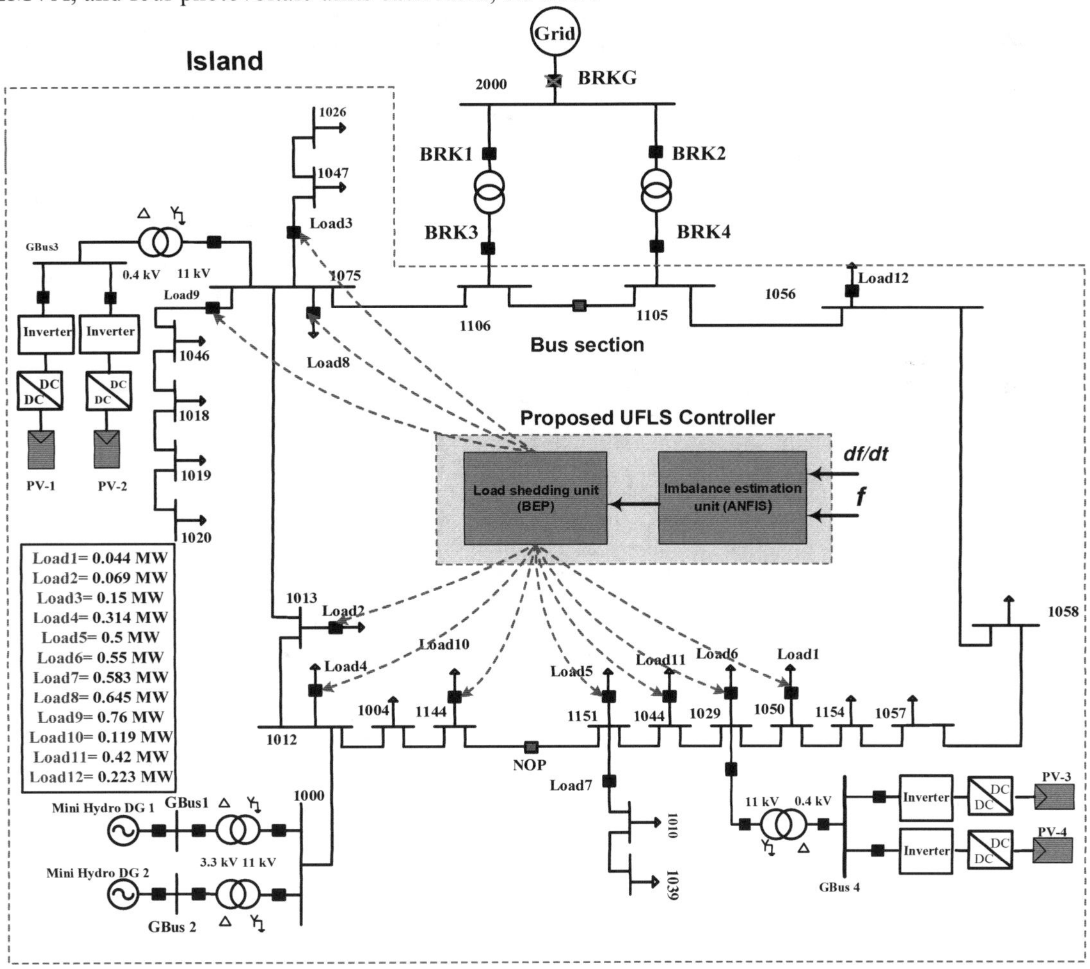


**Figure 1.** The overall test system

IOP Conf. Series: Earth and Environmental Science **164** (2018) 012015 doi:10.1088/1755-1315/164/1/012015

## 3. Methodology of proposed UFLS scheme

The methodology of proposed UFLS scheme is shown in Figure 2. When a disturbance (islanding, load increment, generator trip) occurs in the distribution network, PIEU estimates the amount of load to be shed by using the ANFIS. Since, the PSCAD does not provide ANFIS tool, it is designed and trained in MATLAB software as shown in Figure 3. The ANFIS has two inputs (frequency and rate of change of frequency) and one output power imbalance (ΔP). After calculating the amount of load to be shed, the system frequency is checked to show whether it drop to certain value (49.5 Hz). If this happens, The LSU receive the amount of load to be shed value. Then, if this value is larger than the total random priority loads, the LSU directly shed all random priority loads and start shedding from fixed priority loads. Otherwise, the BEP is initialized to shed the optimal combination of loads.

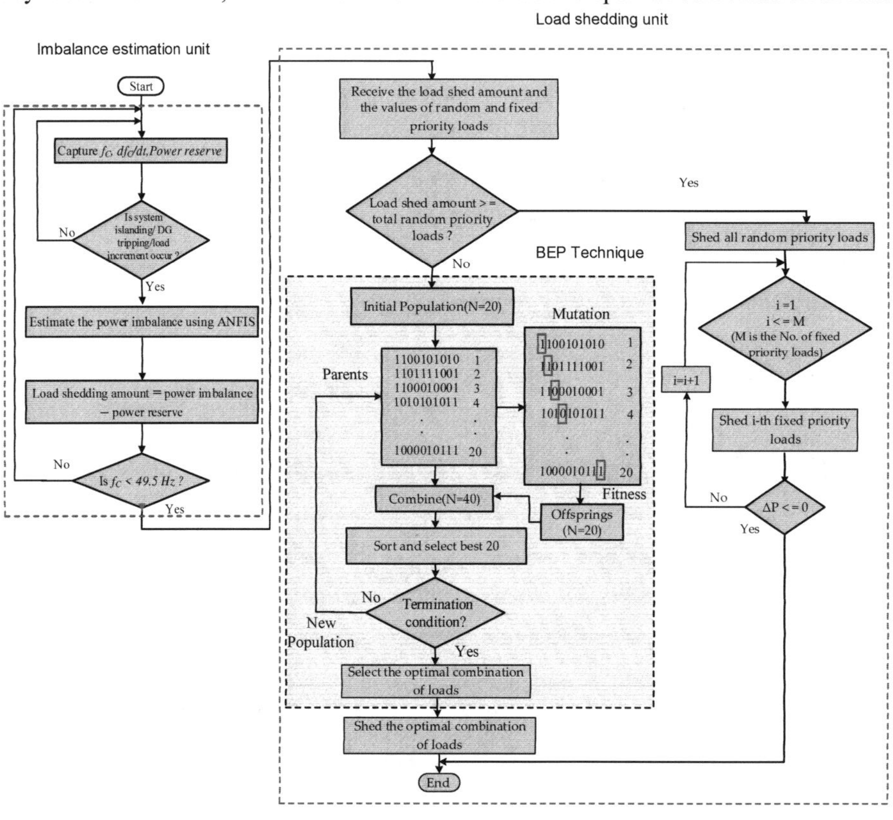


**Figure 2**. Proposed UFLS scheme flow chart

## 4. Results

In this paper, a comparative study is implemented to show the preference of proposed UFLS scheme over the adaptive and conventional UFLS schemes. This comparative study is performed for load increment of 1 MW occurring at 40s after islanding. Immediately after islanding, the system frequency begins to decline in response to excess loads (0.32 MW). Accordingly, the mini-hydro generators use their spinning reserves (0.48 MW) to cover the unbalance power. At 40s after islanding, the total power demand will be 6.68 MW. For this reason, the UFLS automatically activate to stop the frequency declination by shedding the (loads 2,3,8) for proposed UFLS scheme, (loads 1-5) for adaptive and conventional schemes. The frequency responses of three scheme and adaptive UFLS scheme are shown in Figure 3.

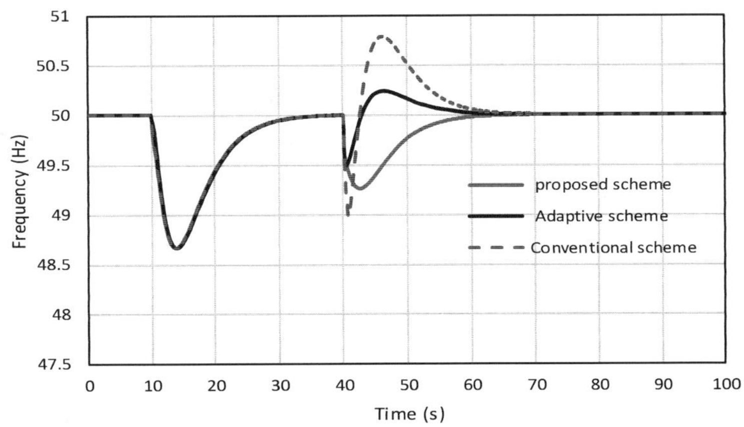


**Figure 3.** The Frequency response for 1-MW load increment

As shown in Figure 3, it is noticed that due to fixed priority of loads, the adaptive and conventional UFLS schemes shed more load (1.07 MW), which leads to overshoot in the system frequency. However, the proposed UFLS scheme shed less load (0.84 MW) and recover the system frequency without any overshooting. To show the advantage of proposed scheme over FRPSL scheme, another comparative study is implemented as shown in Figure 4. Immediately after islanding, the system's frequency begins to decline in response to excess loads (0.32 MW). Accordingly, the mini-hydro generators recover the frequency. At 40s after islanding, a mini-hydro DG of (1.71 MW) is tripped from the islanded microgrid. For this reason, all load shedding techniques are initiated to restore the system frequency by shedding the same amount of power (1.63 MW). However, FRPSL technique fails to prevent the system's frequency from dropping below 47.5 Hz, which leads to total blackout. In fact, the large execution time of this technique is the main reason of operation inability.

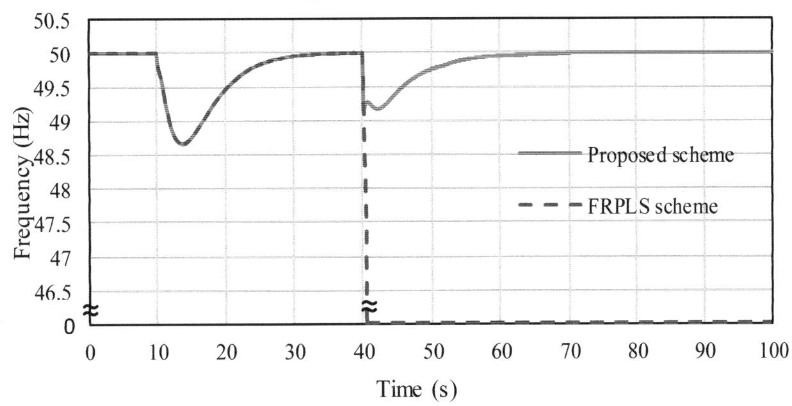


**Figure 4.** Frequency response for mini hydro DG tripping event

## 5. Conclusion

This paper proposed a new UFLS scheme that is suitable for islanding distribution networks. This scheme utilizes the application of ANFIS and BEP techniques to determine the optimal combination of loads needed to be shed from the network. As concluded from the simulation results, the proposed UFLS scheme combines the advantage of high speed response with the ability to separate the optimal loads, therefore, this UFLS scheme can improve the reliability of the distribution network after islanding.

IOP Conf. Series: Earth and Environmental Science **164** (2018) 012015 doi:10.1088/1755-1315/164/1/012015

## References

[1] D. Xu and A. A. Girgis, "Optimal load shedding strategy in power systems with distributed generation," in *Power Engineering Society Winter Meeting, 2001. IEEE*, 2001, pp. 788-793.
[2] J. Laghari, H. Mokhlis, A. Bakar, and H. Mohamad, "Application of computational intelligence techniques for load shedding in power systems: A review," *energy conversion and management,* vol. 75, pp. 130-140, 2013.
[3] J. Tang, J. Liu, F. Ponci, and A. Monti, "Adaptive load shedding based on combined frequency and voltage stability assessment using synchrophasor measurements," *IEEE Transactions on power systems,* vol. 28, pp. 2035-2047, 2013.
[4] C.-T. Hsu, H.-J. Chuang, and C.-S. Chen, "Adaptive load shedding for an industrial petroleum cogeneration system," *Expert Systems with Applications,* vol. 38, pp. 13967-13974, 2011.
[5] R. Hooshmand and M. Moazzami, "Optimal design of adaptive under frequency load shedding using artificial neural networks in isolated power system," *International Journal of Electrical Power & Energy Systems,* vol. 42, pp. 220-228, 2012.
[6] S. Javadian, M.-R. Haghifam, S. Bathaee, and M. F. Firoozabad, "Adaptive centralized protection scheme for distribution systems with DG using risk analysis for protective devices placement," *International Journal of Electrical Power & Energy Systems,* vol. 44, pp. 337-345, 2013.
[7] S. Tso, T. Zhu, Q. Zeng, and K. Lo, "Evaluation of load shedding to prevent dynamic voltage instability based on extended fuzzy reasoning," *IEE Proceedings-Generation, Transmission and Distribution,* vol. 144, pp. 81-86, 1997.
[8] J. Sasikala and M. Ramaswamy, "Fuzzy based load shedding strategies for avoiding voltage collapse," *Applied Soft Computing,* vol. 11, pp. 3179-3185, 2011.
[9] H. Mokhlis, J. Laghari, A. Bakar, and M. Karimi, "A fuzzy based under-frequency load shedding scheme for islanded distribution network connected with DG," *Int Rev Electr Eng,* vol. 7, pp. 4992-5000, 2012.
[10] N. Sadati, T. Amraee, and A. Ranjbar, "A global particle swarm-based-simulated annealing optimization technique for under-voltage load shedding problem," *Applied Soft Computing,* vol. 9, pp. 652-657, 2009.
[11] A. El-Zonkoly, M. Saad, and R. Khalil, "New algorithm based on CLPSO for controlled islanding of distribution systems," *International Journal of Electrical Power & Energy Systems,* vol. 45, pp. 391-403, 2013.
[12] Y.-Y. Hong and P.-H. Chen, "Genetic-based underfrequency load shedding in a stand-alone power system considering fuzzy loads," *IEEE Transactions on Power Delivery,* vol. 27, pp. 87-95, 2012.
[13] W. Luan, M. R. Irving, and J. S. Daniel, "Genetic algorithm for supply restoration and optimal load shedding in power system distribution networks," *IEE Proceedings-Generation, Transmission and Distribution,* vol. 149, pp. 145-151, 2002.
[14] J. Laghari, H. Mokhlis, M. Karimi, A. H. A. Bakar, and H. Mohamad, "A new under-frequency load shedding technique based on combination of fixed and random priority of loads for smart grid applications," *IEEE Transactions on Power Systems,* vol. 30, pp. 2507-2515, 2015.
[15] M. Dreidy, H. Mokhlis, and S. Mekhilef, "Application of Meta-Heuristic Techniques for Optimal Load Shedding in Islanded Distribution Network with High Penetration of Solar PV Generation," *Energies,* vol. 10, p. 150, 2017.

# Effect Method Evaluation of Forecasting Scour by Propeller in Harbours

**P T Ghazvinei[1*], H H Darvishi[1], S Soon[2]**

[1]Department of Civil Engineering, Technical and Engineering College, Ale Taha University, Tehran, Iran / Mailing address Ale Taha University, Koohsar Blv., Shahran, Tehran, Iran
[2]Department of Civil Engineering, Faculty of Engineering, University of Malaya, 50603 Kuala Lumpur, Malaysia

p.taherei@gmail.com

**Abstract.** The scour induced by a propeller wash is experimentally in uniform and non-uniform sand. A series of laboratory experiment with different experimental parameters was conducted in the current research. Moreover, the characteristics of scour hole were studied and compared including depth, length, and width between uniformly and non-uniformly distributed sand. Equations of predicting maximum scour depth, proposed by previous researchers, were evaluated using the current experimental data. Results revealed that there is a need to improve the study in propeller-induced seabed scour in non-uniform sand.

## 1. Introduction

Ship's propeller jet (commonly referred to as propeller wash) is the backwash produced by its rotating propeller, which is usually a high velocity flow used to propel a vessel [1]. Seabed without adequate protection system could be vulnerable to scouring from the wash, particularly in shallow and confined water [2]. Although most damage to structure occur at the flash stormy times, undermining the structure s' foundations in a relatively long duration of time make the foundation weak for the final failure at the flash stormy times conditions [3]. Low under-keel clearances happen in harbour basins, where can cause severe scouring, and consequent jeopardize the structures in harbour by undermining its foundation. Consequently, this leads to the need for costly maintenance and repairing work [4].
A number of case studies relevant to the propulsion action of ships caused serious damage in various harbours and ports, which have been reported over the past three decades [2, 5-12]. The understanding of the extent of seabed scour due to propeller wash is essential to minimize the risk of failure of harbour structures. The interaction between propeller jet and the seabed was also described by Sumer [13] and Gaythwaite [14]. Research on propeller-induced seabed scour was initiated in an unconfined propeller jet. The influences of the rudder [15, 16], hull and the berth geometry to the velocity within the jet were included later on. Johnston, Hamill [15] investigated the scouring action in an unconfined propeller jet. Moreover, Hong, Wu [17] studied the time-dependent maximum scour depth induced by an unconfined jet where Ryan, Hamill [18] carried out an experimental investigation of scour in conjunction with quay walls in harbour using Artificial Neural Networks (ANNs). The review of the literature showed that existing experimental studies of propeller-induced seabed scour by previous

* Corresponding author: p.taherei@gmail.com

researchers were conducted on uniformly-distributed non-cohesive sediments, where practically in ports area, sediments are non-uniformly-distributed. In addition, it was found that the amount of empirical data based on unreliable observations and simulating at the actual conditions of the scour were limited [19]. Non-cohesive sediments were used in the experiments to provide a suitable range of applicability, with one uniformly and one non-uniformly distributed sands. The characteristics of scour hole such as depth, length, and width between uniformly and non-uniformly distributed sand were observed through a series of the laboratory experiments in the current research. The outcomes improve the knowledge of the propeller-induced seabed scour in uniformly and specially in non-uniformly-distributed non-cohesive sediments.

## 2. Experimental setup

The experiments were conducted in a 3 m × 1.2 m × 1 m deep rectangular acrylic tank to observe scour process. A three-bladed, fixed-pitched, open propeller was used to model the scour in the bed. Table 1 shows the propeller characteristics.

**Table 1.** Propeller characteristics.

| Propeller Diameter (mm) | Hub Diameter (mm) | Number of Blades | Coefficient of Thrust | Pitch ratio | Blade Area Ratio | Rudder Angle |
|---|---|---|---|---|---|---|
| 220 | 80 | 3 | 0.4 | 1 | 0.45 | 0 |

To neglect scaling effects of the viscosity, values of the flow and propeller Reynolds Numbers were designed to be larger than 3×103 and 7×104.

Two types of non-cohesive sediments were applied in the experimentations to make available reliable range of applicability, in both uniform (d50 = 0.9 mm) and non-uniform (d50 = 0.75 mm) distribution of the sands with geometric standard deviations equal to 1.4 and 2 for uniform and non-uniform sediments, respectively. Figure 1 shows the schematic layout of the experimental setup. The sediments of 2.1 m long × 1.2 m wide × 0.2 m deep was laid flat in the water tank. The propeller was turn on with two different speed 350 and 500 rpm, for each series of experiments. Propeller was turned off at time = 0.5, 1, 2, 4, 8, 16, 32, 64, and 128 hours after the commencement of each test without any disturbance. Scour depth was measured using digital point gauge on a 100 mm x 100 mm grid, with depth measure along the scour hole.

## 3. Methods

The evolution of scouring profile along the longitudinal direction (downstream of the propeller) of the current study showed similarity and good agreement to Hong et al. (2013)'s experiments resulted a small scour hole lower than the propeller, meanwhile a deposition downstream was next to the primary scour hole within the existence of four stages of scouring process. A primary scour and deposition with ripples occurred downstream of the scour hole was observed downstream side of the propeller. The size of main scour hole increased with the increasing time during developing stage. A small scour hole is formed below the propeller as the propeller drains fluid from behind the propeller. The experiments were continued and ended after 128 hours to ensure an asymptotic condition.

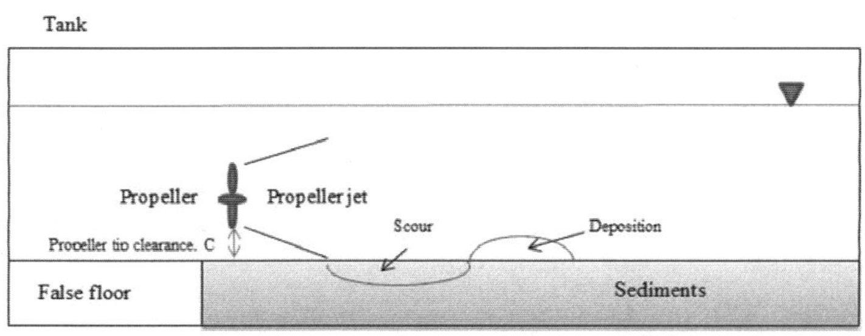


**Figure 1.** Side view layout of experimental setup.

## 4. Results and discussion

### *4.1 Scour profiles under asymptotic condition*

Figure 2 compares the asymptotic dimensional scour profile induced by propeller in uniformly-distributed and non-uniformly distributed sand for the experiments within different offset height and rotational speed. It was found that the scour hole prompted by propeller in uniform sand is deeper than the one in non-uniform sand with the similar clearance and rotational speed. Furthermore, deposition formed in uniform sand was slightly thicker than non-uniform.

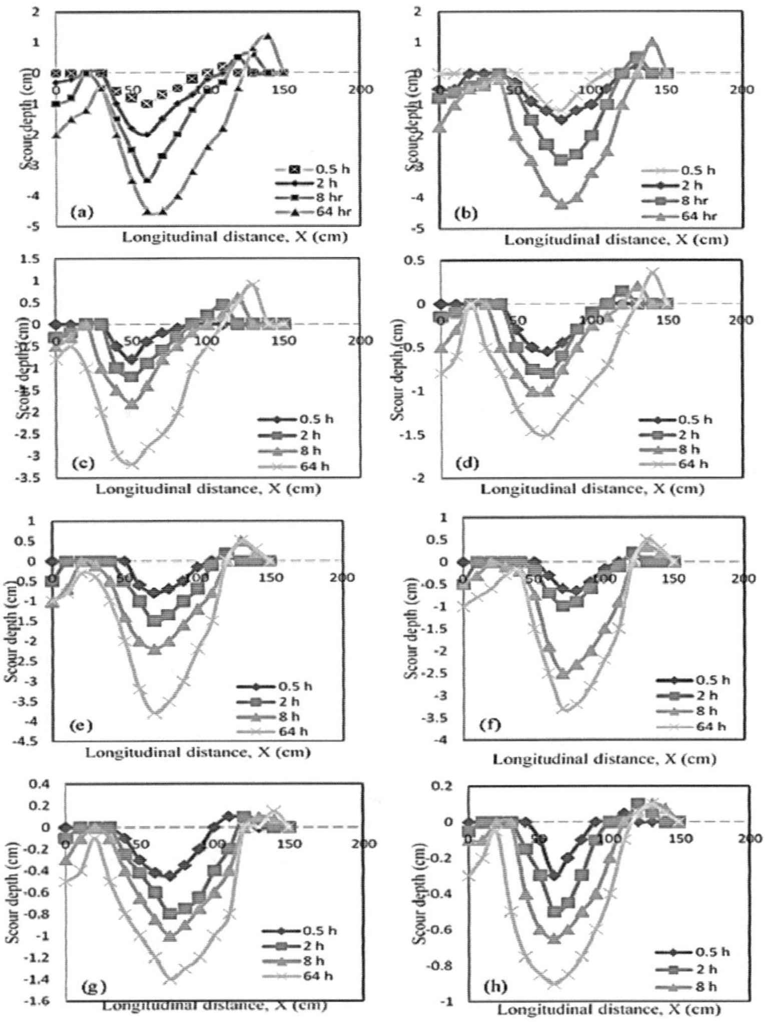


**Figure 2.** Typical scour profiles induced by propeller jet in uniform sand (a) 500rpm, 0.5Dp (b) 500rpm, 1Dp (c) 350rpm, 0.5Dp (d) 350rpm, 1Dp; typical scour profiles induced by propeller jet in non-uniform sand at four different stages for: (e) 500rpm, 0.5Dp (f) 500rpm, 1Dp (g) 350rpm, 0.5Dp (h) 350rpm, 1Dp.

### *4.2 Evaluating the existing methods in prediction maximum scour depth*

Ryan and Hamill (2011) suggested an equations to calculate the erosion depth in a combined existence of berth structures and a rudder. The equation was offered $\varepsilon_{max}$ as the maximum scour depth for the associated $\Omega$ and $\Gamma$ parameters as below:

$$\varepsilon_{max} = \Omega\,[\ln(t)]^{\Gamma} \qquad (1)$$

Table 2 shows the range of applicable and current experimental parameters using the above equations. Compare with the current experimental characteristics, equations. (1) to (3) are not suitable to be used in the current experiments. Furthermore, Hong et al. (2013) suggested a semi-empirical equation to

IOP Conf. Series: Earth and Environmental Science **164** (2018) 012016 doi:10.1088/1755-1315/164/1/012016

predict the time-dependent maximum depth of the scour. The dimensionless equation for the estimation of time-dependent scour depth induced by a propeller wash was:

$$\frac{d_{s,t}}{D_p} = k_1 \left[ log_{10}\left(\frac{U_0 t}{D_p}\right) - k_2 \right]^{k_3} \qquad (2)$$

$$k_1 = 0.014\, F_o^{1.120} \left(\frac{y_o}{D_p}\right)^{-1.740} \left(\frac{y_o}{d_{50}}\right)^{-0.170} \qquad (3)$$

$$k_2 = 1.882\, F_o^{-0.009} \left(\frac{y_o}{D_p}\right)^{2.302} \left(\frac{y_o}{d_{50}}\right)^{-0.441} \qquad (4)$$

$$k_3 = 2.477\, F_o^{-0.073} \left(\frac{y_o}{D_p}\right)^{0.53} \left(\frac{y_o}{d_{50}}\right)^{-0.045} \qquad (5)$$

where, ds,t is the time-dependent maximum scour depth, Uo is the efflux velocity of propeller jet.

**Table 2.** Range of experimental parameters

| | Range of applicable experimental parameters | Range of current experimental parameters |
|---|---|---|
| $F_o$ | 4.87 - 11.20 | 11.72 - 16.74 |
| $C/d_{50}$ | 19.40 - 85.53 | 122.22 - 293.33 |
| $D_p/d_{50}$ | 39.10 - 172.37 | 244.44 - 293.33 |
| $Y_w/D_p$ | 1.52 - 3.05 | 2.73 |
| α | 0 to -35° | 0° |

The proposed equations were demonstrated to be valid with good accuracy for scour data in uniform sand, in which the measured scour depth and predicted scour depth were mostly in the zone of acceptable agreement (less than 10% differences compared with the line of perfect agreement). While in non-uniform sand, the difference between the measured and predicted scour depth were almost out of the zone of acceptable agreement (exceeded 10% differences compared with the line of perfect agreement). Figure 3 compares the measured and predicted depth of maximum scour proposed by equation (2) with the line of perfect agreement in both uniformly-distributed and non-uniformly distributed sand. The figure shows that maximum scour depth predicted in non-uniformly distributed sand using equation (2) is deeper than uniformly distributed sand, which contradict to the experimental results. The measured data showed that scour depths induced by propeller wash were deeper in uniform sand, while using the published equation, the scour depths were deeper in non-uniform sand as shown in the figure 3. This implies that the equation (2) has over predicted the maximum scour depth in non-uniform sand. A more accurate equation is therefore needed to improve the study of propeller-induced scour in non-uniform sand.

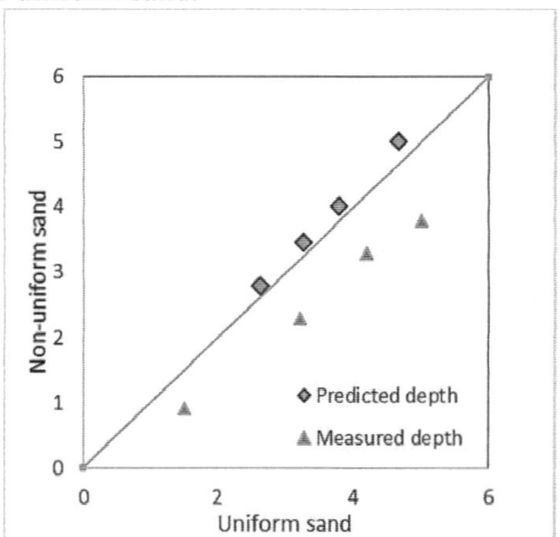

**Figure 3.** The measured and predicted depth of maximum scour proposed by equation (4) with the line of perfect agreement in both uniformly-distributed and non-uniformly distributed sand.

## 5. Conclusions

Experiments on ship's propeller jet-induced seabed scour in uniform and non-uniform sands with two different propeller clearance and two different propeller rotational speeds were conducted. The evolution of scouring profile downstream of the propeller showed a similarity to the previous published results of the literatures. There is a small scour hole below the propeller, a primary scour hole downstream that was followed by deposition. Recent equations of predicting the time-dependent maximum scour depth proposed by Hong, Wu [17] was used in the current study. While these proposed equations have been demonstrated to be applicable with good accuracy with less than 10% differences for predicting scour depth in uniform sand, there has been considerable uncertainty and over prediction when using these equations to predict scour in non-uniformly sand. Therefore, more series of laboratory test in propeller-induced scour in non-uniform sand is conducted to enhance the existing predictive methods of the maximum scour.

## 6. References

[1]. Demirbilek, Z., "Review of Hydrodynamics of High-Speed Marine Vehicles by Odd M. Faltinsen: Cambridge University Press, New York, 2006. Price: $100.00. pp. 454". 2006, American Society of Civil Engineers.

[2]. Evans, G., "Seabed Protection Systems to prevent Scour from High-Speed Ships", No. (2010),

[3]. Ghazvinei, P.T., et al., "Assessment of local scour at bridges abutment", *Research Journal of Applied Sciences, Engineering and Technology*, 8, No. 3, (2014), p. 296-304.

[4]. Hashmi, H.N., "Erosion of a granular bed at a quay wall by a ship's screw wash", No. (1995),

[5]. Bergh, H. and K. Cederwall, Propeller erosion in harbours. Hydraulics Laboratory, Royal Institute of Technology Stockholm, Sweden; 1981.

[6]. Bergh, H. and N. Magnusson, "Propeller erosion and protection methods used in ferry terminals in the port of Stockholm. Bulletin PIANC 58", No. (1987),

[7]. Brewster, P., et al. "A CFD model of a marine propeller wash". in *The Seventh International Offshore and Polar Engineering Conference*. International Society of Offshore and Polar Engineers, (1997), p.

[8]. Chait, S., "Undermining of quay walls at South African ports due to the use of bow thrusters and other propeller units", No. (1987),

[9]. Longe, J., P. Hebert, and R. Byl, "Erosion problems at existing quay constructions due to bow thrusters and main propellers of ships when berthing or leaving", *Bulletin of the Permanent International Association of Navigation Congresses [PIANC]*, No. 58, (1987),

[10]. McKillen, H.G., A Model and Field Study of Ship Propulsion Induced Bed Movements at Berths. Queen's University of Belfast; 1985.

[11]. Robakiewicz, M., "Bottom erosion as an effect of ship propeller action near the harbour quays", *Bulletin of the Permanent International Association of Navigation Congresses [PIANC]*, No. 58, (1987),

[12]. Verhey, H., T. Blokland, and M. Bogaerts, "Experiences in the Netherlands with quay structures subjected to velocities created by bow thrusters and main propellers of mooring and unmooring ships", *Bulletin of the Permanent International Association of Navigation Congresses [PIANC]*, No. 58, (1987),

[13]. Sumer, B.M., The mechanics of scour in the marine environment. World Scientific; 2002.

[14]. Gaythwaite, J.W., "Design of Marine Facilities for Berthing", *Mooring and Repair of*, No. (2004),

[15]. Johnston, H.T., et al., "Influence of a boundary on the development of a propeller wash", *Ocean Engineering*, 61, No. (2013), p. 50-55.

[16]. Svedman, C., et al., "Plasma proteins in a standardised skin mini-erosion (I): permeability changes as a function of time", *BMC dermatology*, 2, No. 1, (2002), p. 3.

[17]. Hong, S., et al., "Microstructure and cavitation–silt erosion behavior of high-velocity oxygen–fuel (HVOF) sprayed Cr 3 C 2–NiCr coating", *Surface and Coatings Technology*, 225, No. (2013), p. 85-91.

[18]. Ryan, D., G.A. Hamill, and H.T. Johnston, "Determining propeller induced erosion alongside quay walls in harbours using Artificial Neural Networks", *Ocean Engineering*, 59, No. (2013), p. 142-151.

[19]. Ghazvinei, P.T., et al. "Contraction scour analysis at protruding bridge abutments". in *Proceedings of the Institution of Civil Engineers-Bridge Engineering*. Thomas Telford Ltd, (2015), p. 39-53.

# Enhanced wind-driven infiltration model by incorporating wind direction for design purpose

**L G Perdamaian[1], Z J Zhai[2]**

[1]Department of Nuclear Engineering and Engineering Physics,
Universitas Gadjah Mada, Jalan Grafika 2 Yogyakarta 55281 – Indonesia
[2]Department of Civil, Environmental, and Architectural Engineering
University of Colorado Boulder, UCB 428, Boulder, CO 80309-0428 – USA

E-mail : [1]laksana_gp@mail.ugm.ac.id, [2]john.zhai@colorado.edu

**Abstract.** Research has been conducted to improve and enhance existing LBNL infiltration model in order to incorporate wind direction effect on flow rate. Existing wind driven infiltration of LBNL model only consider terrain and shielding parameter as correction factor. Supplemental analytical model considered in this research was Swami-Chandra model, which the calculation was based on average wind pressure coefficient (Cp). It was evident that for same wind direction, Cp value will not change regardless of wind velocity. Process of simulation shows that infiltration modelling has to be performed in three-dimensional environment with transient calculation in order to achieve convergent result. Wind direction strongly affects infiltration flow rate. However, as the infiltration hole has relatively small size compared to wind velocity, flow rate variation is marginal for energy-based design. Nevertheless, for pollutant and particulate matter analysis, wind direction will give significant change in term of concentration and flow direction.

## 1. Introduction

Both infiltration and ventilation are vital in determining the indoor air quality. Ventilation, both mechanical and natural are most likely controlled either automatically or manually. Meanwhile, infiltration is unintentional and can introduce unwanted contaminants from outside and either hot air from the outside that increasing cooling load or bring cold air inside and increase the heating load. Study of infiltration has been conducted in decades and researcher has developed many kinds of infiltration model. However, the nature of airflow is very difficult to predict and model, because of the uncertainty in turbulence, vortex, and flow separation, for example. Therefore, from design purpose, the model is highly practical to use but prone to error and inaccuracies. An alternative to those analytical model is using computational fluid dynamics. Usage of CFD has increased along with the improvement of computing capabilities. CFD can model varieties of fluid flow with many physical parameter, therefore better result can be achieved.

## 2. Literature Review

### *2.1. Building Infiltration Model*

Building infiltration model can be divided into single zone and multiple zone model. A zone is defined as fully mixed volume with a constant concentration level of the enclosed gas mixture [1]. Single zone model is used in a building which has virtually no partition, whereas multi zone model is suitable for multi-story building with complex partition.

Single zone infiltration can be used to determine impact of infiltration airflow on whole building energy balance. EnergyPlus used simplified single zone model to calculate infiltration flow rate. Two models are used, Effective Leakage Area which based on LBNL [2] model and Flow Coefficient which based on Walker and Wilson method [3]. Both of the models also can be found on ASHRAE Handbook as Basic and Enhanced method, respectively [4].

One of the infiltration calculation method that widely used is LBL infiltration method. Originally the LBL infiltration method was developed for residential building. LBL infiltration uses several parameters which are leakage area, height of the building, inside-outside temperature difference, terrain class of the structure, and wind speed [5]. Infiltration airflow rate through a leakage can be expressed by following equation.

$$Q = A\sqrt{\frac{2}{\rho}\Delta P} \tag{1}$$

Where

Q = infiltration rate ($m^3/s$)

A = effective leakage area ($m^2$)

ρ = density of air ($kg/m^3$)

ΔP = pressure difference (Pa).

Based on its driving forces, there are two different type of infiltration, which are wind regime and stack regime. Wind regime infiltration dominated by dynamic pressure whereas stack regime dominated by pressure difference. Wind and stack regime infiltration are defined as follows.

$$Q_{wind} = f^*{}_w A_o v' \tag{2}$$

$$Q_{stack} = f^*{}_s A_o \sqrt{\Delta T} \tag{3}$$

Where

$Q_{wind}$ = infiltration in the wind regime ($m^3/s$)

$Q_{stack}$ = infiltration in the stack regime ($m^3/s$)

$A_o$ = total leakage area of the structure ($m^2$)

$f^*_w$ = reduced wind parameter

$f^*_s$ = reduced stack parameter

v' = weather tower wind speed (m/s)

ΔT = inside-outside temperature difference (K)

The 'reduced' term on reduced wind parameter is used to distinguish it from the original wind parameter ($f_w$) which does not incorporate shielding coefficient [2]. For reduced stack parameter, the case is the same with reduced wind parameter, however the term 'reduced' is not because of shielding coefficient, as stack infiltration is caused by temperature difference and has little to do with the terrain or geometry of surrounding environment. Rather, the reduced stack parameter includes ceiling-floor leakage difference. After $Q_{stack}$ and $Q_{wind}$ has been determined, the next step is combine two values using the following equation.

$$Q = \sqrt{Q^2{}_{stack} + Q^2{}_{wind}} \tag{4}$$

*2.2. Airflow around Buildings*

Airflow around building also can exhibit variation of pressure as the result of fluctuation of wind direction and velocity. Pressure exerted by prevailing wind on a surface is called wind pressure ($p_v$), which is a dynamic pressure. The value of wind pressure for every surface is different depending the orientation of the surface and the surrounding environment. Hence a factor should be added to correct the actual pressure exerted by the airflow to the surface. This factor is called wind pressure coefficient ($C_p$). $C_p$ can be defined as follows [6].

$$C_p = \frac{P_x - P_0}{P_v} \tag{5}$$

Where
$C_p$ = wind pressure coefficient
$P_x$ = static pressure at given point on the building façade (Pa)
$P_0$ = static reference pressure (Pa)
$P_v$ = wind pressure (Pa)

The value of $C_p$ depends on many factors such as building geometry, façade, surface orientation, degree of exposure/sheltering, wind speed and wind direction. $C_p$ data can be obtained from primary or secondary sources [6]. Primary sources usually in the form of experiment or measurement, whereas secondary source typically uses calculation and analytical method. Secondary sources can come in the form of database or analytical model. Database is prepared by research organization such as the Air Infiltration and Ventilation Centre (AIVC) and ASHRAE. Analytical model uses calculations to find Cp based on experiment using wind tunnel or full scale measurement. Using regressing technique, a set of formula can be generated. Example of analytical model include Swami and Chandra (1988) [7] and Walker and Wilson (1994) [8] methods.

Meanwhile, there is third method which is using computational fluid dynamics (CFD). CFD has been used in fluid flow modeling for decades. In the recent times, as computing performance improved, CFD is widely recognized as a valuable and reliable tool. However there are several challenges in using CFD to determine Cp value, which are include numerical accuracy, boundary condition, and turbulence models (Stathopoulos, 1997 in Cóstola, et. al, 2009) [6].

*2.3. Importance of Wind Direction and Wind Pressure on Infiltration Calculation*

Wind pressure distribution as an important parameter in infiltration calculation is incorporate in several airflow network (AFN) simulation program in the form of Cp value as mentioned by Cóstola et. al (2009) [6]. Majority of these simulation program, however, using average Cp value, which means there is only one Cp value for each surface of the building envelope, instead of distributed or local Cp. Nevertheless, the incorporation of Cp in some AFN programs reflects the importance of wind direction in infiltration calculation, which plays a major role in determining Cp value.

Energy based infiltration models are formulated with the focus of energy efficient design i.e. heat transfer by the means of infiltration airflow. Air infiltration models for energy analysis purpose assume that the input values calculation are annual average. However, for mass transportation analysis of infiltration, such as for contaminant transport and indoor air quality analysis, instantaneous calculation is very important as the nature of airflow fluctuation caused by wind.

COMIS, an international collaboration of multizone airflow and contaminant transport modeling, incorporates the importance of wind pressure distribution on building envelope to correctly determine the contaminant distribution inside the building [9]. The wind pressure distribution on building envelope is calculated by COMIS using pressure coefficient as follows [9].

$$c_p(x, y, z) = \frac{p(x, y, z) - p_o(z)}{p_{dyn}(z)} \tag{6}$$

with

$$p_{dyn}(z) = \frac{1}{2}\rho_{out}(z)v^2(z) \tag{7}$$

where

$c_p(x,y,z)$ = pressure coefficient at coordinates x, y, and z, respectively [nondimensional]
$p(x,y,z)$ = surface pressure at coordinates x, y, z [Pa]
$p_o(z)$ = atmospheric pressure at height z [Pa]
$p_{dyn}(z)$ = dynamic pressure in the undisturbed flow at height z [Pa]
$\rho_{out}(z)$ = density of outside air at height z [kg/m$^3$]
$v^2$ = wind speed at height z [m/s]

The practice of using average pressure distribution represented by one Cp value in AFN program caused uncertainties compared to multiple local Cp values that form a wind pressure distribution on a particular surface.

### *2.4. Improvement of Existing Air Infiltration Model*

L.C. Ng, et.al. (2015) [10] have proposed a model to improve the accuracy of building air infiltration calculation on building energy analysis. Improved infiltration model was intended to be implemented on building energy simulation, namely EnergyPlus. The Strategy implemented was coupling and incorporating EnergyPlus with other software specialized in multiple airflow modeling and calculation, which is CONTAM. Research was focused to improve the formulation used in EnergyPlus Result shows that each building has unique value of empirical factors.

## 3. Methodology

### *3.1. CFD Case Setup*

In this simulation, a simple, square blockage is considered as a model. Study will be focused on the infiltration through the building. Any flow outside the building surface will not be considered but will be valuable for verifying factor. Two dimensional simulation will be considered in this research. This is because for low rise building, the difference of wind between the base and the top of the building is very small, therefore for simplicity and practicality, two dimensional simulation is adequate to provide reasonable result that can be used in preliminary design purpose. The terrain roughness category is comparable to that of LBNL model, therefore direct comparison can be done. The roughness is set at its lowest level, a representative of open sea which has the roughness of 0.0002 m.

### *3.2. Study of $C_p$ around Building*

The first step of the research is conducted a numerical experiment to study the influence of varying wind speed and direction on Cp value on each surfaces. With the advantage of CFD, study of Cp value can be conducted based on small area (dA) rather than surface average, which the experiment shows there are unique pressure and Cp distribution based on measurement location. The size of the area is of course depends on the size of cells near the wall. Numerical experiment will be conducted with a wind speed of 1, 2, 3, and 4 m/s and the wind direction will be varied from north (0 degrees) clockwise to the east (90). Experiments will only cover a quarter of a full circle because the model is square in shape and will exhibit symmetricity.

### *3.3. Study of Air Infiltration through a Lump Sum Hole*

On the second step, the experiment will start incorporating infiltration feature which is leakage in the wall. This is in accordance with existing infiltration model, which based on lump sum leakage or effective leakage area, particularly LBNL model and its derivative in EnergyPlus and ASHRAE. As the research is intended the existing LBNL model, this step is very important although it is not realistic in most cases. Implementation of the numerical experiment is by adding a hole in two opposite surfaces to let the air going through. The size of the hole is set at 1/20 or 0.05 of building length for each two surface.

IOP Conf. Series: Earth and Environmental Science **164** (2018) 012017 doi:10.1088/1755-1315/164/1/012017

### *3.4. Study of Infiltration through Multiple Small Hole*

The purpose of the third step is to study the effect of individual and distributed Cp around the surface. Another purpose of this step also useful to study the influence of Cp, wind speed and wind direction flow rate of infiltration air either incoming or outgoing which is important factors in formulating improvement of existing building infiltration model. Numerical simulation (CFD) case setup will be identical with previous step. The only difference is now the hole will be distributed but total area is constant. Size of distributed hole depend on the cell size used in the case. The smaller the cell size the smaller possible size of the hole and more information can be obtained.

### *3.5. Proposed Improved LBNL-Based Infiltration Model*

To incorporate the effects of Cp factor into existing LBNL model, this research will utilize regression method. Ideally, the difference between variations of observed variables, in this case wind velocity, wind direction, Cp factor, and infiltration flow rate, is as small as possible so that the trendline can be smooth and accurate regression can be obtained.As CFD can only handle one case at a time only a handful of representative value is observed and then be regressed. The regression analysis result is then used as a correction factor in existing LBNL model. As a complement the Cp factor will be added to accommodate effect of wind direction to infiltration flow rate.

## 4. Study of $C_p$ around Building

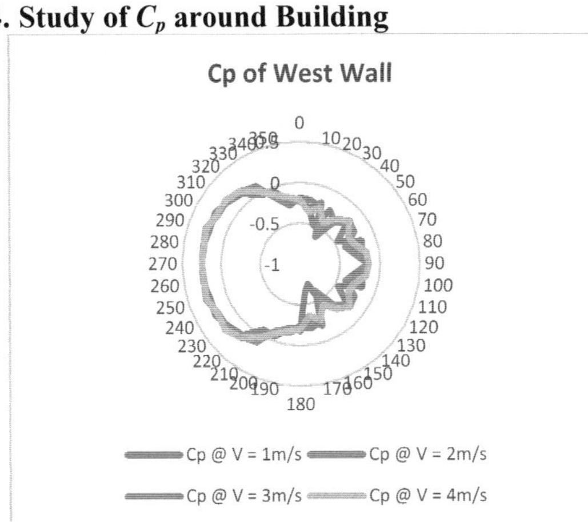


**Figure 1.** Cp of west-facing wall with various wind velocity.

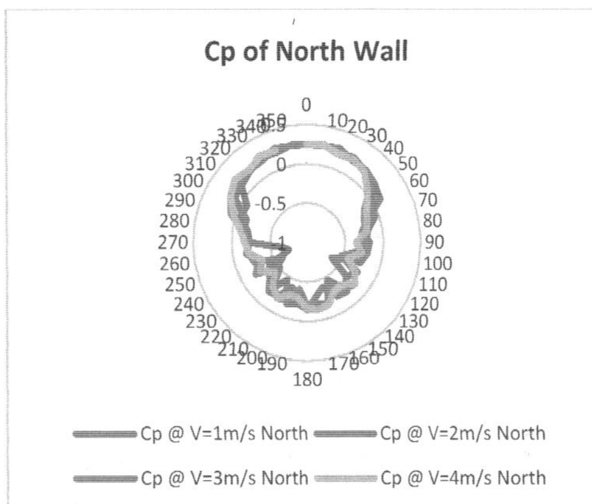


**Figure 2.** Cp of north-facing wall with various wind velocity.

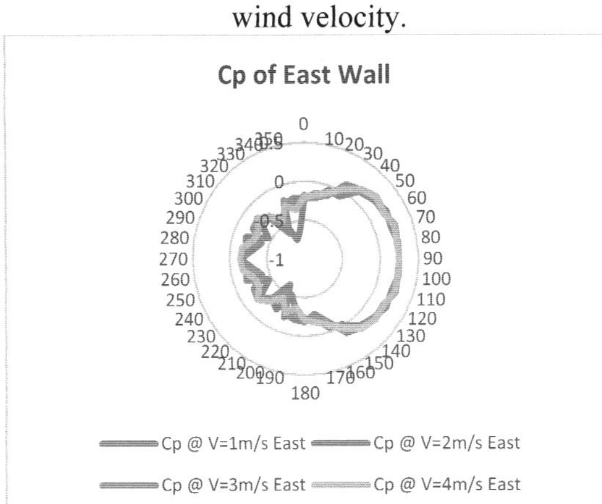


**Figure 3.** Cp of east-facing wall with various wind velocity.

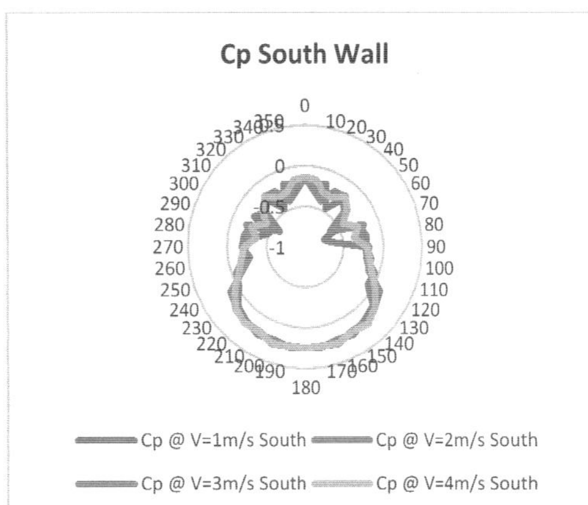


**Figure 4.** Cp of north-facing wall with various wind velocity.

Figure 1 through 4 show the relationship between velocity, wind direction, and surface averaged coefficient of pressure (Cp). For each four surfaces, the pattern is just rotated 90° or a quarter of full circle. This is because the model is in the form of quarter shape. Also, there is no obstruction whatsoever in every direction. In real cases, the shape usually come not in the form of square and there are variety of obstruction and blockages around the building, therefore the surface averaged Cp will be different.

Experiment results show that there is little or no impact of velocity of the value of Cp. This is because Cp is basically a ratio between pressure at the surface of the building and to that of prevailing wind without obstruction. Increasing wind velocity will increase both wind pressure at the surface (Pw) and wind pressure without obstruction (Pv) hence the ratio, which is Cp, will stays the same. The relatively small variation of Cp value for the same wind direction with different wind velocity is caused by the profile of the flow itself.

## 5. Study of Air Infiltration through a Lump Sum Hole

Simulation was conducted using wind direction variation from 0° to 90° with increment every 5° and fixed wind velocity of 3 m/s. The largest Y-velocity occurred when the wind is blowing from 0 to 10°. At this wind direction, the north surface act as the windward side. The resultant of velocity vector is maximum since the angle between the velocity vector and normal axis of the surface is zero.

As the wind approach the building, wind vector will be deflected and turbulence is created. This creates the resistance around the building. When the air is inside the building, it will be trapped. The pressure builds up until it reaches some point when it can overcome outside resistance and drive the air outward. The phenomenon shows the importance of analysing velocity component separately from the resultant velocity. The resultant velocity will always positive and in the case of wind blowing parallel with the opening, the velocity can be large and use it directly to determine inflow or outflow will be misleading.

To obtain the correct infiltration airflow rate using Swami-Chandra model, a series of steps have to be followed, which are depending on the case. For this case, the building has two openings, one at the north surface and the other at the south surface. The opening is a representative of infiltration leakage accumulated in one large opening. Before calculating the infiltration airflow through the opening, Cp value for the surfaces has to be calculated. The result of Cp calculation for wind direction of 0° to 90° for both north and south wall is presented by Figure 5.

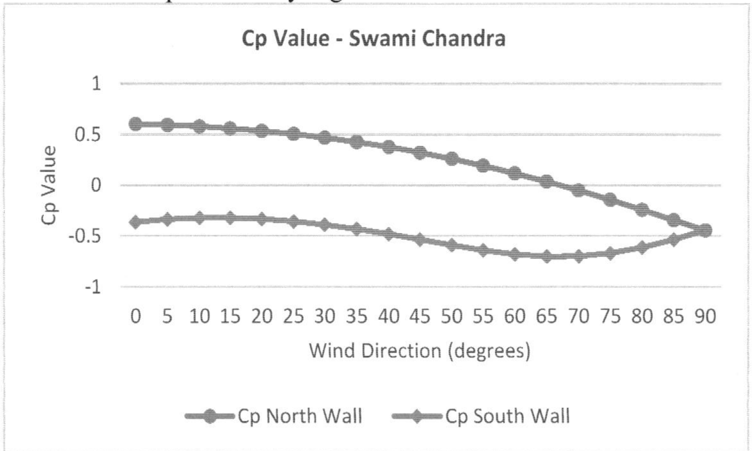


**Figure 5.** Cp Value for North and South Wall.

Figure 5 shows that for the north surface, which is windward side, the Cp is decreasing when the angle of the wind is getting larger relative to the normal of north wall. Cp for north wall is going negative after the wind direction is about 70°. As for the south wall, which in this case is leeward side, the Cp is always negative for wind direction from 0° to 90°.

After the Cp, or to be more precise, the average Cp has been determined, the next step is selecting the method of calculation based on window or opening configuration of the building. As mentioned earlier, the building leakage and crack is assumed only can be found in north and south wall. The cracks and leakage are represented by one large opening each at the north and south wall. Size of both opening is 0.1 m in width and 0.1 m in height, which is also the height of the building. As the height of the building is relatively short, the building can be classified as low-rise building.

The result of calculation is as following, which is relatively small compared to the CFD result or LBNL infiltration model. Figure 6 shows that the flow rate relatively stable and unaffected by the change of wind direction. However, after 60° of wind direction, the flow rate start decreasing with larger wind direction. Eventually, the flow rate becomes zero when the wind is blowing parallel with the north and south opening.

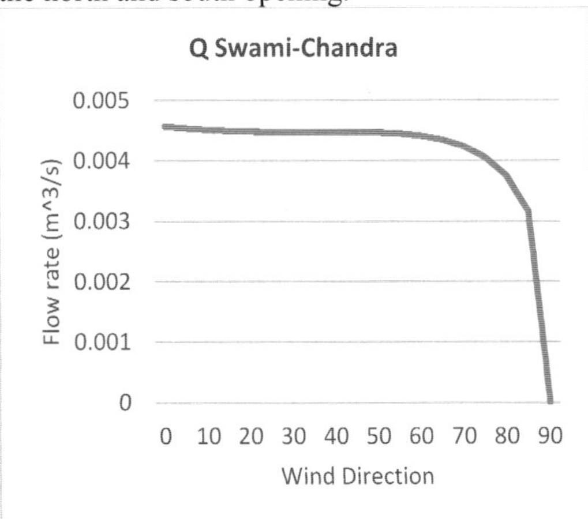


**Figure 6.** Result of infiltration flow rate calculation based on Swami-Chandra model.

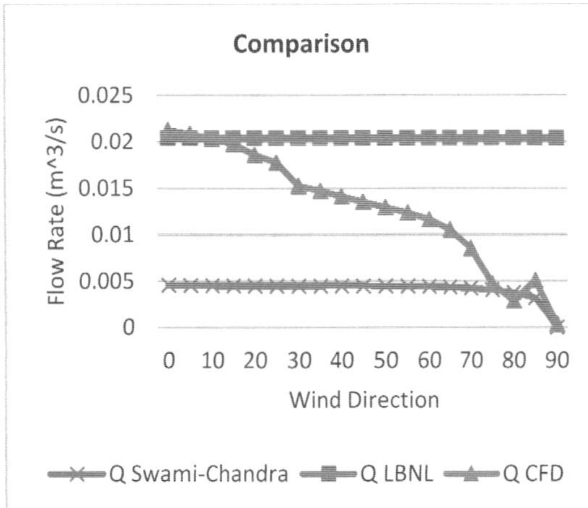


**Figure 7.** Comparison of three different methods.

The reason of the unaffected flow rate until 60° of wind direction is that the calculation of Swami-Chandra model is based on the value of ΔCp. The value of ΔCp itself relatively constant until the wind direction is larger than 60°. After 60° the Cp value of north surface is close to zero and becomes negative after 70°. Meanwhile, the Cp value of south wall is relatively constant at 0.3 but increasing after 45° and reaches its largest magnitude (0.7) at around 65° then decreasing again.

The small flow rate produced by calculation of Swami-Chandra model is because there are several correction factor applied in the calculation. There are discharge coefficient, flow coefficient, and effective window area. The calculation is as such because of the nature of application of Swami-Chandra model which is for ventilation via the large opening. As opposite, infiltration flow rate will always incorporating small opening. Therefore, the correction factors introduced in the calculation might not suitable for calculation of infiltration flow rate.

Calculation using LBNL infiltration model for lump-sum opening case is straightforward. However, as the LBNL model does not incorporate wind direction and made no distinction between inlet and outlet opening, some adjustment should be made. Instead, the LBNL focusing on location of the leakage, i.e. at the floor or at the ceiling. This is will be irrelevant with this case, which is a two-dimensional case. Therefore, the effect of leakage location will be eliminated in the calculation. As expected, Figure 7 show that the flow rate is constant and unaffected by wind direction change, as the formula itself does not include wind direction term.

The aforementioned three different methods will be compared to evaluate and determine their characteristics. Figure 7 shows that the infiltration flow rate from Swami-Chandra model is relatively small compared to two other models. Meanwhile at the wind direction of zero degree (wind blowing from the north), there is small difference between LBNL model and CFD simulation. This is indicating that the LBNL model is basically calculate the infiltration airflow of wind blowing perpendicular to the opening. Based on correlation evaluation, this is a good start because the LBNL model is ready to be corrected by including some terms that will correlate with the change of wind direction.

Swami-Chandra model is an innovative method because of the incorporation of Cp terms. However, the calculation method should be adjusted for infiltration calculation, since correction factors applied diminish the infiltration flow rate.

## 6. Study of Infiltration Flow Rate from Multiple Holes

### *6.1 Distributed Holes at Front and Rear Sides (North and South)*

In this case, wind direction variation was from 0° to 90° with 5° difference. In all cases, air pressure inside the building is positive and relatively larger compared to ambient temperature. This made air velocity vector is pointing outward from the building. However, mass flow rate analysis showed that there were air mass inflow as well as outflow.

Figure 8 shows the mass flow rate pattern of air flowing through distributed holes located at the front and rear. It is evident that the highest mass flow rate is achieved around 45 to 55 degrees, which is diagonal wind direction. Meanwhile, the lowest mass flow rate is achieved at 0°. This is because the size of the hole is far too small compared to the velocity of the wind. The difference of order between hole size (1.25 × 10-3 m width) and wind velocity (3 m/s) is more than 1000. Small hole makes resistance, as a result the entering air is partially blocked and makes the flow rate small.

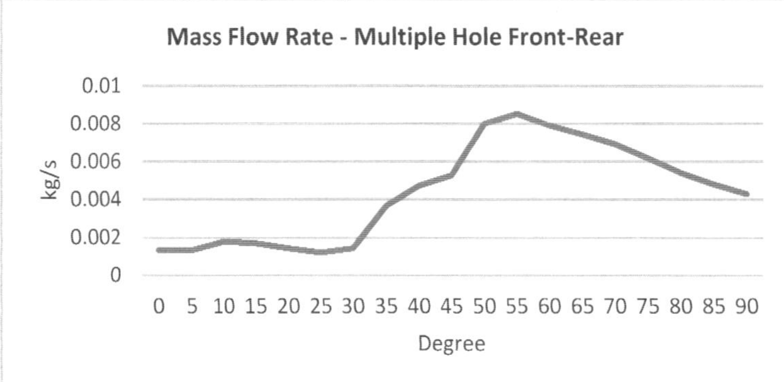


**Figure 8.** Mass flow rate of multiple located at the front and rear of the building.

On the other hand, at the diagonal wind direction, the direction of the airflow is not perpendicular but rather parallel with the airflow inside the building, which tends to flow in circular pattern. Based on the Bernoulli law, the difference between air velocity flowing around the building wall and air velocity inside the building produce pressure difference. As a result, airflow is forced to flow.

### *6.2 Distributed Holes at All Four Sides*

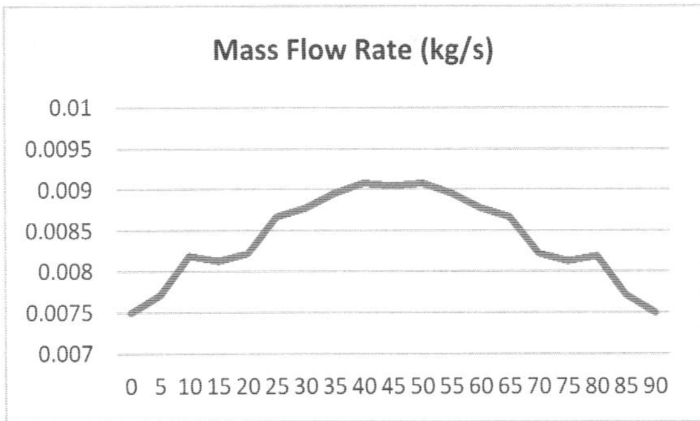


**Figure 9.** Mass flow rate of multiple located at all four sides of the building.

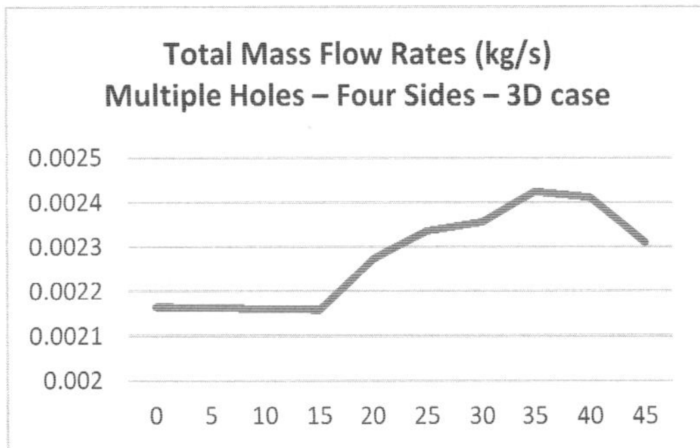


**Figure 10.** Mass flow rate of multiple located at all four sides of the building simulated in 3D.

Figure 9 shows the mass flow rate pattern of air flowing through distributed holes located at the front and rear. Again, almost similar with the previous front and rear case, it is evident that the highest mass flow rate is achieved around 45 degrees, which is diagonal wind direction. Meanwhile, the lowest mass flow rate is achieved at 0°. This is because the size of the hole is far too small compared to the velocity of the wind. The difference of order between hole size (1.25 × 10-3 m width) and wind velocity (3 m/s) is more than 1000. Small hole makes resistance, as a result the entering air is partially blocked and makes the flow rate small.

On the other hand, at the diagonal wind direction, the direction of the airflow is not perpendicular but rather parallel with the airflow inside the building, which tends to flow in circular pattern. Based on the Bernoulli law, the difference between air velocity flowing around the building wall and air velocity inside the building produce pressure difference. As a result, airflow is forced to flow.

The difference between the four direction and front-rear cases is that the mass flow rate difference between the lowest and highest value is rather smaller. This is obviously because the nature of the hole location. More distributed hole means more airflow can flow inside. Although the size of the holes is smaller, the location has a significant effect in introducing air. As explained in the front-rear case, hole faced directly to the wind tend to has less airflow compared to hole located not facing directly.

*6.3 Three-Dimensional Simulation*

Figure 10 shows the mass flow rate of wind direction from 0° to 15° is almost constant, about 0.00216 kg/s. The change of mass flow rate because of wind direction is almost negligible. Mass flow rate start to increase when the wind is coming from 15° and steadily moving up until the wind is coming from 35°. At wind direction of 40° the magnitude of mass flow rate is somewhat decreasing and decreasing further in 45° wind direction.

It is interesting that the 3D case provide lower mass flow rate compared to 2D cases. This is because in 2D simulation, convergence is very hard to achieve, mainly because when the airflow 'hit' the hole and blockage around it, it should flows to other direction, including Z direction. As Z-direction is not available in 2D case, the flow profile is somewhat inaccurate.

From the simulation it is can be seen that wind direction do have a role in determining mass flow rate. Similar with previous scenarios, wind direction between 35 to 55° degrees will produce maximum mass flow rate. However, if the mass flow rate is seen as a whole, there is only 12% difference between the lowest (0.00216 kg/s) and the highest (0.00242 kg/s) mass flow rate magnitude. In term of ACH, the infiltration will cause around 1.01 to 1.13 ACH.

One of the interesting phenomenon in the simulation is that the pressure inside the building will always higher than the environment. With the time of simulation is 60 seconds, it is sufficient enough for the wind to sweep the entire domain (80 m x 80 m). Therefore it can be said that in steady condition, the pressure of air inside the building will be higher than the ambient environment. Velocity vector analysis also shows that airflow is tend to flow outward in every direction, however they tend to flow through hole in leeward side rather than through the windward side.

Meanwhile, mass flow rate analysis shows that the airflow as a mass rate always enters through west and south holes, whereas airflow exits through holes located and east and north. This is because the wind direction is moving gradually to the northeast in a clockwise manner.

In order to determine whether placement of the hole in a building will affect mass flow rate or not, a full 1-story building (3 meter height) has been simulated. In the 3-m case, the hole is located in the middle of building height (1.5 meter). The size and configuration remain the same. There is slight difference between the two, with the 3-m result is relatively lower compared to its 0.1-m counterpart

**Table 1.** Comparison between mass flow rate of 0.1 m and 3 m height.

| Wind Direction | Mass Flow Rate (kg/s) | |
|---|---|---|
| | 0.1 m height | 3 m height |
| 0° | 0.00218 | 0.00215 |
| 45° | 0.00227 | 0.00218 |

## 7. Proposed Improved LBNL-Based Infiltration Model

To determine the correction term to be applied in the existing LBNL infiltration model, the first step that has to be done is find the mathematical formula of the flow rate pattern obtained from simulation result. As shown in the previous chapter, the flow rate has a repetitive pattern based on the angle of the wind direction. In mathematics, one equation that exhibit repetitive (cyclic) pattern is sinusoidal wave, which in fact is also based on the angle. Therefore, as the base of the mathematical formula is sinusoidal equation.

The pattern of the flow rate is not purely sinusoidal because there are multiple peak and valley in one cycle, in contrast with a pure sinusoidal wave which there is only one peak and one valley in every cycle. However, this does not mean that mathematical formula cannot be used to track the pattern of the flow rate result. There will be some modification to the sinusoidal wave.

The proposed formula is actually a sum of more than one sinusoidal wave. This is partly inspired by the creation of square wave in Fourier analysis by summation of odd harmonics of original sinusoidal wave equation. In this case, the constituent of the wave will not following nth harmonics as prescribed to create a square wave, because the patter of the flow rate is not a square wave.

To make an improvement on the existing LBNL model, the flow rate value of the experimental model is not important. Rather, the more important thing is the factor determining how the flow rate of experimental model correlated to that of the existing LBNL model. The original LBNL infiltration model only produce one infiltration flow rate value, as the formula does not include wind direction.

The first step is to look at the shape of the pattern. There are several key characteristics of the pattern that is very important to determine the constants of the sinusoid function. Sinusoid function has frequency and amplitude. To determine the frequency, one can look at the wave cycle. The pattern can be divided into two parts, the large and small wave.

Based on the pattern of the wave, the large wave is assumed starts at its lowest value (valley). Sinusoid function that starts at its lowest value (when there is no phase shifting) is negative cosine. The large wave reach its peak on 45°, 135°, 225° and 315°. On the other hand, the large wave will have its lowest value at 0°, 90°, 180° and 270°. Therefore, one cycle of the large wave requires 90°, which means the frequency is four times larger than ordinary cosine function.

Meanwhile, the amplitude can be determined by observe the difference between the largest and smallest value, then divided by two. Thus, the mathematical formula of the large wave is shown below.

$$-0.055\,cos(4\theta) \tag{8}$$

For the small wave, the method to determine the amplitude and the frequency is similar with that of the large wave. The difference is that the small wave is a positive cosine wave rather than the negative one. This is because the small wave is assumed starts at the largest value and then decreases as the wind angle is getting larger.

The small wave is required to adjust the large wave so that it can closely tracks the pattern of correction factor. The small wave is set to complete a full cycle by 30°. Ordinary sinusoid function needs 360°, therefore the small wave has the frequency 12 times larger than ordinary cosine function. Therefore the mathematical formula for the small wave is shown below.

$$-0.03\,cos(8\theta) \tag{9}$$

Finally to obtain the formula for improvement of the LBNL Model, both of the formula of large and small waves have to be summated and added with another value which is 1.36. Therefore, the final formula is shown below.

$$f_d = -0.055\,cos(4\theta) - 0.03 + 1.36 \tag{10}$$

The formula then compared with the pattern of correlation factor to determine the error. Error of the curve fitting is between 0.4 to 3.7%, which is under 10%. Therefore, the final correction factor will be as following.

$$Q_{wind} = f^*{}_w f_d A_o v' \tag{11}$$

where
$Q_{wind}$ = infiltration in the wind regime ($m^3/s$)
$A_o$ = total leakage area of the structure ($m^2$)
$f^*_w$ = reduced wind parameter
$f_d$ = wind direction parameter
v' = weather tower wind speed (m/s)

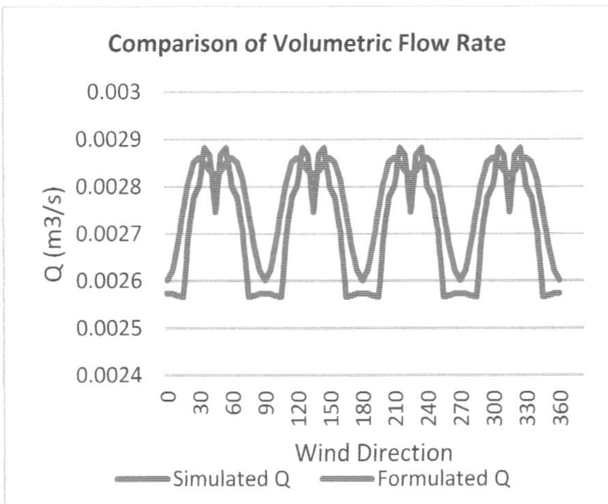


**Figure 11.** *Curve Fitting.*

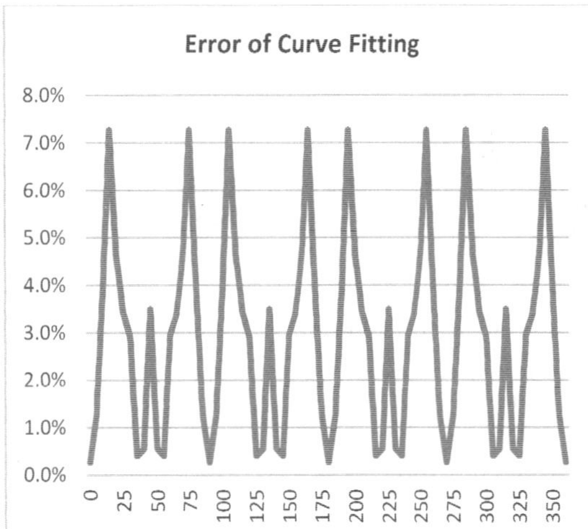


**Figure 12.** *Error of Curve Fitting*

Figure 11 shows the curve fitting between infiltration volumetric flow rate obtained by CFD simulation and volumetric flow rate obtained by adding $f_d$ factor into the simulation. As the value of $f_d$ is obtained by combining two sinusoidal function, the line is rather smooth. The error of the curve fitting is shown by Figure 12. The highest error value is around 7%, which is well under 10% value, therefore the formulated value can be applied for design.

Testing of the improved model is essential that the formulation is ready to be applied in similar condition with different wind speed and wind direction. There will be five test condition, each will use random wind velocity and wind direction. However, all of the terrain parameter will be the same. Result of the simulation will be compared with the calculated result to determine the error.

**Table 2.** Comparison between simulation and enhanced LBNL model

| Wind Speed | theta | $f_w$ | $A_o$ | Q | $f_d$ | $Q_{enhanced}$ | $Q_{CFD}$ | Error |
|---|---|---|---|---|---|---|---|---|
| 3 | 45 | 0.34 | 0.002 | 0.0544 | 1.385 | 0.002825 | 0.002746 | 2.90% |
| 4 | 45 | 0.34 | 0.002 | 0.0544 | 1.385 | 0.003767 | 0.003821 | 1.42% |

## 8. Conclusion

Research has been conducted to improve and enhance existing LBNL infiltration model in order to incorporate wind direction effect on flow rate. From the step one of the research, which is studying wind pressure coefficient (Cp) factor based of numerical simulation result, it is evident that Cp for the same wind direction, Cp value will be the same regardless of wind velocity.

Meanwhile, step two of the research shows that wind direction will affects infiltration flow rate so that wind direction should be considered in infiltration calculation. It is also evident that overall (resultant) of wind velocity is not accurate for infiltration calculation, since it cannot have negative sign, which is very important to determine the flow direction of the airflow.

In distributed hole cases, it is evident that largest flow occurred when the wind direction is 45°. There is huge difference of order between hole size and wind velocity, so that virtually the hole is relatively almost nonexistent. This creates resistance at the hole. Meanwhile, while the outdoor air flow is parallel with the indoor, more flow rate can be generated because of Bernoulli Law.

In term of relationship between wind direction and infiltration flow rate, it is indeed there is strong relationship between the two. Given the nature of infiltration flow rate which is very small, for thermal design purpose it will be marginal. Therefore, for simplicity, it can be ignored. However, in other infiltration flow analysis, particularly in pollutant or particulate matter study, it is very important.

Air inside the building is tend to flow in circular pattern. For particulate matter and pollutant analysis, more investigation should be done, especially in term of pollutant distribution.

For future infiltration analysis, it should be noted that 3D cases is better in term of convergence achievement and obviously closer to reality. More building shape and geometry should also be studied, in order to achieve more understanding.

## 9. Acknowledgement

This paper is first author's master project conducted at University of Colorado Boulder under supervision from the second author. The first author would like to thank Indonesia Endowment Fund for Education (LPDP) for providing funding for this project and the master study as a whole. Also, the first author would like to thank to his wife and family for providing tireless support. Both authors would like to appreciate for the support provided by CDSR - SHERA (Center for Development of Sustainable Region – Sustainable Higher Education Research Alliances).

## References

[1] Feustel, H. E., & Rayner-Hooson, A. 1990. *COMIS Fundamentals*. Report LBL-28560, US Department of Energy, Lawrence Berkeley Laboratory.
[2] Sherman, M.H. and Grimsrud, D.T. 1980b. "Infiltration-Pressurization Correlation: Simplified Physical Modeling", *1980 Annual ASHRAE Conference*, June 22-26, Denver, CO.
[3] Department of Energy. 2016. EnergyPlus™ Version 8.5 Documentation. Engineering Reference.
[4] ASHRAE. 2009. *Fundamentals Handbook*, Ventilation and Infiltration, Chapter 16, Atlanta,

GA: American Society of Heating, Refrigerating, and Air-Conditioning Engineers, Inc.

[5] Sherman, M.H. and Grimsrud, D.T. 1980a. "Measurement of Infiltration Using Fan Pressurization and Weather Data", *1st Symposium of the Air Infiltration Centre on Instrumentation and Measurement Techniques*, October 6-8, Windsor, England.

[6] Cóstola, D, B. Blocken and J.L.M. Hensen. 2009. "Overview of pressure coefficient data in building energy simulation and airflow network programs". *Building and Environment* 44 (2009) 2027–2036.

[7] Swami, M.V. and S. Chandra. 1987. *Procedures for calculating natural ventilation airflow rates in buildings*. Final Report FSEC-CR-163-86. Florida Solar Energy Center, Cape Canaveral.

[8] Walker, I.S. and D.J. Wilson. 1994. "Practical methods for improving estimates of natural ventilation rates". *Proceedings of the 15th IEA Conference of the Air Infiltration and Ventilation Centre*, Buxton, U.K., pp. 517-526.

[9] Feustel, H.E. 1999. "COMIS—an international multizone air-flow and contaminant transport model". *Energy and Buildings* 30 (1999) 3–18.

[10] Ng L.C., Persily A.K., Emmerich S.J. "Improving infiltration modeling in commercial building energy models". *Energy and Buildings* 88 (2015) 316–23.

IOP Conf. Series: Earth and Environmental Science **164** (2018) 012018 doi:10.1088/1755-1315/164/1/012018

# The Mediating Role of Organizational Learning Capability On Green Distribution and Green Packaging Towards Sustainability Performance as A Function Environmental Dynamism: Indonesia and Malaysia Fishery Industries

**A Hutomo[1], M Haizam[2] and O Sinaga[3]**

[1] Business and Management Faculty, Widyatama University, Jalan Cikutra No.204A, Bandung, West Java, Indonesia.
[2] Business and Law Faculty, International University of Malaya Wales, Jalan Tun Ismail, 50480, Kuala Lumpur, Malaysia.
[3] Department of International Relations, Padjadjaran University Jalan Dipati Ukur No. 35, Bandung Indonesia.

arry.hutomo@widyatama.ac.id

**Abstract.** This paper analyses part of green supply chain practices created for fisheries industries to moderate environmental dynamism and to implement role of organizational learning capability adoption as mediating on green distribution and green packaging practices across a supply chain towards sustainability performance. It examines green distribution and green packaging adopted in the supply chain as a result of pressures from primary stakeholders. We derive a conceptual framework with underpinning theory. The resulting hypotheses are tested using fishery industries in Indonesia and Malaysia data of 325 samples utilizing primary and secondary data. Finding reveal, a phenomenon with role of environmental dynamism and organizational learning capability adoption have significant effect to enhancing sustainability performance. We will develop an empirical model that can be applied in other sectors to improve performance. Moreover, our results yield insights to green supply chain management practices in optimizing their supply chain performance.
**Keywords.** Green Distribution; Green Packaging; Organizational Learning Capability; Environmental Dynamism; Sustainability Performance.

## 1. Introduction

Based on the phenomenon of developed countries, it turns out that the key performance level of multinational companies lies in the ability of the company to cooperate with its business partners with environmental awareness[1],[2] and[3]. According to [1], more specifically refers to the balance in the mind-set called green practices mind-set for a harmonious balance between economic growth and environmental sustainability. The tendency of lack of balance, it can be understood the position of Indonesia and Malaysia as a country with abundant natural resources on fisheries industry that sustain rapid economic growth, which in turn ignoring environmentally sustainable situation [4],[5],[6] and [7]. According to [8], Green distribution not only executed through the processes, but also changing the organizational culture and the commitment and continuity in the implementation. [9] said that Indonesia and Malaysia increasingly aware of packaging recycling and its importance to the environment. Green packaging for products has now emerged in develop countries as global warming issues and other issues related to environmental pollution have become a problem for many years [10]. Green Distribution practices among industry players have been noticed as one of the critical issues the which are primarily associated with climate change, water pollution, noise, and other environmental features [11]. Recently, environment issue has created immense attention from academic and industrial people. The two play capabilities executed by means of the principles of 4R1D, reviews those are reduce, reuse, reclaim, recycle and degradable [12]. Sustainable and eco-friendly architecture is one of the main aims that humans for creating the ultimate model for all their activities [2]. With green practices (green distribution and green packaging), companies increase environmental impacts and their influences on supply chain operations [3]. We located that the institutional concept and natural-resource-based totally view (NRBV) are two theories which might be extensively carried out in our research area and are appropriate for our purpose [3],[13]. Dynamic environments are characterised through unpredictable and rapid change, which increases uncertainty for people and companies working within them [14]. Previous research advise that the performance of is incredibly depending on the behaviour of their leaders, and that sustainability overall performance is stricken by both transactional and transformational varieties of management [15]. Furthermore, previous research findings suggest that environmental dynamism moderates the leadership–overall performance relationship such that transactional management is most effective when environmental dynamism is low and that transformational management is most effective when environmental dynamism is excessive. These findings are depicted in a moderator model that extends the transactional–transformational management paradigm and provides to our know-how of ways marketers initiate and grow sustainability[14]–[17]. Based on the potential value of the logistics market, the ability use of green supply chain techniques in Indonesia is very large, especially for commodities are perishable (agriculture, livestock and fisheries) that the level of harm estimated at 40% [4],[7],and[18]. According to [19] said that exceptional of infrastructures still on low-stage compare to Singapore and Malaysia and this impact to transportation quality. Indonesia ought to be enhancing excellent port infrastructure [19]. The implementation of the cold chain, there are numerous troubles on each stages or levels of activities inside the supply chain [4],[7],[18],and [20] that degree of the green distribution and packaging, the problem is the availability of capital and lack of learning organizational capacity of green supply chain practices[4]–[6]. Based on some previous research has described relationships are very strong and significant correlation between green distribution and green packaging to sustainability performance, but researchers have hypothesized that environmental dynamism and organizational learning capacity will further improve the relationship in achieving the degree of sustainability performance for the better results, where researchers investigate the gaps in research studies that have not done in previous research studies. Therefore, in this study, researchers will conduct research to analysis whether any mediating effect of Organizational Learning Capacity (OLC) would affect significantly between Green Distribution (GD) and Green Packaging (GP) towards Sustainability Performance (SP) to be influenced also by using Environmental Dynamism (ENDY) as a function moderation variable.

## 2. Literature Review

Previous research emphasized the relationship between the pressures from the external stakeholders and the extent to which the companies implement environmental-related practices[21]and[22]. Based on underpinning institutional theory, the pressures of external stakeholders of the institutional isomorphic adjustments [23]. According to the literature, consumer pressures, competitor pressures, regulatory pressures, and society pressures [21]–[23]. The natural-resource-based view (NRBV) [23] is an underpinning theory and improving model of the resource-based view (RBV) that relates to the adoption of the strategy. The inexperienced strategies delivered inside the NRBV consist of pollution prevention, product stewardship, and clean technology [24]. Based on this review and evaluation, this research suggests that further application of RBV could enhance the understanding for the manufacturer with regard to the importance of resource allocation and to effectively use the resources in the firm [24],[25]. This research investigates the capabilities of the organization in utilizing its resources optimally by using the integrated environmental dynamism and organizational learning capacity.

### *2.1. Green Distribution*

Sustainable distribution practices consist of those who reduce carbon dioxide, are economically feasible and will result in a better quality of life for the earth's future population. distribution practices green variety from changing the way distribution facilities and vehicles are powered to imposing extra transparency concerning the surroundings and distribution practices [3],[26]. The storage facility is every other essential issue of green distribution [26],[27]. The garage facility should be capable of shop exclusive categories of substances [26]. Green distribution (GD) is consisting of (1) Eco green labelling of products; (2) Environmental development in packaging. (3) Alternate for greater environmentally friendly transportation [3] and [13].

### *2.2. Green Packaging*

Green packaging (GP) is not only a package deal of widespread performance, however additionally with two primary capabilities including protecting the surroundings and renewable resources [3] and [10]. The improvement of green packaging is (1) to enable packaging of light-weight; (2) recyclable, re-use, biodegradable substances, and (3) to prevent the usage of non-ecological substances [12]. To enhance the green packaging, the government can undertake regulation prohibiting positive packaging substances use, organizing of storage refunds gadgets, drawing up the recycling or reuse of legal guidelines, giving a discount or punishment in tax, restricting over-packaging, founding various studies institutions to assess packing materials, law to sell the recent development of packaging substances and using such measures [3],[13], and [26].

### *2.3. Organizational Learning Capability*

According to [28] define organizational learning capability (OLC) because of the managerial and organizational functions or details that facilitate the organizational studying [29]–[32] or inspiring an organization to study [33],[34]. Four underlying dimensions of organizational getting to know capability: (1) experimentation and hazard-taking; (2) interaction with the outside environment, (3) dialogue; and (4) participative choice making[35]. Awareness of the importance of organizational learning in addressing the demands of organizational sustainability, and particularly subject to the triple bottom line (TBL) sustainability [15],[33]. A definition of sustainability TBL is supplied, collectively with an exploration of the sensible issues relevant to adopting organizational learning in addressing it [31],[33],and [34].

### *2.4. Environmental Dynamism*

The third perspective consists of managerial perceptions concerning those environmental attributes such as dynamism, complexity, and hostility [14],[36]. Environmental Dynamism (ENDY) shows how frequently factors in the environment are changing. Environmental Dynamism (ENDY) can be taken into consideration as an idea similar to environmental turbulence or high velocity environment [14,16]. Another sort of environmental dynamism refers to the quantity of unpredictable change in an enterprise's surroundings[37]. Environmental dynamism (ENDY) can be defined by way of the indicator are (1) Heterogeneous; (2) Speed and comprehensiveness and (3) leadership behaviour (transformational or transactional) [36],[37].

### *2.5. Sustainability Performance*

This complex issue has no straight forward solution, especially considering that sustainability is an intention for all to attain as they are continually trying to reach towards it [3],[38],[39]. Researchers have advocated sustainability performance, inclusive of (1) Financial performance, (2) Environmental performance;(3) Operational performance and (4) Social performance as vital performance signs, the proposed version of green distribution and green packaging and sustainability of overall performance in a single observe [40]–[43]. Sustainability overall performance in supply chain management (SCM) and including the green concept to it as a way to emphasize the importance of environmental issues[44]. We conclude by suggesting perspectives for future studies in the which the interaction has observed to lead significantly to overall company performance.

### *2.6. Hypothesis Development*

Previous research has explained that there is a positive relationship between green distribution and green packaging on sustainability performance. Based on the results of reviews from several studies rarely found application in fisheries industry where very rarely discuss about perishable food, while the scope is a problem of supply chain is quite dominant Indonesia and Malaysia, based on lack of research and this gap researcher hypothesized as follows:

**H1:** A Fishery Industries (Indonesia and Malaysia) 's Green Distribution (GD) has a positive impact on its sustainability performance (SP).

**H2:** A Fishery Industries Indonesia and Malaysia Green Packaging(GP) have a positive impact on its sustainability performance (SP).

The improvement of organizational learning capacity and sustainability performance mediation indicators is discussed separately in the green supply chain management practices, especially green distribution and green packaging. Based on previous research that mediation of organizational learning capacity and moderation of environmental dynamism give significant influence on sustainability performance and organizational performance. Therefore, based on lack of research on the above, the researchers hypothesized as follows:

**H3:** A Fishery Industries (Indonesia and Malaysia) 's has a significant effect of Green Distribution (GD) on sustainability performance (SP) mediating by Organizational Learning Capacity (OCL) as a function of Environmental Dynamism (ENDY) in the structural model.

**H4:** A Fishery Industries (Indonesia and Malaysia) 's has a significant effect of Green Packaging (GP) on sustainability performance (SP) mediating by Organizational Learning Capacity (OCL) as a function of Environmental Dynamism (ENDY) in the structural model

## 3. Research Methods

As shown in Fig.1 proposed the final hypothesized measurement model this study examines the extent to which parts of Green Distribution (GD) and Green Packaging (GP) affect sustainability performance (SP). We investigate the impact of mediating effect of organizational learning capacity towards sustainability performance as a function of environmental dynamism with Smart PLS 3.0 Software. there were confirmed 325 usable respondents for data analysis and model validation. Product Indicator Approach with GD, GP, OCL and SP have been implement to structural model as an interaction effect as shown in Fig.2. The main respondents in this research study are fishery industries in Indonesia and Malaysia that have register to feasibility of fish processing from Southeast Asia Fisheries Development Centre (SEAFDEC)® and EMS certified. Systematic random sampling the sampling frame is first divided into a number of segments called intervals.

IOP Conf. Series: Earth and Environmental Science **164** (2018) 012018 doi:10.1088/1755-1315/164/1/012018

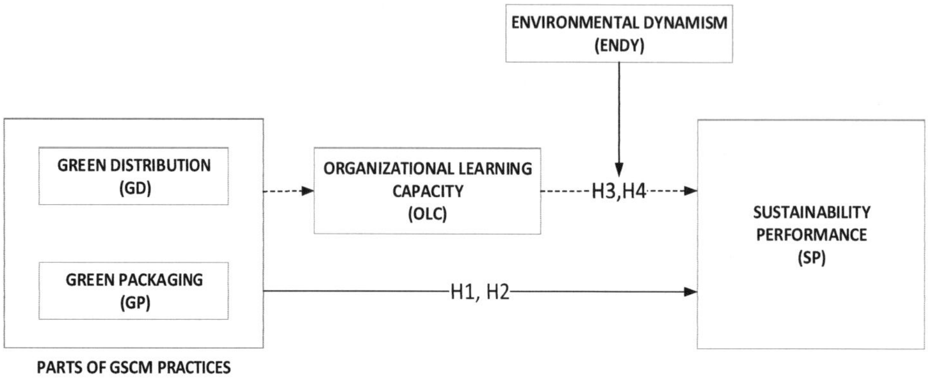


**Figure. 1** Conceptual Framework

Then, from the first interval, using the SRS technique, one element is selected. In this research, a systematic random sample relies on some sort of ordering to choose sample all selected manager from each region fisheries industries. The present study used a seven-point Likert scales for measuring all variables in this study. All of their comments and suggestions regarding the clarity, relevance (content/face validity) and consistencies of the questions incorporated into the survey instrument.

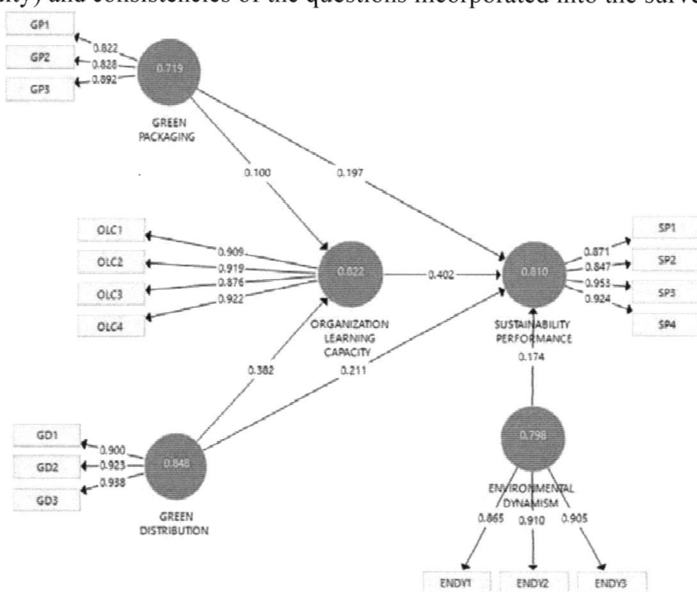


**Figure. 2** Interaction Effect

## 4. Finding, Analysis and Discussion

The researcher evaluates the PLS model based on the prediction orientation which has non parametric properties. The researcher considers that the PLS evaluation model is performed for a "measurement model" or outer model to assess the validity and reliability of the model. The outer model in this study uses reflexive indicators by evaluating through convergent and discriminant validation of latent construct and composite reliability formers and Cronbach alpha for its indicator block [45]. Evaluation structural model or inner model aims to predict the relationship between latent variables. The inner model was evaluated by looking at the percentage of variance described by looking at the R-Square value for endogenous latent construct to test predictive relevance, and average variance extracted [46] for predictiveness by using resampling to obtain stability and estimation.

The discriminant validity test is conducted by the researchers by comparing the square root of AVE for each construct with the correlation value between constructs in the model. Good discriminant validity shown on the square root of AVE for each construct is greater than the correlation between the constructs in the model [46]. Reference [45] expresses the rule of thumb reliability value that the use of Cronbach Alpha to test the reliability of the constructs will give lower value (under estimate) so it is preferable to use composite reliability should be greater than 0.60 - 0.70 is still acceptable for exploratory research adopted from [45]–[47]. Discriminant validity test results from model structure examined that composite that all constructs in the model structure can be considered to have good reliability because they are> 0.70 as shown in Table 1.

**Table 1.** Cronbach'Alpha, Composite Reliability (CR), and AVE

| | Cronbach's Alpha | Composite Reliability | Average Variance Extracted (AVE) |
|---|---|---|---|
| ENDY | 0.875 | 0.922 | 0.798 |
| GD | 0.911 | 0.944 | 0.848 |
| GP | 0.821 | 0.885 | 0.719 |
| OLC | 0.928 | 0.949 | 0.822 |
| SP | 0.921 | 0.944 | 0.810 |

The next step in analysing the structure of the model Rule of thumb for the recommended AVE value must be greater than 0.50 and each cross loading for each variable in the measurement model (> 0.70) means that 50% or more variance of the indicator can be explained [45]–[47]. Researchers noticed that there was a positive influence directly the GD→SP (0.214) and GP→SP (0.188) *We analysed that the following* ***first and second of hypotheses*** *A Fishery Industries (Indonesia and Malaysia)'s green distribution (GD) and green packaging (GP) have a positive impact on its sustainability performance (SP). With analysis data, Fishery Industries Indonesia and Malaysia that Green Distribution(GD) greater than Green Packaging (GP) impact on sustainability performance (SP).*

Researchers have hypothesized that the role of market is a moderating variable because it shows the interaction between exogenous variables (predictors) and moderator variables affecting endogenous variables [48]. Researchers try to avoid measurement errors being inconsistent and biased from the non-linear relationship arising from MRA coefficients by using latent variables. The Role of market (RM) construct is a moderator and a reflexive exogenous construct hence the choice of method to test the interaction effect used by the researcher is Product Indicator Approach [45],[48]. In this study, measurements and modelling were made by making multiplication between exogenous and moderator variable indicators to form interaction constructs. The researcher examined and evaluated the measurement model or outer model for this case by examining valid and reliable. To test the significance to determine the influence between variables based on the literature [48] recommend to use 5000 resampling numbers in the bootstrapping process. From the results of measurement evaluation can be seen the value of R-Square for variable Sustainability Performance (SP) 0.578 which means that included in moderate category. From the results of the path coefficient above can be seen that all variables turned out to have a significant effect with the value of T-Statistics generated for all variables> 1.96. This means that all alternative hypotheses are accepted.

With the product indicator approach, for the variable interaction between Green Distribution (GD) mediating by OCL and moderating by ENDY on Sustainability Performance (SP), GD→OCL→SP obtained by T-Statistic 4.316 (T-Value >1.96) which means that the variable Environmental Dynamism is a moderator variable or can moderate the relationship between variable green distribution, organizational learning capacity and sustainability performance (SP).

With the product indicator approach, for the variable interaction between Green Packaging (GP) mediating by OCL and moderating by ENDY on Sustainability Performance (SP), GP→OCL→SP obtained by T-Statistic 4.986 (T-Value >1.96) which means that the variable Environmental Dynamism is a moderator variable or can moderate the relationship between variable green packaging, organizational learning capacity and sustainability performance (SP).

*We* analysed *the following* ***third and fourth of hypotheses*** *fishery industries (Indonesia and Malaysia)'s Green Distribution and Green Packaging enhancing by organizational learning capacity as mediating variables towards sustainability performance on GSCM Practices has a moderate effect interaction in the structural model by Environmental Dynamism,* means that organizational learning capacity and environmental dynamism as important variable to enhance sustainability performance.

Researchers analysed that GD→SP has a total effect of 0.536 greater than GP of 0.411. Rule of thumb commonly used to assess total effects when the total effect value of the smallest, medium and large is 0.02; 0.15 and 0.35. Based on analysis of data from this evaluation of structural models, the researchers could provide an explanation that the GD→SP (0.536) has a large effect compared GP→SP (0.411) which can be categorized as large effect. Researchers are also trying to analyse the results of the model when the exogenous variables have an effect on the dependent variable, with other exogenous variables, it is said to be an indirect effect. Therefore, the researchers tried specifically reverse logistics indicators to analyse the total effects of exogenous variables specifically by adding direct and indirect effects. One variable may have no direct effect, but it may have an indirect effect as well.

Convergent validity test reflexive indicators with Smart PLS 3.0 program conducted by the researchers to be able to see the value of the loading factor for each indicator constructs. Rule of Thumb is usually used to assess the validity of convergent i.e., the loading factor value (> 0.60 - 0.70) for exploratory research is still acceptable when the Average Variance Extracted (AVE) (> 0.50) [45]–[47]. The entire value of factor loading in this study> 0.70. The highest factor loading value is SP_3 Operational Performance (0.953). From the results of the analysis output can be seen that all the constructs to generate value reflective indicator loading factor> 0.70, which means that all indicators constructs are valid for models studied.

Hence, the researchers conducted an analysis that the interpretation of path coefficient value that existing organization learning capacity and environmental dynamism are significant to increase the independent variable to dependent variable, in this case is sustainability performance. Eco green labelling of products, environmental development in packaging, Alternate for greater environmentally friendly transportation will have a more significant impact when fisheries industry more experimentation and hazard-taking, interaction with the outside environment, dialogue; and participative choice making with attention to some things like heterogeneous, speed and comprehensiveness and leadership behaviour people on doing green practices. This includes implementation of enable packaging of light-weight, recyclable, re-use, biodegradable substances, and prevent the usage of non-ecological substances.

## 5. Conclusions and Recommendations

That is, the Green Distribution and Green Packaging will have a positive on sustainability performance indicators mediation if it can be applied. So it can be concluded that Organizational Learning Capacity and Environmental Dynamism shows the direct influence of independent variable to dependent variable that is sustainability performance. We concluded that the following A Fishery Industries (Indonesia and Malaysia)'s green distribution (GD) and green packaging (GP) have a positive impact on its sustainability performance (SP). With analysis data, Fishery Industries Indonesia and Malaysia that Green Distribution(GD) greater than Green Packaging (GP) impact on sustainability performance (SP) and then, fishery industries (Indonesia and Malaysia)'s Green Distribution and Green Packaging enhancing by organizational learning capacity as mediating variables towards sustainability performance on GSCM Practices has a moderate effect interaction in the structural model by Environmental Dynamism.

This study concluded that alternate for greater environmentally friendly transportation, eco green labelling of product and environmental development in packing had more enhancing sustainability performance not only underlying dimension of organizational capability like experiment, dialogue, participative choice to making but also speed, comprehensiveness and leadership behavior also to enhancing sustainability performance at Fisheries Industries. On the other hand, this study concluded that enable packaging of light-weight; recyclable, re-use, biodegradable substances, and keep to prevent the usage of non-ecological substances had more enhancing sustainability performance not only underlying dimension of organizational capability like experiment, dialogue, participative choice to making but also speed, comprehensiveness and leadership behavior also to enhancing sustainability performance at Fisheries Industries in Indonesia and Malaysia.

This study gives conclusions with this research methodology, the impact of mediating of organizational learning capacity on green distribution and green packaging practices towards sustainability performance as a function on environmental dynamism have significant effect. Researchers suggest policy maker can control sustainability performance on fishery and keep maintain organizational learning capacity. Based on the limitations of the present study, it is proposed that further research should consider following areas or aspects: The present study was conducted among Indonesia and Malaysia's countries. The findings may not reflect the overall situations in the state or government owner enterprises sector. As such, further research should be conducted in other organizational culture and environment to examine the generalizability of the findings of this study. The study employed in the present study was the survey method that used a set of questionnaires as measurement scale. All finding not described green supply chain management practices because this research has limitation focus on investment recovery and reverse logistics. In an attempt to examine the existing sustainability process model, this research will propose a build process model. The proposed green supply chain model will address the issues of sustainability, and efficient forward and green supply chain practices. The process model will examine how the sustainability issues can be deal with in an integrated way using knowledge acquisition and green supply chain linkage in fisheries supply chains. Thus, the proposed research will produce additional knowledge, and contribute significantly on the theoretical aspects of green supply chain practices, sustainability supply chain performance controlling by knowledge acquisition and green bullwhip effect. Government, Fisheries Industries and stakeholders can use the model in their policy analyses by looking into the future behavior of particular inputs and outputs. Input and output values rely on existing circumstances of capacity, investment, constraints and dynamic policy. Measuring future behavior in a timely way for various variables is the key to making decisions for better profitability and sustainability. This integrated model can implement in the fisheries supply chain practices in Indonesia and Malaysia at different levels of the supply chain from harvesting until retailing.

In addition, under limitation this research, researchers suggest recommendation for further research that nuanced investigation of environmental contexts that would reason some of those enablers might additionally need to be investigated. Elaboration of the way close geographical proximity affects sub-suppliers' environmental overall performance and how green distribution and green packaging enforcing techniques evolve is also needed. strength have an effect on asymmetries alongside a couple of tier deliver chains need extra research. this is especially vital in some instances (e.g. perishable and non-perishable food) in which sub-suppliers have significantly extra relative strength than direct providers or even focal groups. basic, the have a look at adds to this inchoate subject of multi-tier green supplier management. the various complexities that exist offer opportunities from technological, organizational, relational, and economic views. This have a look at further identifies the significance and need for multi-tier green supplier management investigations.

**Acknowledgment**
An appreciation is given to the SEAFDEC, Supply Chain Indonesia for their secondary data support and wish to acknowledge of Widyatama Foundation in financial support.

**References**

[1] Little JC, Hester ET, Carey CC. Assessing and Enhancing Environmental Sustainability: A Conceptual Review. Vol. 50, *Environmental Science and Technology*. 2016. p. 6830–45.

[2] Ragheb A, El-Shimy H, Ragheb G. Green Architecture: A Concept of Sustainability. Procedia - *Social Behaviour Science*. 2016; Vol.216(October 2015):p. 778–87.

[3] Chin TA, Tat HH, Sulaiman Z. Green supply chain management, environmental collaboration and sustainability performance. *Procedia CIRP*. 2015;Vol.26:p. 695–9.

[4] Adhuri DS, Rachmawati L, Sofyanto H, Hamilton-Hart N. Green market for small people: Markets and opportunities for upgrading in small-scale fisheries in Indonesia. *Marketing Policy*. 2016;Vol. 63:p. 198–205.

[5] Anderson ZR, Kusters K, McCarthy J, Obidzinski K. Green growth rhetoric versus reality: Insights from Indonesia. *Global Environmental Change*. 2016;Vol.38:p. 30–40.

[6] Chua SC, Oh TH. Green progress and prospect in Malaysia. *Renewable Sustainability Energy Review*. 2011; Vol.15(6):p. 2850–61.

[7] Nielsen M, Ravensbeck L, Nielsen R. Green growth in fisheries. *Marketing Policy*. 2014;Vol.46:p. 43–52.

[8] National Geographic and GlobeScan. Greendex 2014:*Consumer Choice and Environment-A Worldwide Tracking Survey. Vol. 1, www.globescan.com*. 2015.

[9] Shi X. The future of ASEAN energy mix: A SWOT analysis. *Renewable Sustainability Energy Review*. 2016;Vol.53:p. 672–80.

[10] Molina-Besch K. A Supply Chain Perspective on Green Packaging Development-Theory Versus Practice, *Packaging Technology and Science*. 2016.Vol.29: p. 45–63.

[11] Aziz TNAT, Jaafar HS, Tajuddin RM. Green Supply Chain: Awareness of Logistics Industry in Malaysia. *Procedia - Social Behaviour Science*. 2016;Vol.219:p. 121–5.

[12] Zhang G, Zhao Z. Green Packaging Management of Logistics Enterprises. *Phys Procedia*. 2012;Vol.24:p. 900–5.

[13] Masoumik SM, Abdul-Rashid SH, Olugu EU, Ghazilla RAR. A strategic approach to develop green supply chains. *Procedia CIRP*. 2015;Vol.26:p. 670–6.

[14] Ensley MD, Pearce CL, Hmieleski KM. The moderating effect of environmental dynamism on the relationship between entrepreneur leadership behavior and new venture performance. *Journal of Business Venture*. 2006; Vol.21(2):p. 243–63.

[15] Mat a., Razak RC. The Influence of Organizational Learning Capability on Success of Technological Innovation ( Product ) Implementation with Moderating Effect of Knowledge Complexity. *International Journal Business Social Science*. 1991;Vol.2(17):p. 217–25.

[16] Ting H-F. The Moderating Role of Environmental Dynamism on the Influence of Innovation Strategy and Firm Performance. International Journal of Innovation Management Technology. 2012;Vol.3(5):p. 13–6.

[17] Rao P, Holt D. Do green supply chains lead to competitiveness and economic performance?, International Journal of Operation Production Management. 2005;Vol.25(9):p. 898–916.

[18] James SJ, James C. Sustainable Cold Chain. International: *Sustainable Food Processing*. 2013. p. 463–96.

[19] Hutomo A, Fitridayani R. The Impact Of Perceived Environmental Uncertainty, Supply Chain Performance, and Integrated Domestic Economy : Plywood Manufacturer In Indonesia. International Journal of Business Management Study. 2015;Vol.2:p. 198–202.

[20] Lailossa GW. The new paradigm of cold chain management systems and it's logistics on Tuna fishery sector in Indonesia. *AACL Bioflux*. 2015;Vol.8(3):p. 381–9.

[21] Zhu Q, Sarkis J, Lai K. Institutional-based antecedents and performance outcomes of internal and external green supply chain management practices. Journal of Purchasing Supply Management. 2013;Vol.19(2):p.106–17.

[22] Zhu Q, Sarkis J, Geng Y. Green supply chain management in China: pressures, practices and performance. *International Journal of Operational Production Management.*

2005;Vol.25(5):p.449–68.
[23] Zhu Q, Sarkis J, Lai K. Examining the effects of green supply chain management practices and their mediations on performance improvements. *International Journal Production Res.* 2011; Vol.50(5):p.1377–94.
[24] Guang Shi V, Lenny Koh SC, Baldwin J, Cucchiella F. Natural resource based green supply chain management. Supply Chain Management Int J. 2012;Vol.17(1):p.54–67.
[25] Grant R. The resource-based theory of competitive advantage. *California Management Review.* 1991;Vol.33(3):p.114–35.
[26] Mwaura AW. Green Distribution Practices and Competitiveness of Food Manufacturing Firms in Kenya. *International Journal Economy Commercial Management.* 2016;Vol.IV(3):p. 189–207.
[27] Deqqaq H, Abouabdellah A. Design of a green distribution network with multiple tranportation modes. ARPN Journal Engineering Apply Science. 2016;Vol.11(21):p.12354–61.
[28] Goh S, Richards G. Benchmarking the learning capability of organizations. *Europe Management Journal.* 1997;Vol.15(5):p.575–83.
[29] Dibella AJ, Nevis EC, Gould JM. Understanding Organizational Learning Capability. *Journal of Management Study.* 1996;Vol. 33(3):p. 361–79.
[30] Chen G. Management practices and tools for enhancing organizational learning capability. *SAM Advance Management Journal.* 2005;Vol.70(1):4–35.
[31] Yeung ACL, Lai KH, Yee RWY. Organizational learning, innovativeness, and organizational performance: A qualitative investigation. *International Journal Production Res.* 2007;Vol.45(11):p.2459–77.
[32] Goh SC, Elliott C, Quon TK. The relationship between learning capability and organizational performance: A meta-analytic examination. *Learning Organization.* 2012;Vol.19(2):p.92–108.
[33] Smith PAC. The importance of organizational learning for organizational sustainability. *Learning Organization.* 2012;Vol.19(1):p.4–10.
[34] Dibella a J, Nevis EC, Gould JM. Understanding Organizational Learning Capability. *Journal Management Studies.* 1996;Vol.33(3):p. 319–61.
[35] Antonacopoulou E, Chiva R. The Social Complexity of Organizational Learning: The Dynamics of Learning and Organizing. *Management Learn.* 2007;Vol.38(3):p.277–95.
[36] Cingöz A, Akdoğan AA. Strategic Flexibility, Environmental Dynamism, and Innovation Performance: An Empirical Study. *Procedia - Social Behaviour Science.* 2013;Vol.99:p.582–9.
[37] Azadegan A, Patel PC, Zangoueinezhad A, Linderman K. The Effect of environmental complexity and environmental dynamism on lean practices. *Journal of Operational Management.* 2013;Vol.31(4):p.193–212.
[38] Eccles RG, Ioannou I, Serafeim G. The Impact of Corporate Sustainability on Organizational Processes and Performance. *Management of Science.* 2014;Vol.60(11):p.2835–57.
[39] Saufi NAA, Daud S, Hassan H. Green Growth and Corporate Sustainability Performance. Procedia Economy Finance. 2016;Vol.35:p.374–8.
[40] Kusi-sarpong S, Sarkis J, Wang X. Int . J . Production Economics Assessing green supply chain practices in the Ghanaian mining industry : A framework and evaluation. International Journal Production Economy. 2016;Vol.181:p. 325–41.
[41] Chin TA, Tat HH, Sulaiman Z, Muhamad Zainon SNL. Green supply chain management practices and sustainability performance. Adv Sci Lett. 2015;21(5):p. 1359–62.
[42] Cheng TCE, Farahani RZ, Lai K hung, Sarkis J. Sustainability in maritime supply chains: Challenges and opportunities for theory and practice. Transportation Res Part E Logistics. 2015;Vol.78:p. 1–2.
[43] Vijayvargy L, Thakkar J, Agarwal G. Green supply chain management practices and performance: The role of firm-size for emerging economies. *Journal of Manufacturing*

*Technology Management.* 2017;Vol.28(3):p.299–323.

[44] Kafa N, Hani Y, El Mhamedi A. Sustainability performance measurement for green supply chain management Vol. 6, IFAC Proceedings Volumes (IFAC-PapersOnline). 2013.p. 71-78

[45] Chin W. The partial least squares approach to structural equation modeling. *Mod methods Bus Res*. 1998;Vol.295(2):p.295–336.

[46] Fornell C, Larcker DF. Structural Equation Models with Unobservable Variables and Measurement Error: Algebra and Statistics. Journal of Marketing. 1981;Vol.18(3):p. 382.

[47] Hair JF, Sarstedt M, Ringle CM, Mena JA. An assessment of the use of partial least squares structural equation modeling in marketing research. Journal Academy of Marketing Science. 2012;Vol.40(3):p. 414–33.

[48] Vinzi VE, Chin WW, Henseler J, Wang H. Handbook of Partial Least Squares: Concepts, Methods and Applications. *Methods*. 2010. p. 627.

# Increasing public awareness of consuming Indonesian coffee through electronic marketing

**A. Lisdayanti[1] and T.A Anwar[2]**

Faculty of Business and Management, Widyatama University
Bandung - West Java, Indonesia

annisa.lisdayanti@widyatama.ac.id

**Abstract.** Indonesia is one of the five largest coffee producing countries in the world during the 2016-2017 planting season. Indonesia itself has various types of famous coffee to foreign countries. With the diversity of existing coffee and also has been known in foreign countries, still the public awareness is still very low on it. This study aims to increase public awareness of Indonesian coffee consumption through electronic service quality (E-service) and e-word of mouth (E-WOM) as has been done by Otten Coffe in Bandung. Otten Coffee is one of the online stores that sell Indonesian coffee products complete with complete coffee maker equipment. With followers in Instragram accounts of 193 thousand and 17,214 youtube subscribed videos, Otten Coffee has sought to raise awareness of Indonesian coffee quality. However, people just look at existing coffee products without making a purchase. This at least shows that the awareness of the community towards the original Coffee of Indonesia is still low. Unlike foreign consumers who also make a visit to Otten Coffee website and immediately make a purchase.

**Keywords:** Indonesia Coffee, E-service quality, E-WOM, Brand Image

## 1. Introduction

The culture of drinking coffee in Indonesia has been growing since the first imposition of forced cultivation by the Dutch government at the time of colonization. Indonesia is one of the countries that are included into the five largest coffee producing countries in the world during the plant season in 2016-2017, this can be seen in the table as follows:

**Table 1** The five largest coffee producing countries in the world 2016-2017

| Ratings | Country | Coffee in Bags @60kg |
|---|---|---|
| 1 | Brazil | 55.000.000 |
| 2 | Vietnam | 25.500.000 |
| 3 | Colombia | 14.500.000 |
| 4 | Indonesia | 11.491.000 |
| 5 | Ethiopia | 6.600.000 |

Source : www.indonesia-investments.com (2017)

In Indonesia itself has various types of coffee known to other countries such as Aceh Gayo, Mandailing, Lampung, Java coffee, Kintamani, Toraja, Bawaja, Wamena and also one of coffee with flavor and aroma typical of civet coffee. Variety and magnitude of coffee produced by Indonesia is inseparable from the history of its development. Coffee became one of the mainstay commodities in the early 1900s.

Product quality is an important factor in a business interaction, both products especially national coffee products. Currently the awareness of Indonesian people is still low to consume national coffee, whereas the coffee produced has better value for health than instant coffee. Kotler and Keller (2012: 121) states that product quality is the ability of a product to perform its functions, including durability, reliability, accuracy, ease of operation and improvement and other valuable attributes [1].

Otten Coffee is one of the online store that sells Indonesian Coffee products with complete coffee maker. With the followers in the account as many as 193 thousand instagram and video subscribe on youtube as much as 17.214, Otten Coffee seeks to raise public awareness of the national coffee in general and in particular increase sales. According to El-Gohary (2010: 210), e-marketing can be viewed as a new philosophy and modern business practice involved with the marketing of goods, services, information and ideas via the internet and other electronics [2]. However, even though Otten Coffee has increased promotional activities through electronics there are still many Indonesians who do not know Otten Coffee. This is not in line with the condition that Otten Coffee is exporting coffee to Africa, Australia, Malaysia, Singapore and Japan. E-marketing activities conducted from 2013 until now have been done with various as join in Google Q Words. Kotler and Keller (2016: 330) state that "brand image describes the extrinsic properties of the product or service, including the ways in which the consumer attempts to meet customers' psychological or social needs" [3]. Brand image becomes one of the main priorities that are used as a reference for consumers before making a purchase. On the basis of the above, the researcher is interested to conduct research with the title **"Increasing Indonesian Consciousness Consuming Indonesian Coffee through Electronic Marketing".**

## 2. Literature Review

### *2.1. E-Service Quality*

According to Kotler and Keller (2009: 474); e-marketing describes each efforts made by the company to inform buyers, communicate, promote, and sell its products and services through internet [4]. And the concept of service quality refers to some scope the definitions of some of which are within the scope of industry, output or supply, processes, and systems. Tjiptono (2007) defines service quality as a measure of how both the level of service provided the company is able to match expectations in the minds of customers [5]. According to Lee and Lin (2005: 162), e-service quality can be interpreted as customer responses to the company as a whole and assessment regarding the superiority and quality in delivering e-service in the market virtual [6]. Lee and Lin (2005: 163) also explains the dimensions of e-service quality within a website adapted from the service quality dimension to measure how service delivery quality through the website is divided into 4 dimensions as follows: (1) Website Design, (2) Reliability, (3) Responsiveness and (4) Trust [6].

### *2.2. E-Word of Mouth*

According to **Kotler & Keller (2009: 512)** word of mouth marketing is marketing activities through intermediaries of people to people whether orally, written, or electronic communication tools related to the experience of purchasing services or the experience of using a product or service [4]. **Kotler & Armstrong (2012: 139)** reveals that word of mouth has a great power that can impact consumer buying behavior. Recommendations from trusted friends, associations, and other consumers have the potential to be more reliable than commercial sources, such as advertising and salespeople [7]. In his research, Goyette, et al., (2010) divides the eWOM into three dimensions:

1. Intensity
2. Valence of Opinion
3. Content

### *2.3. Brand Image*

**Keller (2013: 432)** states that, one important vital aspect of the brand is the image [8]. Meanwhile, according to **Kotler (2012: 235)** defines brand image as a set of beliefs, ideas, and impressions that are owned by a person to a brand. Therefore, the attitude and actions of consumers to a brand is determined by the brand image, Kotler added that brand image is a requirement of a strong brand [9]. Keller (2008: 56) explains, there are three dimensions in forming a brand image (brand image) in relation to the association perceived by consumers against a particular brand:

1. Strength of brand association
2. Favorability of brand association
3. Uniqueness of brand association [10].

## 3. Paradigm Research and Hypotheses

Based on the previous theory, then made a paradigm of research and hypothesis can be seen below:

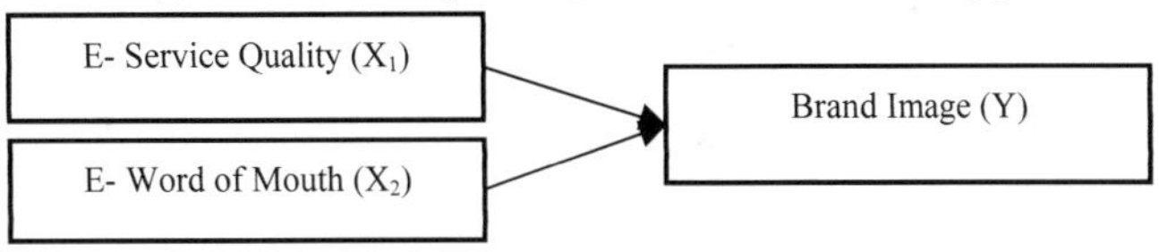


Figure 1 : Research Framework

Based on the research framework, the following hypotheses were developed :

H1: There is the effect of e-service quality on brand image at Otten Coffee
H2: There is the effect of e-wom on brand image at Otten Coffee
H3: There is the effect of e-service quality and e-wom on brand image Otten Coffee simultaneously

## 4. Research Methods

Data analysis method used in this research is descriptive and verifikatif. Descriptive research is a study that has a purpose to describe something and usually characteristics or functions. And verifikatif research is used to know the influence of e-service quality and e-wom to brand image either, simultaneously or partially. The research model used is multiple linear regression.

**Sampling Techniques**

While in the determination of sample size that will be used for this research is calculated using iteration method. According to Sitepu (1994: 108-109) the sample determination can be done through three stages of calculation, namely by determining the smallest correlation coefficient (ρ) value between independent variables with dependent variable, determining the real level (α) and test power (1-β) and determine the sample size [11]. The calculation of the sample by using the calculation of three times iteration so that the minimum sample used in this study as many as 115 respondents. The criteria specified for sampling in this study were Otten Coffee potential consumers.

## 5. Finding Research

The research data was collected by distributing questionnaires with statements about e-service quality and e-word of mouth spread to 115 consumers of Otten Coffee.

Table 2. Recapitulation of E-Service Quality

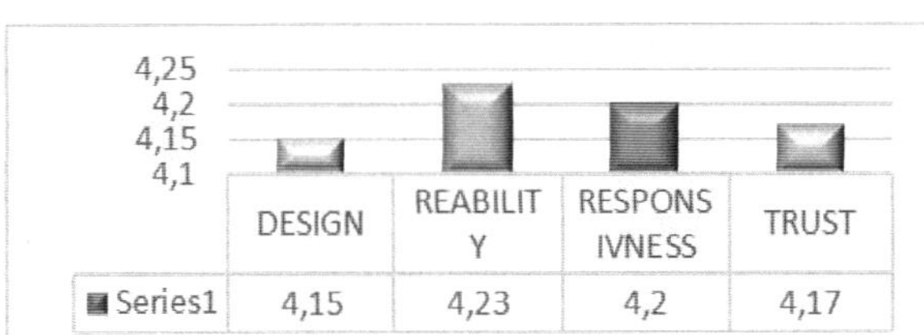


Source : data processed (2017)

Based on the table above can be seen that from 4 (four) dimensions of e-service quality, it can be explained that the highest value is assessed reliability. This relates to the quality of information, the service system to the consumer as well as the accurate representation of product visualization depicted in the website or instragram.

Table 3. Recapitulation of E-WOM

Source : data processed (2017)

Based on the table above can be seen that from 3 (three) dimensions of e-wom, it can be seen that the highest value that is assessed content. This is related to the positive comments from website users or an instragram about Otten Coffee, the use of hastags to recommend products and view other consumer comments

Table 4. Recapitulation of Brand Image

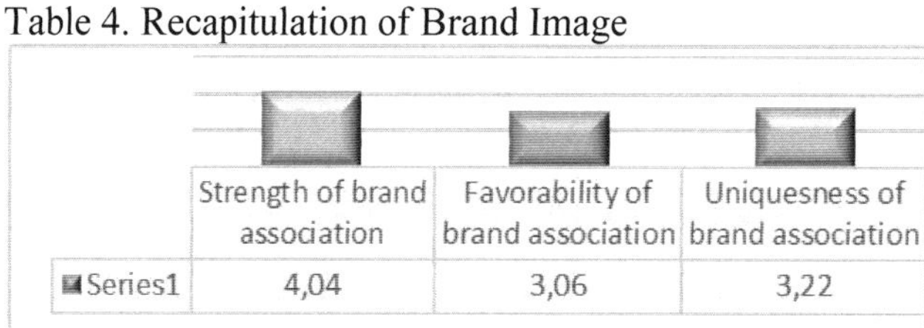


Source : data processed (2017)

Based on the table above can be seen that from 3 (three) dimensions of the brand image, it can be explained that the highest value is assessed strength of brand association. This relates to the quality of coffee offered, the general quality of storage, the variety of products and the credibility of Otten Coffee. Based on the results of data processing, the following is an explanation of the influence of each relationship that exists in this study.

**Table 5. Correlations**

| | $X_1$ | $X_2$ | Y |
|---|---|---|---|
| $X_1$ | 1,00 | 0,889 | 0,971 |
| $X_2$ | | 1,00 | 0,918 |
| *Y* | | | 1,00 |

Source : data processed (2017)

**E-Service Quality to Brand Image**

The results stated that e-service quality (X1) has a positive and significant effect on brand image (Y). With a relationship value of 0.971 then the interpretation of the magnitude of the relationship of both is very strong. Based on the value of the relationship, the value of influence between e-service quality on brand image is 94.28%.

**E-WOM on Brand Image**

The results stated that E-WOM (X2) has a positive and significant effect on brand image (Y). With a relationship value of 0.918 then the interpretation of the magnitude of the relationship of both is very strong. Based on the value of the relationship then obtained value of influence between E-WOM on brand image is equal to 84.42%.

**The Effect of E-Service Quality and E-WOM on Brand Image**

**Tabel 6. Model Summary[b]**

| Model | R | R Square | Adjusted R Square | Std. Error of the Estimate | Change Statistics | | | | |
|---|---|---|---|---|---|---|---|---|---|
| | | | | | R Square Change | F Change | df1 | df2 | Sig. F Change |
| 1 | ,978[a] | ,957 | ,956 | ,13726 | ,957 | 1242,556 | 2 | 112 | ,000 |

a. Predictors: (Constant), X2, X1
b. Dependent Variable: Y

Based on the results of data processing can be known also the influence of together between E-Service Quality and E-WOM on Brand Image has a relationship of 0.954 with very strong relationship interpretation and the influence between these variables is 95.4%. Based on the previous sub-structural can be seen that from the second (2) independent variables studied, the first independent variable is E-Service Quality becomes the largest variable that can affect Brand Image in Otten Coffee.

## 6. Conclusion

Based on the results of research conducted on Otten Coffee on E-service quality, E-Wom and brand image, then obtained the following conclusions:

1. Based on the assessment, it can be concluded as follows: (a) E-service quality, which is considered to have the highest by the consumer is reability, such as quality of information, the service system to the consumer as well as accurate representation of product visualization depicted in the website or instragram. (b) The highest value of the E-WOM assessed by the respondent is the website or instragram content. This is related to positive comments from website or instragram users, hastag usage to recommend products and view other consumer comments. (c) The highest value of the greatest brand image is the strength of the brand association. This relates to the quality of coffee offered, the general quality of storage, the variety of products and the credibility of Otten Coffee.
2. To find out how much influence of E-service quality variable to brand image on Otten Coffee either partially or simultaneously, based on research result indicate variable E-service quality affect the brand image with the influence value of 94.28% and E-WOM affect the brand image with a value of 84.42%. and the results showed that the two independent variables give a high effect that is equal to 95.7%.
3. To see which effect is greatest in improving brand image, it can be concluded that E-service quality has greater influence compared to other independent variables. So companies need to pay attention to website design, reliability of E-service quality, responsiveness to consumers and more able to increase consumer confidence.

## 7. Suggestion

1. Company

   To increase public awareness of existence of Indonesia coffee, Otten Coffee needs to improve design of the website to make it look more attractive and provide clear information about the difference each type of Indonesian Coffee.

2. Researcher

   For next researcher, this research can be cultivated by addition of other variable that can be studied with the aim of developing public awareness of the existence of Indonesia coffee. The addition of variable such as marketing mix may have a big influence on public awareness in Indonesia coffee.

## References

[1] Kotler, Philip, and Kevin Lane Keller. (2012). Marketing Management, 14th Edition United States of America : Pearson

[2] El-Gohary, Hatem. (2010). E-Marketing- A literature Review from a Small Businesses Perspective. Vol. 1, No. 1, pp 214-244. United States of America: International Journal of Business and Social Science.Goyette 2010

[3] Kotler, Philip, and Kevin Lane Keller. 2016. *Marketing Management,* 15th Edition, Pearson Education, Inc.

[4] Kotler, Philip, and Kevin Lane Keller. 2009. *Marketing management.* Upper Saddle River, N.J.: Pearson Prentice Hall.

[5] Tjiptono, Fandy. 2007. Strategi Pemasaran. Edisi Kedua. Yogyakarta : Andi.

[6] Lee, G.-G.,dan Lin, H.-F. 2005. Customer perceptions of e-service quality in online shopping. International Journal of Retail dan Distribution Management.161-176.

[7] Kotler, Philip and Armstrong, Gary. 2012. Priciples of Marketing, 4th Edition. United States of America: Person.

[8] Keller, Kevin Lane. 2013. Strategic Brand Management, 4th ed., Upper Saddle River, NJ: Pearson Prentice-Hall

[9] Kotler, Philip. 2012. Marketing Management. 14th Edition. United States of America: Person.

[10] Keller, Kevin Lane,. 2008, Strategic Brand Management, 3rd ed., Upper Saddle River, NJ: Pearson Prentice-Hall.

[11] Nirwana SK. Sitepu, 1994, Analisis Jalur (Path Analysis),Unit Pelayanan Statitika Jurusan Statistika, Bandung : FMIPA UNPAD.

# Environmental Management in Central Siberia (Based On Analysis of the Krasnoyarsk Territory)

**N P Koptseva**[1,2], **A A Sitnikova**[1]

[1] Humanitarian Institute, Siberian Federal University, Krasnoyarsk, Russia
[2] Research Department, Krasnoyarsk State Pedagogical University named after Astafiev

E-mail: decanka@mail.ru

**Abstract.** The article presents a structural description of the Krasnoyarsk territory's environmental management system as the territory representing Central Siberia. The study is largely based on key provisions of the "Environmental policy concept of the Krasnoyarsk Territory until the year 2030". Certain structural elements of the region's environmental management are analysed. These are nature protection activities of large industrial enterprises operating in the Krasnoyarsk Territory (United Company Rusal, Nornickel, The Mining and Chemical Combine of Rosatom, Siberian Generating Company), the transport system, and waste processing and recycling in the Krasnoyarsk Territory.

## 1. Introduction.

The studies in the environmental management field in Central Siberia based on the analysis of the Krasnoyarsk Territory are extremely important for a number of reasons. According to numerous statistical surveys and environmental ratings, the Krasnoyarsk Territory and its capital, the city of Krasnoyarsk, are considered as among the most environmentally unsatisfactory regions in Russia (in the "Environmental rating of 100 Russian cities according to the opinions of local residents, 2016", compiled by Domofond.ru, Krasnoyarsk takes 93rd place). The term "black sky" was introduced by mass media in Krasnoyarsk to denote the high levels of smog in the city and excessive industrial emissions of harmful substances into the air. The power stations that currently supply most of the region's energy requires the use of coal for fuel, which is extremely harmful to the body. Coal burning is the leading cause of smog that negatively affects the atmosphere, the population's health, etc. For this reason, the use of coal is rapidly declining in developed countries. In order to tackle unfavourable environmental conditions in the Krasnoyarsk Territory, the "Environmental policy concept of the Krasnoyarsk Territory until the year 2030" was adopted. It states the following: "According to the total mass of pollutant emissions from stationary sources (2,516.8 thousand tons in 2011), the Krasnoyarsk Territory ranks first among other federal subjects of the Russian Federation and it is far ahead of all other regions in terms of specific weight (average mass of emissions per source).
The territory's five largest industrial enterprises emit almost 90% (2,210 thousand tons) of total emissions from all stationary sources in the region into the atmosphere. The industrial centres of the territory, which are the cities of Krasnoyarsk, Norilsk, Achinsk, Lesosibirsk, and Minusinsk, are included in the priority list of cities in the Russian Federation with the highest level of air pollution". Western scientists became interested in social and economic environmental studies a few decades ago. In Russian Siberian studies, however, ecology has long remained a subject of natural sciences. This study aims to describe the environmental management situation in the Krasnoyarsk Territory from the

IOP Conf. Series: Earth and Environmental Science **164** (2018) 012020 doi:10.1088/1755-1315/164/1/012020

social and economic sciences perspective, which will help determine the prospects for studying environmental management problems within the framework of socio-economic disciplines.

The term "environmental management" was brought into scientific use roughly in the 1990s. At the present time, it's gradually establishing itself in the Russian scholarly world. For this reason, it is important to continue to study the environmental management system in various regions of Russia, in particular, in the Krasnoyarsk Territory – one of the most ecologically unfavourable regions.

The main purpose of this article is to clarify the extent and nature of all significant participants' involvement that are causing an environmental crisis in the Krasnoyarsk Territory and to identify promising areas in environmental management research in the Krasnoyarsk Territory.

## 2. Literature Review

Western scientists have been actively engaged in environmental management issues since the late XX century [1, 2, 3, 4, 5, 6, 7]. Nowadays, hundreds of studies are being devoted to various aspects of environmental management. Scholars attempt to assign a definition to the concept of the environmental management and determine whether it refers only to enterprise management systems or can be understood as a system for managing the relationship between the environment and society in general [8, 9]. There are also works presenting analysis of certain environmental management aspects, such as the ecological outcome of implementing the ISO 14000 standard in enterprises [10]. Some specialists focus on environmental management policies in different regions [11, 12, 13, 14, 15].

Environmental management is a scientific trend that is also under active development in Russia. Most of its publications on this topic are devoted to describing environmental management technologies undergoing large-scale production. The ecological management of socio-economic orientation (human resources management for environmental protection), however, is understudied.

Despite the fact that there are few works examining environmental management in the Krasnoyarsk Territory [16], the problem of the need to develop a special environmental policy and to introduce environmental management in the region's enterprises at the international standards level has not yet been properly addressed in existing studies.

## 3. Materials and methods

This research was carried out on the basis of qualitative content analysis of official websites of large industrial enterprises operating in the Krasnoyarsk Territory - RUSAL, NORNICHEL, Siberian Generation Company, Rosatom. Reports and statistics related to the activities of the largest financial and industrial groups operating in Central Siberia were analyzed.

## 4. Results

In the broadest sense, the notion of environmental management is regarded as management of society-based interactions and the impact society has on the environment. In this respect, one must consider diverse aspects of human interaction with the environment within the framework of environmental management: government and business organisations' environmental policy, management of both domestic and industrial waste, administrative measures for reducing harmful human impact on the environment in everyday life (use of transport, energy consumption, etc.), system methods for cleaning, maintaining, and restoring biosystems (water and land systems, animal and plant care), environmental education programmes at levels ranging from kindergarten all the way through university. In a narrower sense, environmental management is an area of enterprise management that ensures a balance between economic benefits and environmental costs. In 1996, ISO 14000 was developed, the universal international standard for environmental management. It has been revised and supplemented in accordance with the current environmental and technological situation and is now used to check the degree of environmental responsibility that large industrial enterprises have. The standard assumes continuous improvement of a particular organisation's technical equipment for the purpose of reducing its harmful effects on the environment, internal monitoring of the environmental situation, as well as external auditing of environmental indicators. It also requires enterprises to be open about their impact on the environment by publishing reports with statistical indicators, environmental policy programmes, etc. In this article, we try to provide some data on what modern environmental management requirements, both in a broad and in a narrow sense, are being implemented in the Krasnoyarsk Territory.

To begin with, as previously noted, the environmental policy conception until the year 2030 is developed in the Krasnoyarsk Territory, which is the first important step towards effective improvement of the environment in order to cope with such current problems as climate change and harmful anthropogenic impact on the environment. According to this concept, improvement is planned of the waste management situation in the Krasnoyarsk Territory. There is not enough specialised space for municipal solid waste storage at the moment. Furthermore, there are burial places of hazardous nuclear waste that require constant improvement of its safe storage system. Nevertheless, the most necessary waste management measure for the region today is the implementation of waste recycling programmes since there is an extremely limited quantity of sorting and further waste processing points (for tires, paper, glass); the waste sorting system has not been introduced into citizens' everyday life (non-indifferent citizens must look for the waste sorting points themselves and then carry out the waste removal); there are not enough enlightening activities promoting the idea of waste sorting (environmental art festivals, public petitions). The world practice shows that the introduction of the idea of waste sorting into society requires special educational campaigns on television and in schools.

The Krasnoyarsk Territory is also among the first in Russia in terms of a harmful impact on road transport on the atmosphere. The environmental policy concept of the Krasnoyarsk Territory implies complete switching of public transport to the fuel that meets international standards Euro 5 and Euro 6 by the year 2030. However, there still remains an issue concerning the quantity of personal vehicles and the lack of alternative transportation means. The problem has been discussed year in and year out, but has not yet been solved.

One of the central issues of the environmental policy of the Krasnoyarsk Territory is the restoration of natural resources, forests, and territories inhabited by the indigenous small-numbered peoples. Large industrial enterprises that operate in areas inhabited by the indigenous small-numbered peoples bear a certain share of social responsibility for these territories. It is expected that their environmental policy entails a commitment to the "conservation and restoration of biodiversity".

The key point of this study is a qualitative content analysis of the public information of several large industrial enterprises located in the Krasnoyarsk Territory. According to the "Environmental policy concep of the Krasnoyarsk Territory until the year 2030", the organisations that bear primary responsibility for the current ecological situation in the region are such large enterprises as Rusal, Siberian Generating Company, Limited Liability Company "Krasnoyarsk Cement", Nornickel, Public Joint-stock Company "Polyus", Limited Liability Company "RN-Vankorneft", Mining and Chemical Combine, Rosatom, and others.

According to Rusal's official website, the company's environmental policy is documented in compliance with international standards and involves such aspects as risk management, conformity with the international and Russian environmental law, technological minimization of harmful anthropogenic impact on the environment, personnel training programs on environmental management, cooperation with partners in compliance with environmental standards, regular assessment of the enterprise's environmental impact (both external and internal), and public reports on the results of such assessments. The company "Rusal" is engaged in its environmental modernisation. The official website provides information on planned modernization for 2012; further information is not presented. At the moment, the official reports on environmental protection indicate that, thus far, the company only plans to move onto the international environmental management standards ISO 14000 by the year 2020. In terms of environmental management, as one of its achievements, the company publicly reports that it has managed to reduce greenhouse gas emissions by 50% since 1990. Moreover, it compensates society by implementing several social investment programs that support public initiatives to improve the local environment and allocate individual grants for the development of human capital.

Siberian Generating Company supplies the Krasnoyarsk Territory and certain other regions of Russia with heat and energy power. It produces heating by burning coal, one the most environmentally harmful fuels, and, therefore, entails the most adverse impacts on the environment. Despite the fact that the company website postulates environmentally friendly principles, there is no information on whether the company complies with the environmental standard ISO 14000 or at least plans to move on to this standard in the foreseeable future. Siberian Generating Company receives a lot of criticism from environmental activists and citizens of Krasnoyarsk who care about this issue and is blamed for the city's "black sky". The news media have even covered some "stories" on the company, including one stating that the owner of the company, Andrei Melnichenko, is living in Monaco while residents of Krasnoyarsk breathe poisoned air because he generates heating by burning coal for the sake of cheaper production. In 2016, environmental activists of Krasnoyarsk sent a petition, signed by 1,600 people, to the Prince of Monaco asking him to forbid A. Melnichenko to reside on the territory of the state due to the crime he committed.

The environmental policy of the mining and smelting company "Norilsk Nickel" entails such an aspect as compliance with the environmental standard ISO 14001. This implies external audits of the company's activities for compliance with the international environmental protection requirements. Nornickel took 4th place (out of 33) in the environmental responsibility rankings of Russian mining companies published by the World Wide Fund for Nature (WWF). The following parameters were evaluated based on these rankings: the certified environmental management system that complies with Russian and international standards ISO 14000, the availability of documents on environmental policy, policy on cooperation with the indigenous small-numbered peoples of the North, Siberia, and the Far East; transparency in the company's environmental impact data (statistical reports), greenhouse gas emission indicators, and programs for reducing energy consumption; biodiversity conservation programs on the territory where the company operates; environmental monitoring of the contractors' activities; and environmental risk insurance.

Thus, through the example of environmental policy analysis of the three large enterprises (Rusal, Nornickel, and Siberian Generating Company), it becomes clear that there are organisations with different levels of environmental responsibility in the Krasnoyarsk Territory. Some of them meet the international environmental management standards, while others do not implement enough technologies in their activity to minimize their environmental impact.

## 5. Discussion and Conclusion

The following conclusions can be drawn from this study:

1) Environmental management is quite a new phenomenon for the administrative sector of the Krasnoyarsk Territory (the environmental policy concept was adopted only in 2013); therefore, many aspects of environmental policy (introducing environmental education programs, involving citizens in the processes of reducing harmful anthropogenic impact on the environment, monitoring enterprises' activities and their environmental protection strategies, etc.) are on the project development stage and are planned for implementation in the foreseeable future.

2) In order for their solution, each of the designated environmental policy problems of the Krasnoyarsk Territory requires the involvement of humanities specialists for analysing public opinion on the willingness of the population to participate in environmental protection activities, reviewing international successful practices in reducing the harmful anthropogenic impact on the environment, developing enlightening scientific projects devoted to ecological problems, etc.

Large industrial enterprises located in the Krasnoyarsk Territory bear the greatest responsibility for the deterioration of the ecological situation in the region, although they are aware of the need to comply with the international environmental management standards. Some enterprises conform to the standard ISO 14000, have a developed environmental policy concept, and provide public access to the complete data on their environmental impact. The activity of others, on the contrary, does not meet the international environmental management standard requirements at all.

## 6. References

[1] Bartolomeo M, Bennett M, Bouma J, Heydkamp P, James P and Wolters T 1990 Environmental management accounting in Europe: current practice and future potential *European Accounting Review* **9**(1) 31

[2] Bebbington J, Gray R 2001 An account of sustainability: failure, success and a reconceptualization *Critical perspectives on accounting* **12**(5) 557

[3] Born S, Sonzogni W 1995 Integrated environmental management: strengthening the conceptualization *Environmental management* **19**(2) 167
[4] Margerum R 1999 Integrated environmental management: the foundations for successful practice *Environmental management* **24**(2) 151
[5] Schlosser I 1990 Environmental variation, life history attributes, and community structure in stream fishes: implications for environmental management and assessment *Environmental Management* **14**(5) 621
[6] Selin S, Chevez D 1995 Developing a collaborative model for environmental planning and management *Environmental management* **19**(2) 189
[7] Pearce D, Atkinson G 1993 Capital theory and the measurement of sustainable development: an indicator of "weak" sustainability *Ecological economics* **8**(2) 103
[8] Antunes P, Santos R 1999 Integrated environmental management of the oceans *Ecological Economics* **31**(2) 215
[9] Christmann P 2000 Effects of "best practices" of environmental management on cost advantage: The role of complementary assets *Academy of Management journal* **43**(4) 663
[10] Miles M, Russell G 1997 ISO 14000 total quality environmental management: the integration of environmental marketing, total quality management, and corporate environmental policy *Journal of Quality Management* **2**(1) 151
[11] Dash S, Pradhan M 2014 Environmental Challenges, Climate Change & Development *Journal of Siberian Federal University. Humanities & Social Sciences* **7**(3) 370
[12] Fliervoet J, Geerling G, Mostert E, Smits A 2016 Analyzing collaborative governance through social network analysis: a case study of river management along the Waal River in The Netherlands *Environmental management* **57** 355
[13] Koptseva N 2014 Expert analysis of the main trends of Northern Siberia's indigenous small-numbered people's economic development *Economic Annals-XXI* **11-12** 93
[14] Koptseva N, Kirko V 2014 Specificity of ethnogeny indigenous peoples by Central Siberia in the transition from the traditional type of society to modern society *Life Sci* J **11**(7) 409
[15] Tao Y, Li F, Crittenden J, Lu Z, Sun X 2016 Environmental impacts of China's urbanization from 2000 to 2010 and management implications *Environmental management* **57**(2) 498
[16] Kobalinskii M, Sibgatulin V 2016 Environmental Economic Zoning as the Basis of the Sustainable Development of Resource Territories (Based on the Example of Krasnoyarsk Krai) *Journal of Siberian Federal University. Humanities & Social Sciences* **9**(11) 2616

**Acknowledgements**
The reported study was funded by Krasnoyarsk Region Science and Technology Support Fund according to the research project: « Decorative and Applied Art of Indigenous Peoples of the Krasnoyarsk Territory: Current Status, Prospects for Development.
The reported study was funded by Russian Foundation for Basic Research, Government of Krasnoyarsk Territory, Krasnoyarsk Region Science and Technology Support Fund to the research project № 17-16-24601.

IOP Conf. Series: Earth and Environmental Science **164** (2018) 012021 doi:10.1088/1755-1315/164/1/012021

# A Study on the Evaluation Indicators of the Regeneration of Brownfields in Taiwan - A Case Study of Formosa Chemicals & Fibre Corporation (Changhua Plant)

**S J Feng[1,2], N C Chao[2] and P T Li[2]**

[1] Department of Architecture, Chaoyang University of Technology, 168, Jifeng E. Rd., Wufeng District, Taichung, 41349 Taiwan
[2] Department and Graduate Institute of Architecture, Chaoyang University of Technology, 168, Jifeng E. Rd., Wufeng District, Taichung, 41349 Taiwan
*Corresponding author: fengshyhren@yahoo.com.tw

**Abstract.** Industries in Taiwan have developed for up to 60 years, but many industrial zones, pieces of industrial land or other related resources have gradually shown signs of decline, and manufacturers and factories have even gone bankrupt or transferred their production sites to other countries, which has caused the redundancy and neglect of the industrial land. Most of these sites were originally located in the urban fringe, but are, due to the cities' expansion and alteration, now gradually adjacent to them. Civilians and environmental groups are considerably concerned about the impact of "brownfield" on their sphere of life. Accordingly, the study summarizes the relevant research and data regarding brownfield, and constructs the evaluation indicators of the regeneration of brownfields through empirical research of FORMOSA CHEMICALS & FIBRE CORPORATION. The purpose of the study is to offer indicators, in multiple aspects, for assessment and inspection, in hope of being a reference to the relevant policy and studies of the assessment for activation and regeneration of brownfields, so that the brownfields in cities can play its due value.

## 1. Introduction

While Taiwan has experienced decades of high economic growth and urban development up to the present time, the industry structure has changed. Nevertheless, the land has been continuously developed in the pursuit of economic progress, causing urban growth to be out of control. Constant development toward suburban area also results in the decline of city central parts[1]. Notably, the rapid growth of Taiwanese industries in the 1950s to 1980 has led to the construction of numerous pieces of land for industrial purpose. According to the statistics in 2012 by Bureau of Industry, Ministry of Economic Affairs, the current land area of urban planning industrial zones is 21,999.23 hectares, and that of non-urban D-class construction land is 22,284.94 hectares. However, the industries in Taiwan has continually developed to the present time, many of which have gradually declined and transferred their production sites from Taiwan to other countries, as the times and industrial structure change. In addition, policies have been focused on economic development in the past, neglecting the importance of environmental protection, for which industrial development and its alteration have even caused land pollution, or conditions such as redundancy and neglect or abandonment. Nonetheless, opinions and thoughts about development of brownfield are still not be concerned or adopted in Taiwan. Brownfield development can only apply the laws and approaches relevant to urban planning and urban renewal, but relevant pollution remediation falls within the responsibility of environmental protection authorities. Moreover, other bottlenecks exist, to the implementation of the regeneration of brownfield, such as lack of large amounts of money and technology, the prolonged process, etc. For this reason, the study has reviewed relevant domestic and foreign literatures and related cases, and summarized the environmental, social and economic problems of industrial zones in Taiwan, attempting to apply the thought of brownfield regeneration to the cases in Taiwan. The purpose of this study is to construct assessment indicators for regeneration of brownfield, through the empirical case study on Changhua Plant of FORMOSA CHEMICALS & FIBRE CORPORATION (hereinafter referred to as FCFC plant), which has reviewed the current condition of the brownfield in that plant, hoping to offer a new thinking direction for regeneration of brownfield, in consideration of environmental protection, socio-economic factors, and sustainable development.

## 2. Literature review

### *2.1 Formation and definition of brownfield*

Most brownfields are the deserted fields after industrial economic activities. They began to appear in the first stage of industrialization, and in Europe, they emerged in the United Kingdom in 1800-1914, in Germany in 1870-1940, in most of Eastern Europe and Southern Europe In 1900-1970[2], but in the United States in the period between 1877 and 1980. Along with the general urban development trend, brownfields are involved in the urban sprawl, and are also the result of the influence of urban economy and culture. Namely, brownfields are the irreversible consequence of the land use for extensive industrial development in the early stage. The problem of early brownfields did not attract much attention from all sectors of the community. In recent years, with the awareness of environmental protection enhanced, the problem of brownfield is increasingly noticed. The government and regeneration of brownfields can not only provide local authorities with financial income and employment opportunities, but also comprehensively promote the development of society, economy, and environmental health, leading to win-win benefits in the economic and environmental aspects[3].

However, the formation of brownfields is often accompanied by changes in urban structure, industrial restructuring, environmental pollution or other factors. Moreover, urban development is even severely affected by the redundancy and neglect due to the outward shift of industries, and low or improper use and under-utilization of brownfield. A number of pollution incidents have broken out and pollution sites have emerged in Taiwan over the past few decades, along with the process of the active development of secondary and tertiary industries. Considering the nature of industrial development in Taiwan and the characteristics of Taiwanese brownfields, the most significant discrepancies between Taiwan and foreign brownfields are their scopes and locations. An entire metropolis is likely to be regarded as a potential foreign brownfield site, but in terms of scale[4], a Taiwanese brownfield is usually hidden in an individual factory or on an idle piece of land, which is situated in the fringe of urban planning zones.

Therefore, according to the above analysis, brownfields are the deserted land and are transformed into dilapidated sites, after the industrial economic activities that utilized them, and they may be associated with pollution, waste, redundancy and neglect, low use of land, or other issues. Moreover, the term "brownfield" is profoundly discussed and explored in Germany, European Union, China and other countries, while in Taiwan, in-depth discussion or research is not common, but is only in the form of a certain initial concept or an exploration.

### *2.2 Regeneration of brownfield*

The emergence of brownfield not only results in the shattering urban space, and the disappearance of urban functions, but also wastes the resources of urban development, enhances crime rates, and causes economic depression, environmental pollution or other problems. Therefore, the development of brownfield shall be considered to be a sustainable development concept, and its goals are to mend the developed space and integrate functions in urban areas, to improve the urban public facilities and enhance disaster prevention as well as other functions, meanwhile meeting the economic and social needs of current and future generations. "European Urban Environment Brownfield Sustainable Development Plan" demonstrates that the sustainable development of brownfield is an approach to managing and mending land use, and to ensuring that the present and future needs of human beings will be satisfied, within the areas which are environmentally sensitive, economically and socially acceptable where social systems are healthy. Moreover, the benefits of regeneration of brownfield include alleviation for the pressure of land use, intensive and economical use of land, stimulation on economic growth, facilitation in the improvement of the environmental quality and economic development of the whole society, and a positive effect on the ecological environment and human health[5]. For these reasons, this study summarizes the perceptions and opinions of many scholars on the regeneration of brownfield as shown in figure 1.

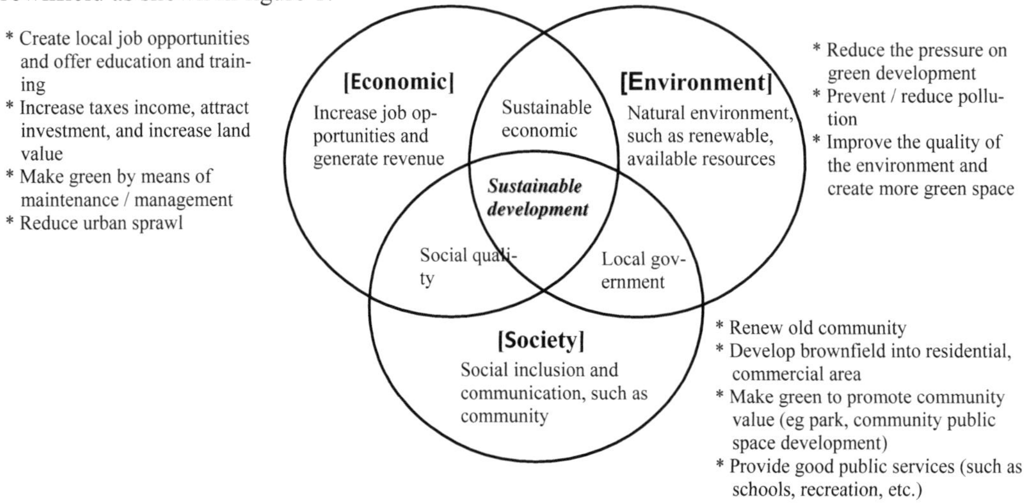


**Figure 1.** Regeneration of brownfield.

## 3. Research method

This study, through literature review, interviews, case evaluation and analysis, constructs the reviewed regeneration evaluation frameworks of brownfield, and applies Fuzzy Delphi Method for induction on reviewed indicators, and afterward analyzes and probes into the relative weights between the indicators, for the reference of the follow-up study on construction of evaluation indicators for brownfield regeneration. Furthermore, the study, through the in-depth interviews with experts in various relevant fields, explores their opinions and advice on brownfields regeneration policy, and collects the first-hand information from their abundance of knowledge to enhance the depth and breadth of this research.

## 4. Industrial structure changes of FCFC plant and relevant issues

### *4.1 Changes in industrial structure*

FCFC plant set up factory in Changhua, Taiwan in 1965, which led to local economic development and offered abundant employment opportunities. More than 6,000 jobs were created during the period of its greatest prosperity, and that was also important for Taiwan in the economic take-off stage. However, changes in industrial structure and the transfer of production sites cause development bottlenecks of textile industries in Taiwan, especially staple fiber business. It is estimated that sales of related products will not grow within next 50 years, and Vietnam and the countries in Southeast Asia are increasingly developing their textile industries with their cheap labor, so that FCFC plant has lost its competitive advantage. With that disadvantage, the company, after being struggled by environmental groups, eventually shut down its factory in 2016, and is currently planning how to transition.

### *4.2 Influence of brownfield regeneration on the local area*

The location of FCFC plant is located on the borderland between Changhua Old City and the area in the Extended Urban Plan, and in the surrounding area are many settlements and residents. However, according to many relevant reports and the opinions of the environmental groups struggling against

FCFC plant, the factory has a considerable negative impact on the local area, and the affection of air pollution is notably tremendous, so the factory had been constantly requested to relocate. However, it is undeniable that FCFC plant has made a considerable contribution to the local economy. An interviewee, during an in-depth interview, expressed that the development of a city cannot proceed without economic support. Therefore, it is an important issue of how a consensus can be reached among the local residents to resolve the negative conflict, so as to guide the impact of FCFC plant on the local area toward a positive direction.

### *4.3 Mechanism and approach of brownfield regeneration*

Through the analysis of above information and the compilation of the interview data, the domestic operation mechanism of and approach to brownfield regeneration is quite immature. Additionally, in the case that FCFC plant is private properties, those issues, which have not been settled, need considerable discussion and negotiation, such as how brownfield regeneration is operated in the future, how feedback is made or other relevant regeneration mechanism. Accordingly, the major issue on brownfield regeneration is to establish a set of comprehensible and definite mechanism, regulations or others.

### *4.4 Impact of development of brownfield from the past to the present on urban structure, environment, society and economy*

FCFC plant was located in the fringe area of Changhua City in the past, and with the expansion of the city, is now adjacent to the urban area, for which FCFC plant has an impact in multiple respects. In the urban plan on expansion and rezoning of Changhua City, the plant plays an important role in the urban structure of the entire city. In the scope of the renewed old town of Changhua City, FCFC plant is situated in the core area, so the transformation and the regeneration of the factory are bound to have a significant influence on the future development of Changhua City, the local environment, economy and society.

## 5. Construction of brownfield regeneration indicators

The study has developed evaluation indicators for brownfield regeneration, and concludes that the assessment must proceed in consideration of six levels, after studying and researching on the relevant literatures and materials, and after open interviews with experts and scholars in restrictive and objective conditions. The brownfield regeneration indicators description as shown in table 1.

**Table 1.** Brownfield regeneration indicators description

| Goal Layer | Layer 2 | Description |
|---|---|---|
| Brownfield Regeneration evaluation indicators | Environmental | The formation of brownfield is due to the past development of the industry, resulting in the pollution of industrial manufacture procedure to the environment. Accordingly, the evaluation for regeneration must take into account the environmental conditions, any concern regarding pollution, and the extent of contribution to the environment. |
| | Society | The long-term impact of the brownfield on the local area is tremendous, so the regeneration evaluation should be associated with the rights and interests, development or other social factors concerning the local area, so as to estimate the contribution to the local community. |
| | Economic | To assess the contribution of regeneration to the economy, taken into consideration and explored are the benefits of economic development, and the industrial operation model of regeneration, industrial capacity or other economic factors. |
| | Policy Implementationt | As brownfield regeneration is realized through the cooperation and specialization in various fields, discussed are the mechanism and operation regulations of policy implementation, and aware of are the laws and regulations, systems, implementation approaches, or other policy-relevant factors concerning brownfield regeneration. |

| | |
|---|---|
| Residence and Life | Taken into consideration are of brownfield regeneration, the influence on residence and life, and the extent of contribution to the quality and the need of life. |
| Health and Safety | Taken into consideration are the extent of planned contribution to the safety of citizens, and the factors regarding assessment and control of the impact of brownfield regeneration, on citizens' health. |

According to the need of this research, questionnaires of the study are distributed mainly to experts, officials and scholars, who have acquired the fundamental and relevant knowledge regarding brownfield and urban re-development. Thirteen questionnaires have been issued and distributed, and ten have been collected, with two invalid ones included. After the arithmetic mean is calculated of the data collected in the study, the research threshold of the study is set as 7.0, which is within the commonly accepted range of 6-8, and the data with $G^i$ value below 7.0 have been eliminated. Consequently, the recommendations are produced on the suitability and importance of and "indicators for evaluation of brownfield regeneration." as shown in table 2.

**Table 2.** Fuzzy delphi method questionnaires result.

| Evaluation Indicators | Ci | | Oi | | a | | Gi | Eliminated |
|---|---|---|---|---|---|---|---|---|
| | min | max | min | max | min | max | | |
| Assimilative capacity of environment | 6 | 8 | 9 | 10 | 8 | 9 | 8.20 | |
| Improvement extent of environmental quality | 5 | 8 | 8 | 10 | 7 | 9 | 7.60 | |
| Investigation, evaluation, and remediation of hazardous substance | 5 | 9 | 9 | 10 | 7 | 10 | 8.50 | |
| Ecological protection and rehabilitation planning | 4 | 9 | 7 | 10 | 6 | 10 | 9.00 | |
| Growth and decline of biodiversity | 3 | 7 | 7 | 9 | 4 | 8 | 6.20 | v |
| Changes to green space | 4 | 7 | 6 | 10 | 5 | 9 | 7.00 | |
| Growth and decline of landscape functions in the environment | 4 | 8 | 6 | 10 | 5 | 9 | 6.49 | v |
| Employment opportunities for the local populace | 3 | 8 | 7 | 10 | 4 | 9 | 6.61 | v |
| Preservation of cultural assets | 2 | 7 | 6 | 10 | 4 | 10 | 7.39 | |
| Enhancement of social security and welfare | 2 | 7 | 6 | 9 | 5 | 8 | 6.54 | v |
| Increase the area of public facilities | 5 | 7 | 7 | 10 | 7 | 9 | 7.50 | |
| Care for underprivileged Groups | 1 | 7 | 6 | 9 | 5 | 8 | 7.00 | |
| Community development and identification | 4 | 8 | 8 | 10 | 7 | 9 | 8.00 | |
| Promote population dynamics to facilitate local development | 2 | 7 | 6 | 9 | 5 | 9 | 8.00 | |
| Diversification in industrial structure | 2 | 7 | 6 | 10 | 5 | 9 | 6.58 | v |
| Increased industrial output value | 4 | 7 | 7 | 9 | 6 | 8 | 6.50 | v |
| Benefits from promotion of the local economy | 5 | 8 | 7 | 10 | 7 | 9 | 7.00 | |
| Price change rate of the surrounding real estate | 3 | 8 | 6 | 10 | 5 | 9 | 7.66 | |
| Percentage of site land value added | 2 | 7 | 5 | 9 | 4 | 9 | 6.77 | v |
| Unemployment rate and employment rate | 4 | 7 | 6 | 10 | 5 | 9 | 6.04 | v |
| The extent to which the government implements the project | 5 | 8 | 7 | 10 | 6 | 9 | 6.42 | v |
| Implementation approaches, resource allocation and planning on public - private partnerships | 6 | 8 | 9 | 10 | 8 | 9 | 8.70 | |
| Prevention of land speculation | 5 | 8 | 7 | 10 | 6 | 9 | 8.32 | |
| Establishment of incentives and feedback system | 5 | 9 | 9 | 10 | 7 | 9 | 8.20 | |
| Sound legitimate basis | 3 | 9 | 7 | 10 | 6 | 10 | 7.80 | |
| Coordination to land use plan | 3 | 7 | 6 | 10 | 5 | 10 | 7.88 | |
| Public safety measures planning | 5 | 7 | 8 | 10 | 7 | 9 | 6.95 | v |

| | | | | | | | | |
|---|---|---|---|---|---|---|---|---|
| Environmental green area ratio and per capita area of open space in green space | 4 | 7 | 6 | 10 | 6 | 9 | 7.35 | |
| Integrity of human-oriented traffic planning | 4 | 8 | 7 | 10 | 6 | 9 | 6.59 | v |
| Mobility of public transport | 3 | 8 | 6 | 10 | 5 | 9 | 7.34 | |
| Industrial transport line planning | 4 | 7 | 6 | 9 | 5 | 8 | 6.90 | v |
| Impact of site use | 4 | 7 | 6 | 10 | 5 | 9 | 6.48 | v |
| Operational approaches to management and monitoring of environment | 5 | 9 | 8 | 10 | 7 | 9 | 6.32 | v |
| Disaster prevention base and facility planning | 4 | 9 | 8 | 10 | 6 | 10 | 8.41 | |
| Escape shelters and evacuation routes planning | 4 | 9 | 7 | 10 | 6 | 9 | 8.35 | |
| Green energy use and reuse planning | 3 | 8 | 6 | 10 | 5 | 9 | 7.85 | |
| The possibility of healthy lesions | 4 | 8 | 7 | 10 | 5 | 10 | 6.83 | v |
| Waste management of industrial development | 4 | 8 | 6 | 10 | 5 | 9 | 7.62 | |

## 6. Conclusion

### *6.1 Impact on Taiwan of brownfield regeneration*

This study, by literature analysis and expert interviews, probes into the origin and development of the brownfield in Taiwan, and then further explores the impact of brownfield regeneration on Taiwan. It is concluded that the area is considerably finite of land in Taiwan, but in the industrial development, numerous plants and industrial zones have been built in Taiwan, which indeed, has created an economic miracle for Taiwan in the past, and also laid the current industrial basis of Taiwan. However, with the progress of time and industrial structure changes, many industries are facing the dilemma of elimination or migration, identical to FCFC plant, the case in this study, but this study assumes that the approach to future development is to enhance product value through transformation and redevelopment, adapted to the international trends of environmental protection and low pollution. Nevertheless, in the expansion and alteration of urban structure, the location and regeneration of brownfields affect the development of the cities which the sites are situated in. For instance, the brownfield such as FCFC plant site, the area of which is so immense that it should be taken into consideration in the urban planning process. As the experts proposed that the city should have had a comprehensive strategic plan, in which each single piece of land plays its due role and function. Accordingly, the impact on Taiwan of brownfield regeneration includes not only the improvement of environmental quality and solution to pollution, but also the enhancement of its economic power and its influence on society, through innovative thinking regarding regeneration.

### *6.2 Evaluation indicators for brownfield regeneration in Taiwan*

The study has summarized that the fields on brownfield regeneration in Taiwan are very complicated and professional, after the analysis of brownfield regeneration and the expert interviews. Therefore, it is expected that the developed indicators can be applied to the ecological restoration of brownfield and the basis of probe into brownfield regeneration, so that the study may realize its due value, meaning that the regeneration is not just a single land development that becomes a price creation. However, due to the unique characteristics and differences of other brownfields, it is proposed to supplement additionally critical assessment items for a single regeneration case, and to adapt the indicator groups to the distinct nature of the brownfield. The evaluation indicators are summarized as follows table 3.

**Table 3.** Brownfield evaluation indicators summarized.

| Aspect | Evaluation Indicators | Explanation |
|---|---|---|
| Environmental Aspect | Assimilative capacity of environment | Apply the evaluation items stipulated by Environment Protection Administration (hereinafter referred to as EPA), including soil, geology, topography, surface water and groundwater resources, water quality, air quality, noise, meteorological or other main factors, to as- |

| | | |
|---|---|---|
| | | sess the contribution of brownfield regeneration to environmental carrying capacity, and to monitor its quality and maintain the performance to meet the Standards, so as to achieve the goal of environmental cleaning and protection. |
| | Investigation, evaluation, and remediation of hazardous substance | Investigate, rectify and monitor contaminants according to the EPA Pollution Control Project, such as wastes, toxic chemicals and radioactive materials, which may be remained by the past industries. Notably, any factors must be described that affect the safety of residence. |
| | Ecological protection and rehabilitation planning | Assess the contribution of brownfield regeneration to the areas of natural ecosystem, including land ecology, water ecology, natural ecological landscape, habitat and other major areas, so as to achieve the goal of ecological protection and rehabilitation. |
| | Changes to green space | Assess the contribution of brownfield regeneration to the growth rates of major green space, including the growth rates of tree planting and green area in natural environment. |
| Society Aspect | Preservation of cultural assets | As the value of historical and cultural assets lies on the base to understand the important meaning of human history, not only should those declared historical resources with cultural assets be preserved, but their potential resources also be described, such as historical sites, historical buildings, cultural assets, industrial landscape, relics, settlement culture, cultural activities, local cultural characteristics, landscape recreation resources and other major areas. |
| | Increase the area of public facilities | Public facilities are constructions serving the public, including transportation, and facilities for people's livelihood and leisure, education and cultural activities. With the increase in the area of public facilities, the number of served people by the base is accordingly increased. |
| | Care for underprivileged Groups | Assess the contribution extent of brownfield regeneration to the care of underprivileged groups and blighted communities in the surroundings. |
| | Community development and identification | As the long-term industrial development on the brownfield has led to the stigma attached to and the sense of disagreement with the locality, assess the contribution extent of brownfield regeneration to the overall improvement of the community and the recognition of the people to the local area. |
| | Promote population dynamics to facilitate local development | Population dynamics has always been the power source of urban development, the brownfield regeneration facilitates the increase in the local population and causes the regional population inflow, driving the growth of human resources that urban development needs. |
| Economic Aspect | Benefits from promotion of the local economy | Assess the contribution extent of brownfield regeneration to the local industry stability, industry-driven effect, local industrial development and economic interests. |
| | Price change rate of the surrounding real estate | Assess the contribution extent of brownfield regeneration to the real estate in the peripheral areas. |
| Policy Implementation Aspect | Implementation approaches, resource allocation and planning on public - private partner- | Assess the coordination extent on operation among the government, citizens, and the third sector, in which approaches are proposed to incentive system, resource allocation, and policy implementation. |

| | | |
|---|---|---|
| | ships | |
| | Prevention of land speculation | Land speculation is a major issue of urban development dilemma. For the smooth operation of urban functions, prevention of land speculation can curb improper land use, and land prices speculation, facilitating the development of urban functions. |
| | Establishment of incentives and feedback system | Current urban renewal incentives: building dimension rewards, building dimension transfer, reduction of or exemption from tax and others; feedback means that the actor of land change is responsible for the cost of such alteration, according to the philosophy of internalization of external cost, and that feedback is required from the value - added benefits of such alteration, in accordance with the philosophy of accruement of land value increments to the public. Feedback contribution items include land, cash equivalents, planned road land, building a planned road, construction of park square, floor area, and others. |
| | Sound legitimate basis | Brownfield regeneration involves a number of professional fields, and the extent of its complexity is considerably high. To improve the functions, effective management and operation of various competent authorities, it is required to develop sound laws and regulations in relevant fields to promote brownfield regeneration. |
| | Coordination to land use plan | Land use plan associates with specific codes, regulations, use intensity, and other directions. The optimal use of land requires prevention of the disorder of land development or excessive land use, and reduction of inconsistent or incompatible land use. |
| Residence and Life Aspect | Environmental green area ratio and per capita area of open space in green space | Parks and green space are not only landscape buffer zones, but also leisure places, which have the effect of reducing the noise, air pollution and high temperature effect. Environmental afforestation is an important approach to the alteration of urban environment and the improvement in the quality of citizens' lives. |
| | Mobility of public transport | Increasing the mobility of urban mass transport system will accordingly enhance the utilization rate of the mass transit system, strengthen the function of urban transportation and circulation, reduce pollution to the environment and waste to energy, and improve the convenience and safety of transportation, reaching the goal of sustainable development. |
| | Improvement extent of environmental quality | Assess the percentage of improvement extent to the status of monitoring indicators for the brownfield site, which did not meet the required standard, such as soil quality, water quality, air quality and others. |
| Health and Safety Aspect | Disaster prevention base and facility planning | Planning of the disaster prevention base can reduce the loss of the disaster, precluding the deprivation of citizens' lives and enhancing the health and safety mechanism of the city. |
| | Escape shelters and evacuation routes planning | Proper planning and design on escape and evacuation route can, in case of an occurrence of disaster, enhance the efficiency of escape, evacuation, and salvage, and increase the effectiveness of the rescue, to facilitate the urban security mechanism. |
| | Green energy use and reuse planning | Energy reuse can both reduce the amount of waste disposal, and promote the re-use of resources, reflecting the degree of recognition of the environmental protection and resource use efficiency in an area. |
| | Waste management of industrial development | The sustainability of brownfield regeneration requires the mechanism to properly handle the various wastes. |

## 7. References

[1] Shiu J Y, Research of Policies of Industrial Lands and Production Efficiency in Taiwan, 2004, National Chengchi University.
[2] Hsu Y C, A study on the development of brownfield - the new direction of urban industrial zones renewal in urban planning, 2002, National Taiwan University.
[3] Guo P, Analysis of American Brownfields Cleanup and It's Revelation of China, Environmental Protection Science, vol. 36, no. 3, Jun 2010, pp. 73-76.
[4] Cheng Y H, A Study of the Industrial Heritage Regeneration as Local Revitalization: An Application on Taiwan Cement Factory in Kaohsiung, 2008, National University of Kaohsiung.
[5] Teng Y C, The Evaluation and Potential Analysis of Turning Brownfield into Urban Habitat—A Case Study of Yungkang City, 2006, National Cheng Kung University.

# Predicting Fish Ecological As Indicator of River Pollution Using Decision Tree Technique

**Che-Yu Hsu [1], Sheng-Jung Ou [2] & Wei-Fan Hsieh[3]**

1.Adjunct Assistant Professor, Department of Landscape and Urban Design, Chaoyang University of Technology, 168, Jifeng E. Rd., Wufeng District, Taichung, 41349 Taiwan
2.Professor, Department of Landscape and Urban Design, Chaoyang University of Technology, 168, Jifeng E. Rd., Wufeng District, Taichung, 41349 Taiwan
3. Assistant Professor, Department of Landscape and Urban Design, Chaoyang University of Technology, 168, Jifeng E. Rd., Wufeng District, Taichung, 41349 Taiwan

sjou@nchu.edu.tw, tony428tw@hotmail.com (Corresponding author).

**Abstract.** The Goal of the research is to introducing the principle of Decision Tree that is being used to forecast river pollution, it provides a new method to evaluate the river pollution based on its water quality. We collected monthly monitoring data of water quality from Dezikou River basin of Yilan County, and the data of fish ecology obtained from ecological survey and report, in which to build a water quality and ecology resources database through an actual field investigation. By using data mining software, IBM SPSS Modeler 14.1's decision tree, conducting the River Pollution Index. Shannon-Weaver diversity, Pielou's Evenness Index, Margalef's Species Richness Index, Fish Tolerant Index and Simpson's Index of Diversity′ s classification and prediction, to build a model for river pollution prediction, and to compare this with the Multiple Logistic Regression Analysis. The results showed that the model for river pollution prediction built under the Decision Tree can obtain a better forecast result. The following are the accurate rates of Decision Tree: 88% for CART, 90% for CHAID, 91.67% for C5.0, and 86.11% for Multiple Logistic Regression Analysis. Therefore, the Decision Tree's algorithm shows a better result in forecasting than the Multiple Logistic Regression Analysis.

## 1. Introduction

Rivers serve many societal functions and belong to the most intensively human influenced ecosystems worldwide [1]. By monitoring the fish in the river, it not only reflects the water quality of the river by the presence of the migratory fish and the fish biology indicators, but also it can further reflect the changes in the environment underwater by RPI. Fishes have a number of advantages as biological-integrity indicators of a watershed system [2]. Some species of fish are very sensitive to changes in water chemistry, such as pH or dissolved oxygen, which may be caused by pollutants. After the monitoring outcome of the success of restructure the river in Singapore and Guyahoga River in America, it is not just to pay attention to the improvement of the water quality, but also it is to concern the effectiveness of the environmental ecology in rivers, meaning fish is one of the main targets for monitoring. This research applies the Shannon-Weaver diversity, Pielou's Evenness Index,

Margalef's Species Richness Index, Fish Tolerant Index and Simpson's Index of Diversity to evaluate the analysis. IBM SPSS Modeler is a data mining and text analytics software application built by IBM. Classification and regression trees are ideally suited for the analysis of complex ecological data [3]. An ensemble algorithm of data mining decision tree (DT)-based CHi-squared Automatic Interaction Detection (CHAID) is widely used for prediction analysis in variety of applications. [4]. The main goals of this study were to introduce the principle of Decision Tree that is being used to forecast river pollution, it provides a new method to evaluate the river pollution based on its water quality. The model for river pollution prediction built under the Decision Tree can obtain a better forecast result.

## 2. Research method

### *2.1 Study area*

Dezikou River is on the northern side of Yilan County, meanders for 19.3 km through a basin area of 113.17 square km, Toucheng Town. Dezikou River is a typical urban river. The importance of the area is derived from the presence of different habitat types and the numerous flora and fauna species. However, human activities in recent years, such as land use change and urbanization, deteriorated the water quality and the ecosystem of Dezikou River. At present, the species that found are most tolerant species, but its number is still not large enough.

### *2.2 The Decision Tree's algorithm*

Decision tree learning uses a decision tree (as a predictive model). It is one of the predictive modelling approaches used in statistics, data mining and machine learning. At present, academia and practitioners use Decision Tree's algorithm including CART, CHAID, and C5.0 etc, due to its research target's variation to be that of ordering type of variation, it cannot use QUEST to undergo analysis.

### *2.3 Data Resource*

*2.3.1 Water quality data information of investigation* This research collects relevant monitoring data information on water quality monthly from the Yuan EPA National Environmental Quality Monitoring Information Network providing information which it has collected in a total five year from 2009 to 2013, in which the information was obtained from Yilan County Water Pollution Control Inspection and basin management plan. Every monthly water quality monitoring data builds relevant data, and builds a water quality database [5] [6].

*2.3.2 Ecological survey* Survey collected fish ecology information between 2009 and 2013 from Yilan County Council EPA It records fish' category, species, individual number, feeding characteristic, movement characteristic, pollution resistance characteristic, habitat characteristic and pollution resistance factor to build fish ecological database by its actual field investigation. Furthermore, its ecology factor is calculated through Shannon-Weaver diversity, Margalef's Species Richness Index, Fish Tolerant Index, and Simpson's Index of Diversity.

## 3. Result & Discussion

The research is to adopt sample database to build a river pollution forecast model, and to evaluate the model.

### *3.1 Classification and Regression Analysis CART Algorithm*

Through CART algorithm, we can find that if the species diversity is lower or extremely low and if there is less or lesser abundant species, then the RPI has 75% chance to be Severely-polluted; if species diversity is lower or extremely low and if the species abundance is low, higher, and extremely high, then the RPI has 96.88% chance to be Moderately-polluted; if there is a higher level of species diversity, then the RPI has 57.14% chance to be lightly-polluted. If there is an extremely high level of species diversity, then the RPI has 100% chance to be Non (Slightly)-polluted (see Figure 1).

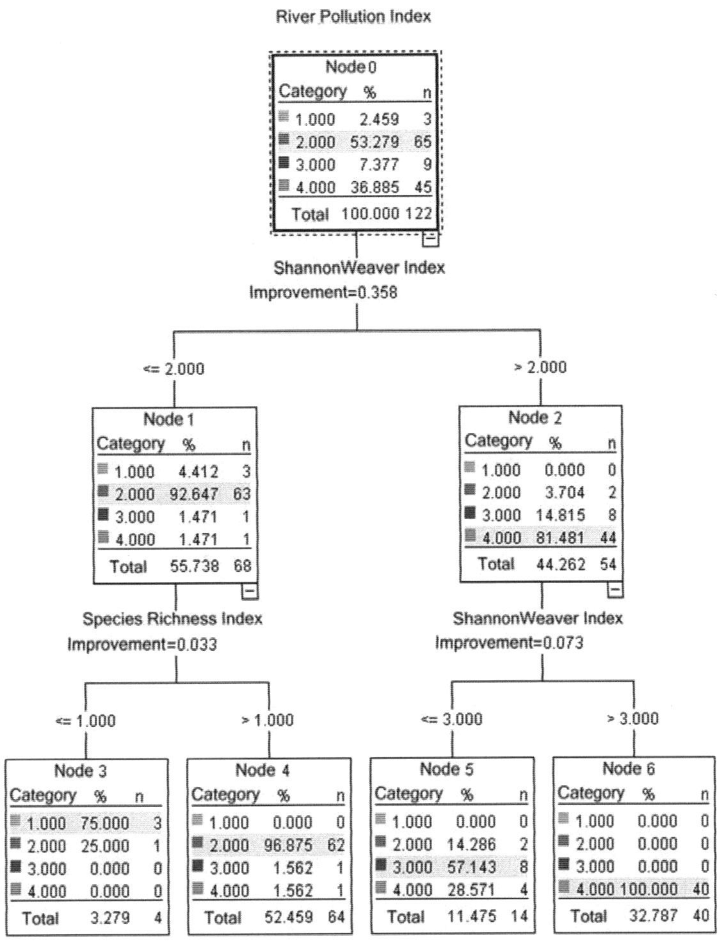


**Figure 1**. Classification and Regression Analysis (CART) Algorithm

*3.2 Chi-Square Automatic Interaction Detection , (CHAID Algorithm*

Via CHAID, we can find that if it there are more fish species with severely-polluted indicator, then the RPI has 50% chance to be Severely-polluted; if there are more fish species with moderately polluted indicators and if there is a low or lower level of low fish diversity, then the RPI has 97.62% chance to be Moderately-polluted; if there are more fish species with lightly-moderated indicators, then RPI has 54.5% chance to be lightly-polluted. If there are more fish species with non- (slightly-) polluted indicators, then then the RPI has 98.44% chance to be non- (slightly)-polluted (see Figure 2).

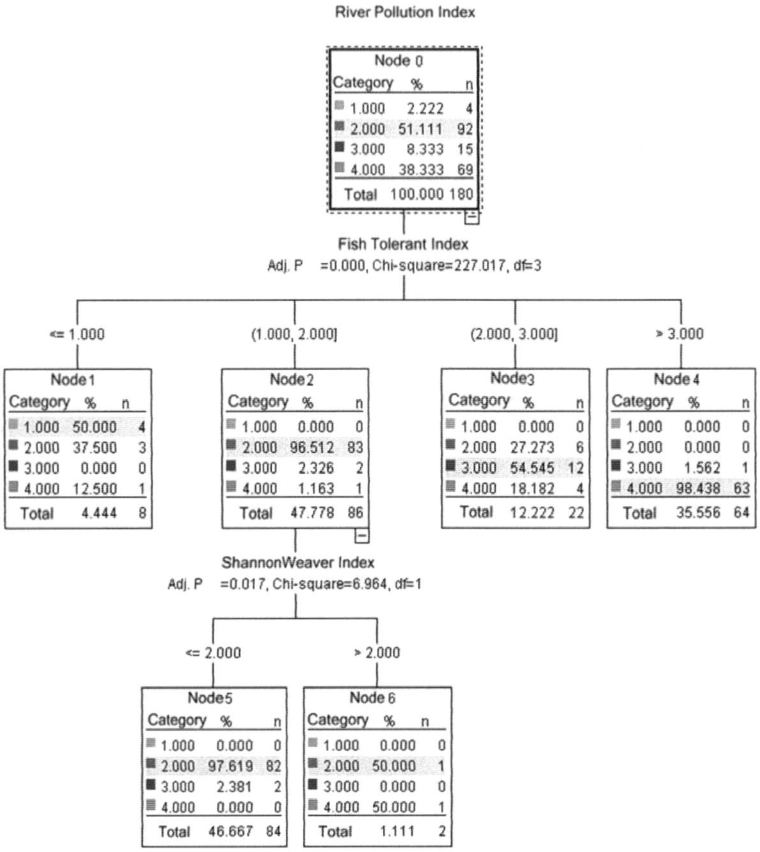


**Figure 2**. Chi-Square Automatic Interaction Detection , CHAID Algorithm

### *3.3 Decision Tree C5.0 Algorithm*

By C5.0 algorithm, we find that if it is severely, moderately and lightly polluted has more fish species and the level of diversity among fish species is low / lower and not abundant fish species then RPI has 66.67% rate to be Severely-polluted; if severely, moderately, and lightly polluted has more fish species and low / lower level of fish diversity among fish species and not abundant fish species , then RPI has 95.56% rate to be Moderately-polluted; if severely, moderately and lightly polluted has more fish species and high / higher level of diversity among fish, then RPI has 60% rate to be lightly-polluted; if non(slightly) polluted has more fish species then RPI has 98.44% rate to be non(slightly)-polluted (see Figure 3).

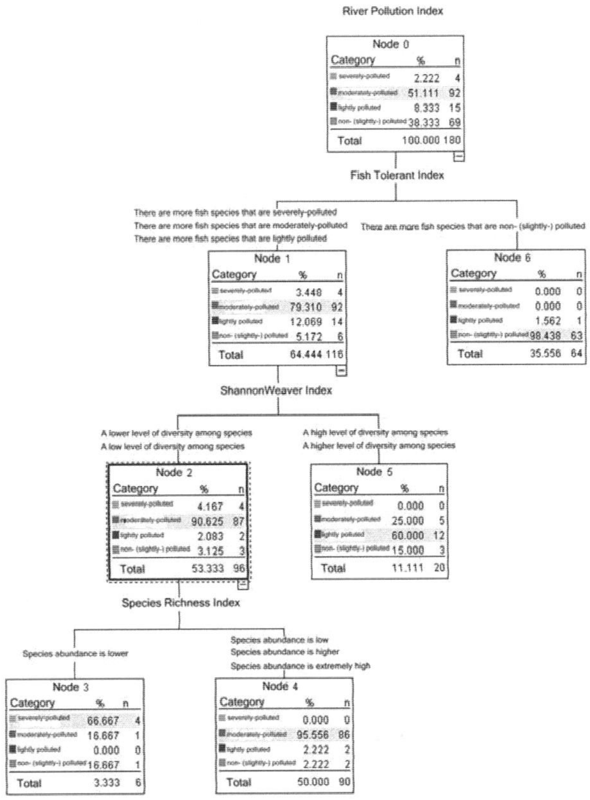


**Figure 3**. Decision tree C5.0 diagram

### *3.4 Multiple logistic regression analysis – building river pollution forecast model*

First test to see how match the variables are within the multivariate logistic regression models. Its test result will show model's P = 0.000, so the variables (independent variable) overall can explain its dependent variables to have significance of the statistics (see Table 1). Their classification and actual categories the accurate rate is 86.11%.

**Table 1.** Multiple logistic regression models all samples goodness-of-fit test analysis

| Method | -2 Log Likelihood | Chi-Square | Sig. |
|---|---|---|---|
| Forward Stepwise | 768.579 | 755.763 | 0.000*** |

*P < 0.05; **P < 0.01; ***P < 0.001

## 4 Conclusion

This research has tried to conduct RPI based on Decision Tree (CART, C5.0, CHAID ), Shannon-Weaver diversity, Pielou's Evenness Index, Margalef's Species Richness Index, Fish Tolerant Index and Simpson's Index of Diversity to categorize and reasoning/evaluate, to obtain the river pollution data of Dezikou River basin and build an ecology forecast model for obtaining a better forecast effect. On evaluation of model side, the accurate rate of forecast is that CART is 88.33%, CHAID is 90%, C5.0 is 91.67% and multiple logistic regression analysis is 86.11%.

Although Decision Tree C5.0 algorithm has the most accurate rate, CHAID algorithm can use only two sets of variables of Fish Tolerant Index and Shannon-Weaver Index to achieve the same result with that of Decision Tree C5.0, which includes three sets of variables; Fish Tolerant Index, Shannon–Weaver Index, Margalef's Species Richness Index. Moreover, Decision Tree model is simpler. Therefore, a simple and easy to understand the Decision Tree's algorithm has better forecast outcome than traditional multiple logistic regression analysis (see Table 2).

**Table 2.** Decision Tree and Multiple logistic regression analysis to predict the accuracy

| | | | |
|---|---|---|---|
| Classification and Regression Analysis | Correct | 159 | 88.33% |
| | Error | 21 | 11.67% |
| | Total | 180 | |
| Chi-Square Automatic Interaction Detection | Correct | 162 | 90% |
| | Error | 18 | 10% |
| | Total | 180 | |
| Decision Tree C5.0 | Correct | 165 | 91.67% |
| | Error | 15 | 8.33% |
| | Total | 180 | |
| Multiple logistic regression analysis | Correct | 155 | 86.11 |
| | Error | 25 | 13.89 |
| | Total | 180 | |

This research's purpose is to provide a simple method to forecast water quality of rivers, based on the survey of the current fish species to substitute complex water quality examination. Due to sample and subject condition limitation, we cannot conduct sample division testing and verify further. One can apply this method to proceed to more specific evaluation when all the data has been provided.

## References

[1] Naftulin IS, Rebrova OY 2010 Application of C&RT, CHAID, C4.5 and WizWhy algorithms for Omar F. Althuwaynee, Biswajeet Pradhan, Hyuck-Jin Park & Jung Hyun Lee, Landslides, 2014, 11(6) 1063–1078..stroke type diagnosis, Artificial Intelligence and Soft Computing. Springer, 651–656.

[2] Plafkin, JL, MT. Barbour, KD. Porter, SK. Gross, & R M. Hughes. 1989. Rapid bioassessmentprotocols for use in streams and rivers: Benthic macroinvertebrates and Fish EPA/440/4-89/001. Washington, DC: U.S. Environmental Protection Agency.

[3] De'Ath G and Fabricius E 2000 Classification and regression trees: a powerful yet simple technique for ecological data analysis. Ecology 81(11): 3178-3192.

[4] Omar F. Althuwaynee, Biswajeet Pradhan, Hyuck-Jin Park & Jung Hyun Lee, Landslides, 2014, 11(6) 1063–1078..

[5] Environmental Protection Administration Executive Yuan, R.O.C. (Taiwan) 2009∽ 2013 River Water Quality History Data, Environmental Water Quality Information,

[6] Environmental Protection Bureau Yilan County, River ecology species survey, http://works.ilepb.gov.tw/01001_W_01/p3_2_1.html#river1

# Investigation of Haloacetic acids (HAAs) levels in water supply and its correlated HAAs formation

**P Saengchut[1], S Karuchit[1] and P Pentamwa[2,3]**

[1] School of Environmental Engineering, Institute of Engineering, Suranaree University of Technology, Nakhon Ratchasima, Thailand
[2] School of Environmental Health, Institute of Public Health, Suranaree University of Technology, Nakhon Ratchasima, Thailand

[3] Corresponding author e-mail: prapat@sut.ac.th

**Abstract.** Haloacetic acids (HAAs) are usually found as disinfecting by-products in water supply. This study aims to investigate the concentration of $HAA_5$ levels and to predict the $HAA_5$ formation by multiple linear regression equations. $HAA_5$ in the university water supply and the vicinity were analyzed by the mass detector (GC-MS) method with the use of gas chromatography. It was found that HAAs in 8 of 19 water samples were higher than US.EPA standard of 60 µg/L. The concentration HAA5 were ranged from 60 to 300 µg/L. The residual chlorine in the water samples were found between 1.2 - 1.8 mg/L which was above the standard set of 0.2 - 1.0 mg/L. The $HAA_5$, pH, DOC, Free $Cl_2$ variables were used to predict the correlation formation by multiple linear regression analysis at the 0.05 level of significance. It should be noted that correlation coefficient among target variables was high at the 0.91 level of statistically significant.

## 1. Introduction

Clean water is very important to support living thing on the earth. In the past, not many problems of water quality was not found for water consumption. The simple water treatment methods which were appropriate to purposes of using such as mixing with alum or chlorination to kill bacteria and germs could be used. At present, in general, natural water resources are contaminated. As a result, the water quality was deviation low. Improving the water quality by water irrigation process is important. The water with good quality must meet the standard of cleanliness and safe for consumption. In other words, there must be no contamination of toxic substances and photogenic microorganism. Water must be safe for the consumers' health for better quality of life.

According to several studies, it was found that water from surface water sources in the treatment process was contaminated by Natural Organic Matters (NOMs) in inorganic and organic substances. In the water treatment process, color, turbidity, inorganic and organic substances and suspended solids are removed by the coagulation-flocculation process to reduce the contamination of microorganisms in the water [1], which may be harmful to the consumers. In the water treatment process, chlorine has been added in water for disinfection. However, the reaction between chlorine and NOMs using this method may caused residues in water. The residues which have the potential to cause cancer was halogenated organic by-products. The large group of chemical substance which has the potential to cause cancer is haloacetic acid (HAAs) [2]. HAAs are colorless compounds and vaporize less.

They can also dissolve in water and rather stable. There are totally 9 types of HAAs but there are 5 types ($HAA_5$) of legislation regulating on drinking water which affect to human health. The 5 types include monochloroacetic acid (MCAA), dichloroacetic acid (DCAA), trichloroacetic acid (TCAA), monobromoacetic acid (MBAA) and dibromoacetic acid (DBAA) [3]. Many results of the studies indicated that HAAs caused cancer such as bladder and colon cancer. It also affected to the pregnant women. If a pregnant woman ingested HAAs in the body, HAAs can cause an abnormal fetus. The standard of HAAs in water should not more than 0.060 mg/L or 60 µg/L [4]. Despite the legislation restricted the HAAs beyond the standard but the toxicity levels of HAAs nowadays are more. Many studies observed the quality of water supply in Mahasarakham province was examined for the contamination of $HAA_5$ in the range of 22.54 - 39.37 µg/L [5]. In addition, the contamination of HAAs in drinking water in the city of Alacant, Spain, was 50.41 µg/L [6].

This study investigated the contamination level of $HAA_5$ in the water supply at university located in north eastern of Thailand including the water supplies in the vicinities. The level of $HAA_5$ was evaluated based on the drinking water standard. The level of $HAA_5$ contamination was used to predict $HAA_5$ formation. This information will use to investigation including improving the water treatment system in order to decrease the risk of getting the substances that cause cancer.

## 2. Materials and methods

### *2.1. Study area and water sampling*

This study was carried out by collecting water from the water supply plant (UWP) and other locations on campus of university in the northeastern part of Thailand and the water supplies in the vicinities. The area investigation covered 13 university-wide water sampling points namely: 1) the water supply unit of the university (UWP), 2) the administration building, 3) the classroom building complex, 4) the sports complex, 5) the Kasalong canteen, 6) the Khrawthanthaw canteen, 7) the classroom building canteen, 8) the staffs residences buildings, 9) the staff residence houses, 10) the students dormitory no.13, 11) the students dormitory no.15, 12) the students dormitory no.16 and 13) the university hospital (UH). For the community water supply plant near university area, the sampling points covers 6 points namely: 1) the water supply unit of Khok Kruat village, 2) the intermediate pipe of Khok Kruat village, 3) the tap water of Khok Kruat village, 4) the water supply unit of Nongrungka village, 5) the intermediate pipe of Nongrungka village and 6) the tap water of Nongrungka village. The water samples were collected during July-August 2017 from totally 19 locations in the study area. The regular water supply system in the study consists of screening, coagulation-flocculation, sedimentation, filtration and chlorination systems.

### *2.2. Analytical Methods*

#### *2.2.1. Sampling*

The water quality parameter analyzed in this study included pH, dissolved organic carbon (DOC), Free $Cl_2$ and $HAA_5$. Analysis of pH used pH meter, DOC were calculated from equation (1) [7] and residual free chlorine (Free $Cl_2$) was measured by iodometric method [8]. For $HAA_5$ analysis, water supply samples were collected in screw-capped glass bottles of 40 mL with Teflon-faced septa. The bottles added with 0.1 N of $Na_2S_2O_3$ x $5H_2O$ to quench residual chorine reactions. All samples were kept at 4°C until analysis, which was performed no later than 14 days after sampling.

$$DOC = \frac{UV_{254} - 0.07}{0.045} \qquad (1)$$

IOP Conf. Series: Earth and Environmental Science **164** (2018) 012023 doi:10.1088/1755-1315/164/1/012023

*2.2.2. Sample preparation*

The sample preparation procedure of $HAA_5$ was followed according to EPA Method 552.2. [9,10]. After adding 2 ml of concentrated $H_2SO_4$, 16 g of $Na_2SO_4$, and the surrogate standard (2,3-dibromopropionic acid), a 40 ml of water sample was extracted with 4 ml of MTBE spiked with the internal standard (1,2,3-trichloropropane) manually for 2 min and then 3 ml of extract was methylated by adding 1 ml of 10% $H_2SO_4$ methanolic solution and kept at 50°C for 2 h. After cleaning with 4 ml of $Na_2SO_4$ Saturated solution, the extracted was then submitted for GC/MS analysis.

*2.2.3. GC/MS analysis*

The total HAAs compounds analyzed in this study included Monochloroacetic acid (MCAA), Dichloroacetic acid (DCAA), Trichloroacetic acid (TCAA), Monobromoacetic acid (MBAA) and Dibromoacetic acid (DBAA) species. The separations of the $HAA_5$ were quantified by gas chromatography with Mass detector. A one mL of sample was introduced into GC-MS by splitless injection and separated on a 30 m 0.25 mm, 0.25 mm film thickness HP-5MS capillary column. The carrier gas flow rate was 1 ml/min (constant flow). The oven temperature was held isothermally at 35°C for 10 min and then ramped to 75°C at a rate of 5°C /min and held for 15 min, further to 100°C at a rate of 5°C /min for 5 min, 135°C at a rate of 5°C /min for 2 min, and 185°C at a rate of 258 °C/min for 2 min. The injector temperature was 200°C and the transfer line was maintained at 290°C.

*2.3. Statistical analysis*

The correlation between $HAA_5$ and pH, DOC, Free $Cl_2$ was examined through the multiple linear regression analysis using the SPSS software program. The specific variables of pH, DOC and Free $Cl_2$ were used to predict the significance correlation of HAAs levels. The HAA5 levels were used as the independent variable and pH, DOC, Free $Cl_2$ levels were used as the dependent variable at the 0.05 level of significance.

## 3. Results and discussion

*3.1. $HAA_5$ in water supply*

The factors which affected the formation of HAAs directly are DOC and Free $Cl_2$. This study investigated several parameters in the water treatment system at university and the vicinity. DOC with the average value of 2.14 ± 0.31 mg/L and Free $Cl_2$ with the average value of 1.22 ± 0.36 mg/L were found in the water treatment system at university whereas DOC with the average value of 3.16 ± 0.34 mg/L and Free $Cl_2$ with the average value of 1.81 ± 0.12 mg/L found in vicinity, respectively. $HAA_5$ levels in the study are showed in Table 1. According to Table 1, the water from the Kasalong canteen and the students dormitory no. 13 were found at higher US.EPA standard. While the most water from the water treatment systems in the vicinity of the university exhibits the highest levels more than 60 μg/L which is higher than US.EPA standard value set of 60 μg/L.

**Table 1.** Concentration of $HAA_5$ in water at university and vicinity.

| Areas | MCAA (μg/L) | MBAA (μg/L) | DCAA (μg/L) | TCAA (μg/L) | DBAA (μg/L) | $HAA_5$ (μg/L) |
|---|---|---|---|---|---|---|
| University | | | | | | |
| Water supply unit of the university | <1.0 | <1.0 | 35.32 | <1.0 | <1.0 | 35.32 |
| Administration building | <1.0 | <1.0 | <1.0 | <1.0 | 13.89 | 13.89 |
| Classroom building complex | <1.0 | <1.0 | <1.0 | <1.0 | <1.0 | <1.0 |
| Sports complex | <1.0 | <1.0 | <1.0 | <1.0 | <1.0 | <1.0 |
| Kasalong canteen | <1.0 | <1.0 | 53.47 | 27.01 | <1.0 | 80.47 |
| Khrawthanthaw canteen | <1.0 | <1.0 | <1.0 | <1.0 | <1.0 | <1.0 |
| Classroom building canteen | <1.0 | <1.0 | 2.32 | <1.0 | <1.0 | 2.32 |
| Staff residences building | <1.0 | <1.0 | 6.77 | <1.0 | <1.0 | 6.77 |
| Staff residence house | <1.0 | <1.0 | 1.0 | <1.0 | <1.0 | 1.0 |
| Students dormitory no. 13 | <1.0 | <1.0 | 72.82 | 105.40 | <1.0 | 178.23 |
| Students dormitory no. 15 | <1.0 | <1.0 | 26.65 | 12.41 | <1.0 | 39.06 |
| Students dormitory no. 16 | <1.0 | <1.0 | <1.0 | <1.0 | <1.0 | <1.0 |
| University hospital | <1.0 | <1.0 | <1.0 | <1.0 | <1.0 | <1.0 |
| Vicinity around university | | | | | | |
| Water supply unit of Khok Kruat village | <1.0 | <1.0 | 159.65 | 187.69 | <1.0 | 347.34 |
| Intermediate pipe of Khok Kruat village | <1.0 | <1.0 | 42.99 | 25.64 | <1.0 | 68.63 |
| Tap water of Khok Kruat village | <1.0 | <1.0 | 111.96 | 153.53 | <1.0 | 265.49 |
| Water supply unit of Nongrungka village | <1.0 | <1.0 | 108.23 | 115.84 | <1.0 | 224.07 |
| Intermediate pipe of Nongrungka village | <1.0 | <1.0 | 97.35 | 159.14 | <1.0 | 256.50 |
| Tap water of Nongrungka village | <1.0 | <1.0 | 122.25 | 179.55 | <1.0 | 301.81 |

### *3.2. $HAA_5$ correlations*

The correlation between observed $HAA_5$ concentration is based on the data set of university and vicinity samples. In terms of $HAA_5$ concentration in water, the prediction of $HAA_5$ is based on the data set which was calculated by multiple linear regression equation as shown in Table 2. The data was divided into 2 sets. The first set was analyzed on day 1 within 1 hour after the sample water had been collected and the second set was analysed on day 2 collection.

**Table 2.** Multiple linear regression equations of prediction.

| datasets | Multiple linear regression equations | $R^2$ | Significance |
|---|---|---|---|
| 1 | $HAA_5 = 142.560DOC - 20.902pH + 99.232Free\ Cl_2 - 246.488$ | 0.789 | 0.000 |
| | $HAA_5 = 144.086DOC + 99.437Free\ Cl_2 - 413.344$ | 0.788 | 0.000 |
| 2 | $HAA_5 = 83.383DOC + 648.824pH + 44.093Free\ Cl_2 - 5418.098$ | 0.583 | 0.004 |
| | $HAA_5 = 172.176DOC + 54.186Free\ Cl_2 - 370.558$ | 0.450 | 0.008 |

Both sets of data reveals the multiple linear regression analysis at the 0.05 level of significance. The coefficient of determination ($R^2$) was used. It was found that the equation of data set 1 is $HAA_5 = 144.086DOC + 99.437Free\ Cl_2 - 413.344$, $R^2 = 0.788$ which is not different with $R^2 = 0.789$ and $p = 0.000$. The equation data set 2 is $HAA_5 = 83.383DOC + 648.824pH + 44.093Free\ Cl_2 - 5418.098$, $R^2 = 0.583$ and $p = 0.004$.

There is relationship among four variables: $HAA_5$, pH, DOC and Free $Cl_2$ at the 0.05 level of significance. The correlation coefficient between the observed $HAA_5$ and the prediction of $HAA_5$ on day 1 is 0.911 as shown in Figure 1 and the correlation coefficient between the observed $HAA_5$ and the prediction of $HAA_5$ on day 2 is 0.782 as shown in Figure 2.

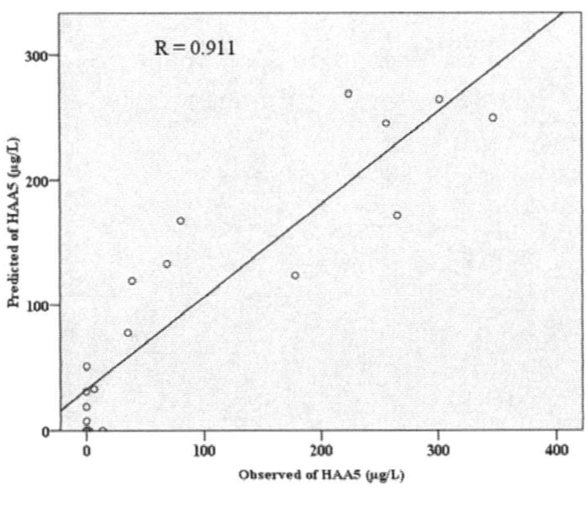


**Figure 1.** Correlation between observed $HAA_5$ and predicted $HAA_5$ on day 1.

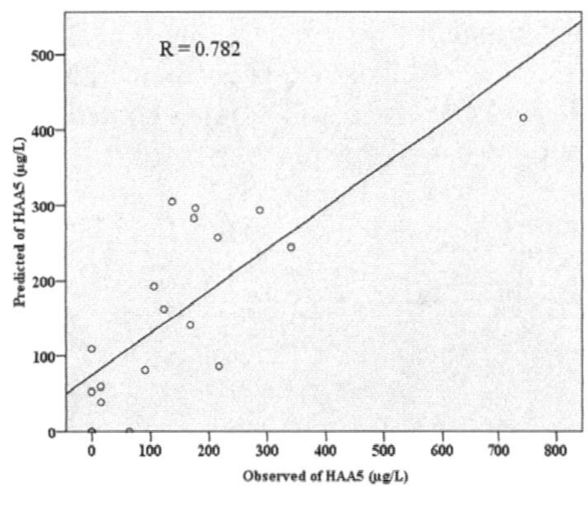


**Figure 2.** Correlation between observed $HAA_5$ and predicted $HAA_5$ on day 2.

## 4. Conclusions

This study shows concentration of $HAA_5$ in water in different areas of university and in the vicinity. The results of the study revealed that the water at the Kasalong canteen and the students dormitory no. 13 as well as included all samples of the water from the vicinity exhibited the highest value which was more than the standard set (60 µg/L). The value of residual chlorine was higher than the standard set (0.2 - 1.0 mg/L)[11] which may caused HAAs formation. Since the parameters can affect the water quality, the parameters which were pH, DOC and Free $Cl_2$ were analyzed by multiple linear regression to predict $HAA_5$ in water. As a result, $HAA_5$ could be decreased to improve the quality of water.

## 5. References

[1] Pentamwa P, Sukton B, Wongklom T, and Pentamwa S 2013 *J. Environmental Science and Development* 538: 538-544
[2] Panyakapo M and Paopuree P 2010 *Occurrence of Trihalomethanes in Chlorinated Drinking Water: A Case Study of The Water Supply System of Nakhon Pathom Municipality* (Thailand: Silpakorn university)
[3] Dmitruk U and Dojlido J 2006 *J.Polish of Environ* 51-56
[4] U.S.EPA. 2012 *The Drinking Water Standards and Health Advisories*. (Washington, DC.)
[5] Uansiri S 2009 *Determination of Haloacetic Acids in Water Samples Using* IonChromatography (Thailand: Mahasarakham University)
[6] Villanuevaa C.M., Kogevinasa M and Grimalt J.O. 2003 *J. Water Research* 37: 953–958
[7] CBCL limited 2011 *Study on Characteristics and Removal of Natural Organic Matter in* Drinking *Water Systems in Newfoundland and Labrador*. (Canada)
[8] AWWA, APHA and WPCF 2005 *Standard Methods for the Examination of Water and Wastewater* 20th ed (USA: Washington, DC.)
[9] US.EPA. 1995 *Determination of haloacetic acids and dalapon in drinking water by liquid-liquid extraction, derivatization and gas choromatography with electron capture* detection *Method 552.2* Rev. 1.0 (Washington, DC.)
[10] Yuefeng X 2001 *J. Water Research* 35: 6
[11] WHO 2006 *Guidelines for drinking-water quality. 3rd ed. Vol. 1* (Switzerland)

**Acknowledgments**
The authors would like to express thanks to Suranaree University of Technology for financial support. Special thanks to the staffs from school of Environmental health, school of Engineering as well as community water supply plant staff for their kindly assistance.

# Estimation of Urban Air Pollutant Levels using AERMOD: A Case Study in Nakhon Ratchasima, Thailand.

**K Katika[1,2] and S Karuchit[1]**

[1] School of Environmental Engineering, Suranaree University of Technology Nakhon Ratchasima, Thailand

[2] Corresponding author e-mail: kamolrat.katika@gmail.com

**Abstract.** This research aims to study the possibility of using AERMOD air quality model to estimate the air pollutant levels of a selected city – Nakhon Ratchasima Municipality, Thailand. Four pollutants were studied: $PM_{10}$, CO, $SO_2$, and $NO_x$. The measurement data were obtained from a PCD automatic pollutant monitoring station for comparison with the model's results. The air pollution sources used were residential, furnace, traffic and industrial sources. Emission factors were used to estimate the concentration of pollutants from the activities of the population in the area. The estimation was based on the Top Down Approach method (TDA). Results shows that the pollutant concentrations obtained from the model were lower than the measurement values in all parameters. The values obtained from the model were 4.14% to 77.88% of the measurement values. The cause may be due to incomplete account of sources. However, the model can be useful for assessing the carrying capacity of the urban area.

## 1. Introduction

The city of Nakhon Ratchasima in north-eastern part of Thailand is a city of significant growth. The expansion of business and industrial sectors have been good for the economy but not so for the environment. One of the problem is air pollution, which is mostly due to human activities. Nakhon Ratchasima Municipality had only one automatic pollutant monitoring station owned and operated by the Pollution Control Departmentt (PCD). Based on recorded measurements, some air pollutants were exceeding standards. However, the fact that the measurement results are available from only one station in the city area making it difficult to gain the understanding of the overall level of pollutants. Since the AERMOD air quality model is popular for use in air quality monitoring, its ability could help solve this problem. Using the AERMOD air quality model [1] with estimated area and line sources representing the city's [2] emission can help predicting the air pollutant concentrate in the form of concentration isopleth. The model application can reduce the time and cost of the actual measurements while providing the visual interpretation of air pollution levels throughout the study area.

The objective of this research is to study the possibility of estimating air pollution concentrations in the city area using the AERMOD air quality model. The study area was selected to be Nakhon Ratchasima Municipality. Estimation of pollutants levels at the PCD monitoring station in the study area were compared with actual measurements. Four pollutants were studied: particulate matters smaller than *10* microns ($PM_{10}$), carbon monoxide (CO), sulphur dioxide ($SO_2$), and oxides of nitrogen ($NO_x$). The results can be used as an example of the application of AERMOD in air quality management of urban areas.

## 2. Materials and methods

### *2.1 Study Area and Pollutant Sources*

The study area, Nakhon Ratchasima municipality, can be divided into residential areas based on the *4* constituencies as shown in Figure 1. The evaluation of air pollution was carried out using appropriate emission factors for 4 major air pollution sources: residential, furnace, traffic, and industrial sources.

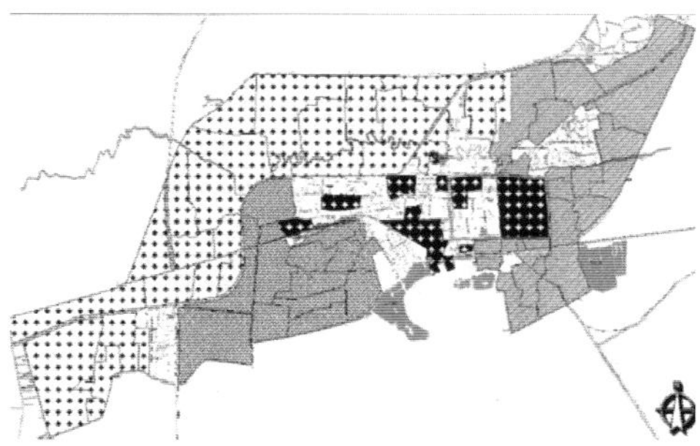

**Figure 1.** The district of Nakhon Ratchasima Municipality

### *2.2 Analysis Methods*

*2.2.1. Instrumentation* The air pollutant measurement data used was secondary data obtained from PCD. The data came from an automatic air quality monitoring station of Nakhon Ratchasima Municipality. Its location is at 14.973785N, 102.085730E, which is approximately at the center of the city. This study used the measurement data in the year 2015 for the comparison with the pollution level estimated from the model. Four principle ambient air pollutants were studied, namely particulate matters smaller than 10 microns ($PM_{10}$), carbon monoxide (CO), sulphur dioxide ($SO_2$), and oxide of nitrogen (Nox).

*2.2.2 Meteorological data* This study uses the meteorological data of the year *2015* obtained from meteorological stations. The raw meteorological data was arranged into the required format of the AERMOD model: the hourly surface data was arranged into the SCRAM format, and upper air data was arranged into the FSL format. The data arrangement resulted in the forms of SFC File (*.SFC) and PFL File (*.PFL). The meteorological data was classified for every hour in the total period of one year for all parameters. The hourly surface data consisted of wind speed, wind direction, ambient temperature, cloud cover, and ceiling height. It was measured at Nakhon Ratchasima meteorological station, coordinate 14.973805, 102.085718. The upper air data consisted of atmospheric pressure, heights, temperature, dew point temperature, wind speed, and wind direction. It was measured at Bangkok Metropolis station, coordinate 13.726389, 100.559722.

*2.2.3 AERMOD Modeling Option* The air quality model used, AERMOD, consists of two sub-programs AERMET and AERMAP. They are programs that prepare meteorological information and geological terrain information, respectively, for the main program AERMOD. The two sub-programs act as pre-processor for data going into the main program. For this study, AERMAP used the terrain map type SRTM3/SRTM1 from WebGIS, the data was based on SRTM3 (Global 90m) and the horizontal datum was WGS 84. AERMET used the meteorological data from monitoring station measurement which was arranged into the SCRAM and FSL formats. The surface station primary tower (anemometer) was 186.6 m. In the AERMOD model, 4 pollution sources were entered: residential, furnace, traffic, and industrial sources. The receptors for the study area were defined by a uniform cartesian grid system, length 12,255.6 × 7,387 square meters. One discrete cartesian receptor was defined for the PCD automatic pollutant monitoring station, coordinate 14.973805,102.085718.

*2.2.4 Emission Factors Determination* Emission factors represent the relationship between pollutant emissions and pollution-related activities. In this study, the emission factors were based on the research on emission inventory in Nakhon Ratchasima Municipality, as shown in Table 1 [3]. The estimation of these emission factors was based on a "Top-Down Approach" (TDA). In order to update the emission values from the year 2012 to 2015, adjustments were made proportional to the change in the numbers of population. Overall, the emission in 2015 were less than in 2012 due to a decreasing trend in the city's population. The estimated emission data was entered into the AERMOD model in the unit of $g.s^{-1}.m^{-2}$ for area and line sources for further computation of the model.

**Table 1.** Emission factors of air pollution sources in Nakhon Ratchasima Municipality

| Source | activities | Emission Factor (kilograms / unit of fuel) | | | |
|---|---|---|---|---|---|
| | | $NO_X$ | $SO_2$ | CO | $PM_{10}$ |
| Area | LPG (Liter) | 0.001810 | 0.000010 | 0.001230 | 0.000100 |
| | Fuel Wood (Kilograms) | 0.000120 | 0.000320 | 0.026400 | 0.011110 |
| | Charcoal (Kilograms) | 0.000030 | 0.000180 | 0.035700 | 0.015690 |
| Temple | Crematory (Kilograms) | 0.115700 | 0.070850 | 0.096200 | - |
| Industrial | Fuel Oil (1000 Liter) | 8.000000 | 19.000000 | 0.600000 | 1.000000 |
| | Diesel (1000 Liter) | 2.400000 | 17.000000 | 0.600000 | 0.240000 |
| | Diesel (1000 Liter) | 2.400000 | 17.000000 | 0.600000 | 0.240000 |
| | Kerosene (1000 Liter) | 2.400000 | 17.000000 | 0.600000 | 0.240000 |
| | LPG (1000 Liter) | 2.280000 | 0.012000 | 0.384000 | 0.072000 |
| | Fuel Wood (1000 Kilograms) | 0.190000 | 0.037500 | 3.300000 | 23.500000 |
| Traffic | Motorcycle (Kilometers.Car) | 0.000240 | 0.000020 | 0.013140 | 0.000086 |
| | Gasoline Engine Car (Kilometers.Car) | 0.000026 | 0.000061 | 0.000513 | 0.000101 |
| | Small Diesel Engine Car (Kilometers.Car) | 0.000469 | 0.000041 | 0.000439 | 0.000042 |
| | Big Diesel Engine Cars (Kilometers.Car) | 0.023120 | 0.000327 | 0.008890 | 0.001150 |

## 3. Results and discussion

*3.1 Pollution levels estimated by AERMOD*

The estimated annual average pollutant concentrations using the AERMOD air quality model are shown in Figure 2-5. The concentrations of $NO_x$, $SO_2$, CO and $PM_{10}$ were found to be higher in the road network. The highest $NO_x$, $SO_2$, CO and $PM_{10}$ concentrations in microgram/m$^3$ were 344.24, 12.53, 414.43 and 30.46, respectively. The maximum concentration of all pollutants was at the same point at coordinate 184246.47, 1657665.85. At the point of the discrete cartesian receptor, which is the location of the PCD air quality measurement station, the levels was found to be 13.78, 27.72, 0.60 and 4.63, respectively. From Figure 2-5, the dispersion of $PM_{10}$ concentrations is the most apparent. The distribution of $SO_2$ concentrations is minimal. From Figure 3, it can be seen that $SO_2$ is highly concentrated near the pollution source.

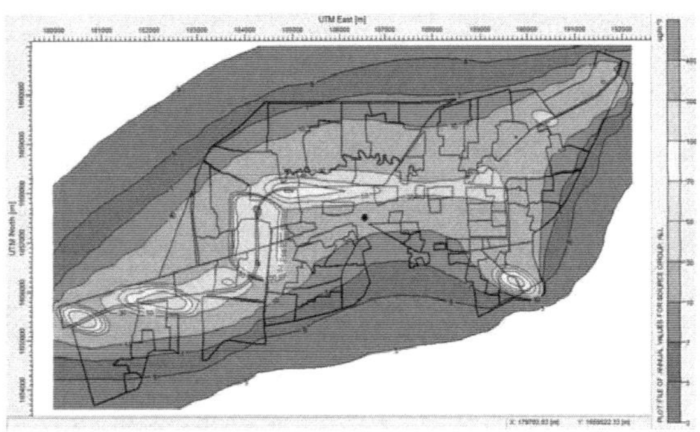

**Figure 2.** Oxide of nitrogen ($NO_x$) concentration estimates by AERMOD, annual averaging.

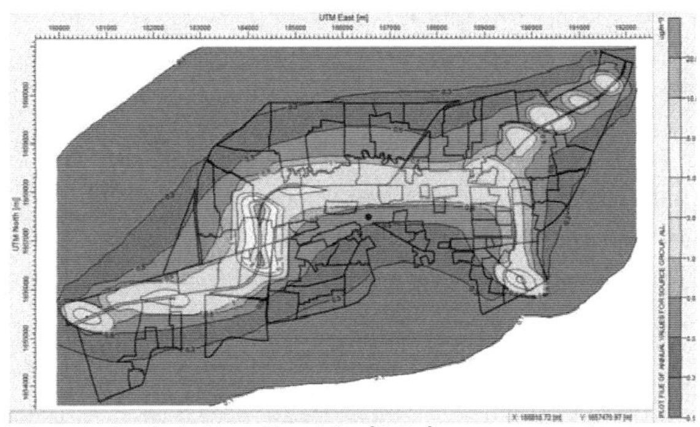

**Figure 3.** Sulphur dioxide ($SO_2$) concentration estimates by AERMOD, annual averaging.

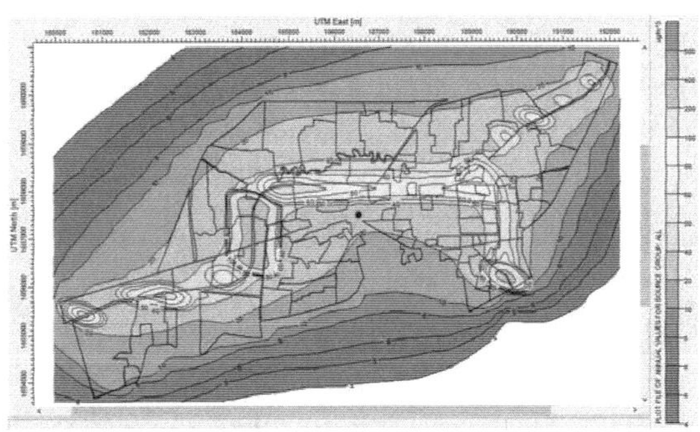

**Figure 4.** Carbon monoxide (CO) concentration estimates by AERMOD, annual averaging.

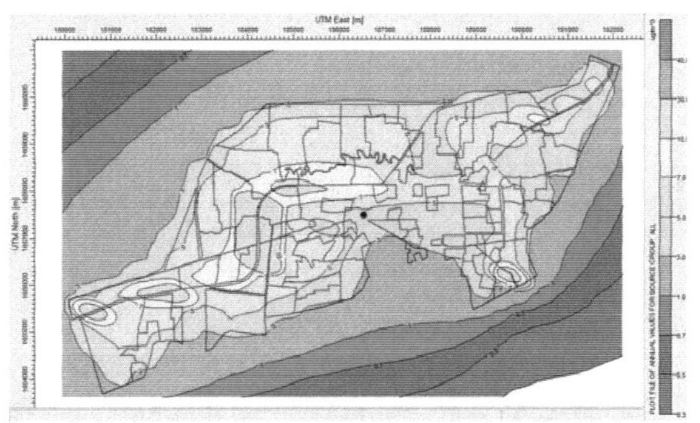

**Figure 5.** Particulate matter <10 microns ($PM_{10}$) concentration estimates by AERMOD, annual averaging.

*3.2 Comparison between AERMOD results and measurement data*

The estimation of annual pollution levels using the air quality model AERMOD were compared to the pollutant concentration measurement at the same location, as shown in Table 2. It was found that the pollutant concentrations derived from the model were lower than the measurement values in all pollutants. The values of $NO_x$, $SO_2$, $PM_{10}$ and CO from AERMOD were 77.88%, 10.59%, 8.98% and 4.14% of those measured at the station, respectively. The reason that the model's estimation was lower than the actual measurement may be due to pollutant sources which are unaccounted for, such as dust resuspension from traffic and ground surfaces and other non-point sources.

**Table 2.** Comparison of measured pollutant concentration with AERMOD results.

| Pollutants | Ambient Standard ($ug/m^3$) | Measurement ($ug/m^3$) | AERMOD ($ug/m^3$) |
|---|---|---|---|
| $NO_x$ | 57 | 17.69 | 13.78 |
| CO | - | 669.94 | 27.72 |
| $SO_2$ | 100 | 5.68 | 0.60 |
| $PM_{10}$ | 50 | 51.58 | 4.63 |

## 4. Conclusions

In this study, we investigated the possibility of using AERMOD to assist in predicting air pollution concentration in a city. Comparison of results with actual measurement at the same coordinate showed that the pollutant concentrations obtained from the model were lower in all parameters. The cause may be due to incomplete sources of pollutant incorporated in the analysis. Overall, the most concentrated area along the boundary road [4]. However, the results showed promising ability for the AERMOD model to be used as a tool for air quality management of a city, possibly for assessing the carrying capacity of the urban area.

## References

[1] Workgroup AI. AERMOD: DESCRIPTION OF MODEL FORMULATION. Office of Air Quality Planning and Standards Emissions Monitoring and Analysis Division Research Triangle Park, North Carolina.: U.S. Environmental Protection Agency; 2004 September 2004. 97 p.

[2] Workgroup AI. AERMOD IMPLEMENTATION GUIDE. Office of Air Quality Planning and Standards, Air Quality Assessment Division Research Triangle Park, North Carolina.: U. S. Environmental Protection Agency; 2015.

[3] Kotula S. Emission Inventory of Air Pollutants in Nakhon Ratchasima Municipality Area. Nakhon Ratchasima, Thailand.: Suranaree University of Technology; 2010.22558-60, 67-70, 58-95, 103-11

[4] Mohammad H.A SHK, Suriya V, Josias Z, Qi Ying. AERMOD for near-road pollutant dispersion: Evaluation of model performance with different emission source representations and low wind options. 2017;7:392-402.

**Acknowledgments**

The authors would like to thank Suranaree University of Technology and Agencies related to Nakhon Ratchasima Municipality for supporting the information for the research.

IOP Conf. Series: Earth and Environmental Science **164** (2018) 012025 doi:10.1088/1755-1315/164/1/012025

# Treatment of $H_2S$ and $NH_3$ from wastewater using a pilot-scale tray biofilter

**S Karuchit[1] and A Rakthaisong [1,2]**

[1] School of Environmental Engineering, Suranaree University of Technology Nakhon Ratchasima, Thailand
[2] Corresponding author e-mail: apisie@hotmail.com

**Abstract.** A tray biofilter was developed and tested for treatment of high volume, low concentration $H_2S$ and $NH_3$ from pig farm wastewater. The combination of 2 bed depth values (0.15 m. and 0.30 m.) and 3 EBRT values (0.6, 1.2, and 1.8 s) resulted in 6 systems for the experiment. The filter media used were compost : chopped coconut shell : dried cow manure : dried wastewater treatment plant sludge at 60:20:10:10 ratio. The treatment efficiency for $H_2S$ and $NH_3$ were in the range of 78% - 89% and 57% - 80%, respectively. Overall, the systems with higher bed depth and/or higher EBRT can treat the two gases with higher efficiency. The mass loading rates, the elimination capacity, and efficiency of this study were comparable to other studies.

## 1. Introduction

Hydrogen sulfide *(*$H_2S$*)* and ammonia *(*$NH_3$) are odorous gases found when anaerobic digestion of organic matters occurs. $H_2S$ has a typical smell of rotten eggs, and $NH_3$ has a strong and repulsive smell of animal urine. Both gases can be detected by human at low concentrations. In Thailand, one of the most significant sources of these gases in communities are livestock production, especially pig farms. The odor mostly comes from the manure and urine of pigs that were collected in the form of wastewater from the farmhouses. Levels of these gases in ambient air can cause significant odor nuisance to the people in nearby areas and may lead to their adverse health impacts – nausea, headaches, or respiratory tract irritation.

With respect to odor from the wastewater, closed-system treatment such as covered lagoons can minimize the odor while also yielding the benefit of biogas. However, $H_2S$ and $NH_3$ are still emitted into the environment via the farmhouse air ventilation system, with no treatment mechanism in place. Among several treatment techniques, compost-based biofilter can be used as a cost-effective method to treat farmhouse emission. It is suitable for low concentration gas flow and has advantages of low operating and maintenance costs [1]. Moreover, most filter media can be obtained and prepared within the farm and the spent media can subsequently be used as fertilizer. The system efficiency, however, is subjected to wide variation depending on the design parameter such as filter media types and mixture ratio, empty bed residence time (EBRT), and mass loading rate (MLR). Typical efficiency of 50-90% were reported in the literature *[*2, 3*]*.

In this research, a simple tray-type biofilter was developed for treatment of high volume, low concentration $H_2S$ and $NH_3$ from pig farmhouses. This paper presents the results from the pilot-scale study focusing on the efficiency and operating conditions.

## 2. Methodology

### *2.1. System design and operation*

Details of the pilot-scale tray biofilter design and operation are presented in a related paper [4] and are summarized as follow. First the design parameters were reveiwed from relevant literatures and the ones that were in accordance with the research objectives were selected. The final design include a filter media box with a screen plate as the bottom of the box. The bed area was 0.4 x 0.5 $m^2$, and the box height was 0.3 m. Two values of bed depth were experimented: 0.15 and 0.30 m., which resulted in the bed volume of 0.03 and 0.06 $m^3$, respectively. The empty bed residence time (EBRT) were selected as 0.6, 1.2, and 1.8 seconds, which were achieved by using the flow rates of 0.017-0.100 $m^3/s$. These bed height and EBRT values were lower than most of those in the literature because they were intended to fit the tray-type configuration which is easy to operate and can handle large gas flow rate.

The combination of 2 bed depth values and 3 EBRT values resulted in 6 systems for the experiment. Each experiment was run and essential parameters were measured continuously for 14 days. New filter media was used every time the new system start its operation. The inlet and outlet concentration of $H_2S$ and $NH_3$ were measured using COMBIMASS serie GA-m and Gasmet serie DX4040 portable gas analysers, respectively.

The tray biofilter system was set-up near a 300-pig housing in the farm in Suranaree University of Technology, Nakhon Ratchasima, Thailand. The average wastewater flow was 21 $m^3/d$. The input gases were collected from a covered wastewater pit via a collection pipe system powered by a 2-HP, 1.5 kW fan with inverter for flow rate variation. The gas was considered a representative of the gas from typical pig farmhouses. The lateral pipes under the media bed were designed to evenly distribute the gas flow upward to the media bed. An automatic water spraying system was attached under the filter's roof to control media bed moisture.

### *2.2. Filter media*

The material type and mixing ratio of the filter media were selected based on optimum cost, ease of operation, and adequate nutrient and porosity. The selection results were compost : chopped coconut shell : dried cow manure : dried wastewater treatment plant sludge at 60:20:10:10 ratio [5]. The filter media were mixed thoroughly with added water, and kept for 2 weeks before loading into the system. The characteristics fo the filter media are shown in table 1.

The microorganism in the filter media was measured by the aerobic plate count method along the operation period. It was in the range of $3.60 \times 10^8$ - $6.30 \times 10^8$ CFU/g, with the average of $5.76 \times 10^8$ CFU/g. The levels were within the recommended range of $10^6$ - $10^{10}$ CFU/g for biofilters [1, 6].

**Table 1.** Characteristics of the filter media in this study.

| Characteristics | Individual Medium | | | | Mixed Media |
|---|---|---|---|---|---|
| | Compost | Chopped Coconut Shell | Dried Cow Manure | Dried WWTP Sludge | |
| Density ($g/cm^3$) | 0.62 | 0.05 | 0.34 | 0.66 | 0.56 |
| Porosity (%) | 54.70 | 78.67 | 75.20 | 40.08 | 55.54 |
| Moisture (%) | 17.46 | 6.58 | 2.38 | 79.78 | 71 |

## 3. Results and Discussion

### *3.1. Treatment Efficiency*

Results from the operations of 6 biofilter systems were analyzed in terms of mass loading rate (MLR), elimination capacity (EC), and efficiency, and the average values were presented in table 2. For $H_2S$, the treatment efficiency was in the range of 78% - 89%, with the thicker bed systems (system 1 - 3) having higher efficiency values. However, system 3 had much higher MLR and EC than system 1 and 2, so it can handle greater concentration and/or gass flow rate. Similar discussion can be made for $NH_3$, although the difference of MLR, EC, and efficiency were less obvious. The $NH_3$ treatment efficiency was in the range of 57% - 80%.

**Table 2.** Mass loading rate (MLR), elimination capacity (EC), and efficiency results.

| System | Bed Depth (m.) | EBRT (s) | Flow Rate ($m^3$/s) | Filtration Velocity (m/s) | $H_2S$ MLR ($g/m^3$.hr) | $H_2S$ EC ($g/m^3$.hr) | $H_2S$ Eff. (%) | $NH_3$ MLR ($g/m^3$.hr) | $NH_3$ EC ($g/m^3$.hr) | $NH_3$ Eff. (%) |
|---|---|---|---|---|---|---|---|---|---|---|
| 1 | 0.3 | 1.8 | 0.033 | 0.17 | 115 | 103 | 89 | 10 | 8 | 80 |
| 2 | 0.3 | 1.2 | 0.050 | 0.25 | 185 | 156 | 85 | 17 | 10 | 61 |
| 3 | 0.3 | 0.6 | 0.100 | 0.50 | 426 | 349 | 81 | 33 | 25 | 74 |
| 4 | 0.15 | 1.8 | 0.017 | 0.08 | 139 | 113 | 78 | 10 | 7 | 66 |
| 5 | 0.15 | 1.2 | 0.025 | 0.13 | 169 | 135 | 79 | 14 | 11 | 77 |
| 6 | 0.15 | 0.6 | 0.050 | 0.25 | 428 | 338 | 80 | 23 | 14 | 57 |

Descriptive summary of the treatment efficiency results during the operation can be presented in box plots as shown in figure 1. Most plots showed distributions that are a little skewed to the left, which means there were occational low values. For $H_2S$, system 1 has considerably higher efficiency than the others, and all systems have approximately the same distribution shapes. For $NH_3$, the differences were rather obscure. Comparison of average efficiency values among the 6 systems were carried out using Tukey's multiple-pair comparison method [7]. For $H_2S$, 5 pairs were found to be significantly different – the mean efficiency value of system 1 was higher than those of system 2, 3, 4, 5 and 6. For $NH_3$, the mean efficiency value of system 1 was significantly higher than system 2 and 6. This suggests that the system with high bed depth and high EBRT can treat the two gases with higher efficiency.

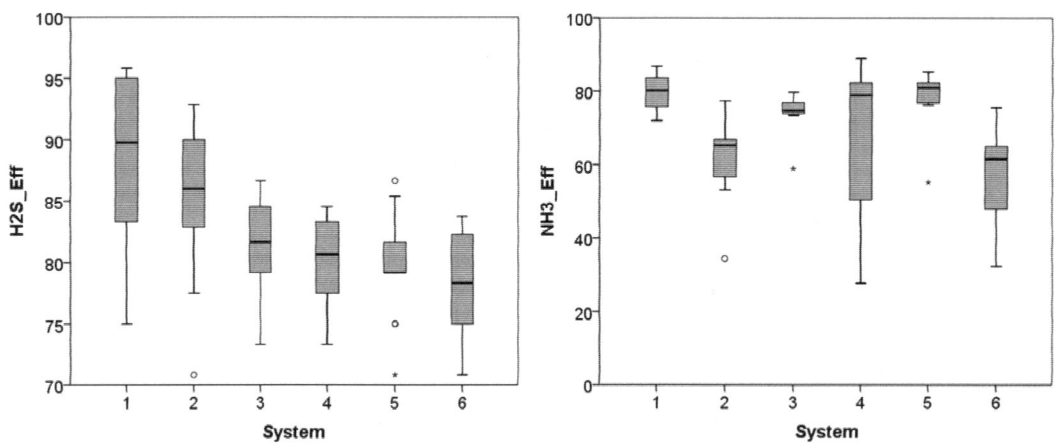


(a) (b)

**Figure 1.** Box plots of treatment efficiency of the 6 systems: (a) $H_2S$ and (b) $NH_3$.

### *3.2. Operating Conditions*

The 4-day moving average efficiency plot of the systems' efficiency during operations are shown in figure 2(a). The efficiency values fluctuated during the first few days but became more stable during the latter part of the operation. Again the systems with high bed depth showed higher efficiency. On the other hand, filter bed compression rates were quite similar for all system, 1.3% - 1.6% per day, as shown by the slopes in figure 2(b). On average, system 1 - 3 compressed 5.5 cm./day and system 4 - 6 compressed 2.7 cm./day.

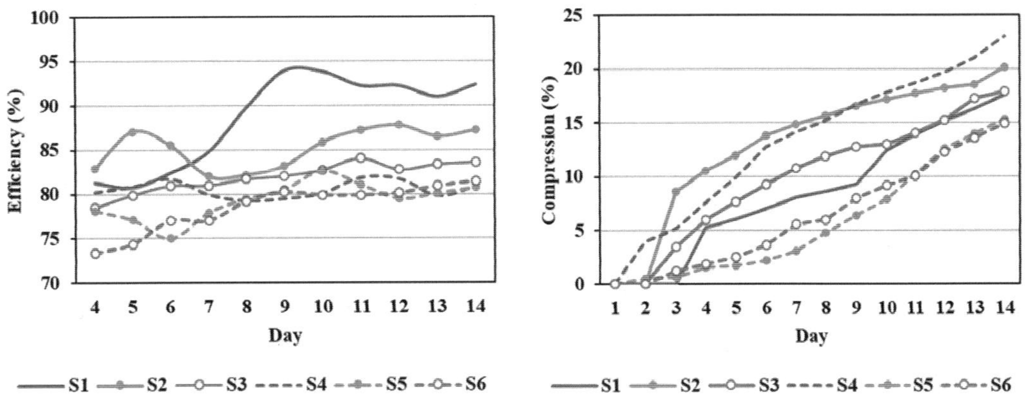


(a) (b)

**Figure 2.** Time series plots of: (a) $H_2S$ efficiency levels (4-day moving average) and (b) filter bed compression.

Correlation analysis was performed on treatment efficiency and essential factos, namely pH, temperature, and moisture, using Spearman's rank correlation coefficient *($r_s$)*. It was found that the $H_2S$ treatment efficiency had significant positive correlation with pH in system 5 *($r_s$ = 0.580)* and 6 *($r_s$ = 0.766)*, and with temperature in system 3 *($r_s$ = 0.624)* *(*figure 3*)*. No significant correlation was found in the case of $NH_3$. Few correlations were anticipated because the factors were controled and have little variations over the operation period.

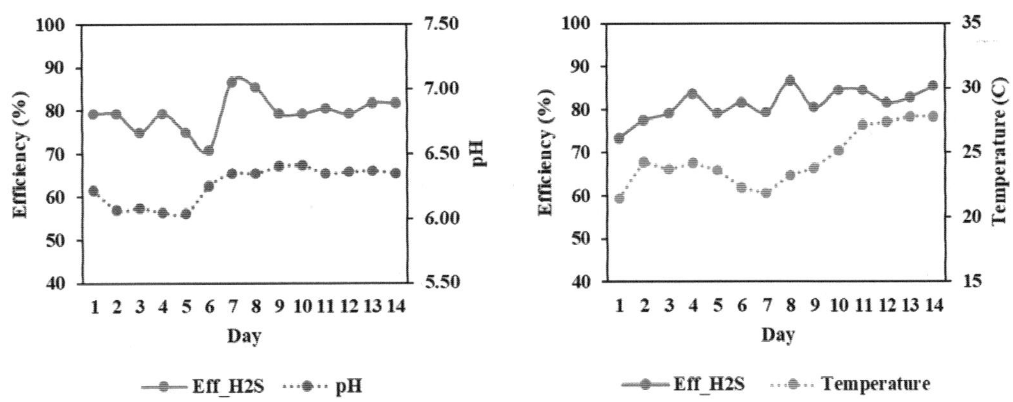


(a) (b)

**Figure 3.** Correlation of $H_2S$ treatment efficiency and other factors: (a) pH in system 5 and (b) temperature in system 3.

Table 3 shows a comparison of operating condition and results of this research with relevant studies. This research used relatively lower inlet concentration of both gases and operated at lower EBRT values. Nevertheless, the mass loading rates, the elimination capacity, and efficiency were relatively better than other studies. Therefore, the tray-type biofilter is a promissing odor treatment alternative for pig farmhouses.

**Table 3.** Comparison of operating condition and results with relevant literatures.

| Gas | Reference | $C_{in}$ ($mg/m^3$) | EBRT (s) | MLR ($g/m^3$.h) | EC ($g/m^3$.hr) | Efficiency (%) |
|---|---|---|---|---|---|---|
| $H_2S$ | [3] | 1.2-1.3 | 0.6 | - | 2.56 | 42.2 |
| | [8] | 32-3,480 | 7-23 | 130 | 250 | 93-100 |
| | [9] | 393 | 45 | 3-34 | 122 | 100 |
| | [10] | 131-1,650 | 32-51 | 0.1-13 | 8 | 45-100 |
| | [11] | 200-1,300 | 60 | 11.7-60 | 58 | 99 |
| | **This Research** | **72-58** | **0.6-1.8** | **115-428** | **103-349** | **79-89** |
| $NH_3$ | [3] | 7-9 | 0.6 | - | 14.4 | 45.8 |
| | [6] | 139-348 | 22.5 | 0.48 -19.06 | - | 99 |
| | [12] | 43-150 | 30-60 | 54 | 52 | 96 |
| | **This Research** | **4-6** | **0.6-1.8** | **10-33** | **7-25** | **57-80** |

## 4. Conclusion

This research yielded a pilot-scale tray biofilter which can treat $H_2S$ and $NH_3$ from pig farm wastewater with 78% - 89% and 57% - 80% efficiency, respectively. The treatment efficiency corresponded to the bed depth and EBRT.

## References

[1] Devinny S, Deshusses A and Webster S 1999 *Biofiltration for Air Pollution Control* New York (Lewis Publishers)

[2] Mudliar S, Giri B, Padoley K, Satpute D, Dixit R, Bhatt P, Pandey R, Juwarkar A and Vaidya A 2010 *J. of Environmental Management* **91** 1039-54

[3] Lim T, Jin Y, Ni J and Heber J 2012 *Biosystems Engineering* **112** 192-201

[4] Rakthaisong A, Karuchit S and Potivichayanon S 2015 *The 4th Int. Sym. on Engineering, Energy and Environment*, 8-10 November 2015, Chon Buri, Thailand

[5] Punapa-amporn S 2008 *Master Thesis* Chulalongkorn University, Bangkok

[6] Sandeep M, Balendu G, Kiran P, Dewanand S, Rashmi D, Praveena B, Ram P, Asha J and Atul V 2010 *J. of Environmental Management* **91** 1039-54

[7] Berthouex P and Brown L 2002 *Statistics for Environmental Engineers 2nd ed.* New York (Lewis Publishers)

[8] Adler P 2001 *Biofiltration: A Primer* )On-line (Available: http//:www.d.umn.edu

[9] Ratanamuk N 2006 *Master Thesis* Chulalongkorn University, Bangkok

[10] Kim J, Rene E and Park H 2008 *Bioresource Technology* **99** 583-8

[11] Omri I, Bouallagui H, Aouidi F, Godon J and Hamdi M 2013 *Saudi J. of Biological Sciences* **20** 169-76

[12] Wu D, Quan X, Zhang Y and Zhao Y 2006 *Biochemical Engineering J.* **32** 84-92

## Acknowledgements

This research was funded by the National Research Council of Thailand and Suranaree University of Technology.

IOP Conf. Series: Earth and Environmental Science **164** (2018) 012026 doi:10.1088/1755-1315/164/1/012026

# The effect of profitability and leverage to the carbon emission disclosure on companies that registered consecutively in sustainability reporting award period 2014-2016

**Khairun Nisak [1a] and Rita Yuniarti [2b]**

[1] Accounting Department of Economics, Widyatama University, Bandung, West Java, Indonesia 40125
[2] Accounting Department of Economics, Widyatama University, Bandung, West Java, Indonesia 40125

[a]khairun.nisak@widyatama.ac.id; [b]rita.yuniarti@widyatama.ac.id

**Abstract.** This study aimed to examine the effect of profitability and leverage to the carbon emission disclosure on Companies that Registered Consecutively in Sustainability Reporting Award Period 2014-2016. The objects in this research are profitability, leverage and carbon emission disclosure. The dependent variable in this study is the carbon emission disclosure and independent variable is profitability and leverage. Research methods used in this research is explanatory. The research results showed that profitability and leverage significantly influental to the carbon emission disclosure. Profitability variable contributes to the influence of 19.05% againts the carbon emission disclosure. Leverage variable contributes to the influnce of 26,47% againts the carbon emission disclosure, while the rest 54.5% is the magnitude of the influence exerted contribution by other factors that not examined in this research for examples exposure media, operational coverage of the company, and environmental performance.

**Keyword.** profitability, leverage, carbon emission disclosure

## 1. Introduction

The issue of tackling global warming is a problem that being aggressively fought in the world. One of the reason is greenhouse gas emissions. Indonesia is one of the largest contributor to greenhouse gas emissions in the world. According to data from the cooperation REDD (Reduction Emissions from Deforestation and Forest Degradation), in 2005 Indonesia contributed for greenhouse gas emissions is 2.05 gigatons. This fact makes Indonesia the third largest emitter of carbon in the world after the United States (5.95 giga tons) and China (5.06 giga tons). Indonesia's carbon emissions predicted to be 3 giga tons of CO2 in 2020.

One of the causes of climate change in the world is the greenhouse gases resulting from human activities. According Carbon Disclosure Project (2013), 50 of the 500 largest companies listed on the world responsible for nearly three-quarters of the 3.6 billion metric tons of greenhouse gases (GHGs). Carbon produced by 50 companies, which mainly operates in the sectors of energy, materials and utilities sectors (materials and utilities sectors). The carbon has increased by 1.65% to 2.54 billion metric tons over the last four years (cdp.net).

Carbon Emission Disclosure in Indonesia still a voluntary disclosure therefore not all companies disclose it information in their report. According to research [9] disclosure practices of greenhouse gas emissions, including carbon emissions is still low to fulfill the ISO 14064-1 guidelines. Although disclosure of carbon emissions in Indonesia is a voluntary disclosure, but companies should pay more attention to it because the last few decades environmental conditions getting worst and claim from citizens to create a livable environment. Carbon emission disclosure have some considerations such as to gain legitimacy from stakeholders, avoiding the threats, especially for companies that produce greenhouse gas to increased operating costs, reduced demand, reputational risk, legal proceedings, and penalties [1][2]. Research [3] finding regarding the control variables allows us to affirm that the volume of information on greenhouse gas emissions significantly depending on the sector of activity in which the company operates. The influence of industry on voluntary disclosure of all types with the aim of reducing the cost of politics. Research [4] found that voluntary disclosure of greenhouse gas emissions are influenced by the presence of an environmental management system, public reporting of the Carbon Disclosure Project (CDP), and the use of the GRI (Global Reporting Initiative). While [5][6] examine the factors that affect to the carbon emission disclosure.

Indonesia made commitments to reduce greenhouse gas emissions. Commitment to reduce greenhouse gas emissions is demonstrated by ratifying the Kyoto Protocol on December 3, 2004 through Law 17/2004 [7]. Based on President Decree 61/2011, 71/2011, 47/2012 explained that industrial sectors can reduce the greenhouse gas emissions and they also have the obligation to carry out Corporate Social Responsibility (CSR), that are mining, agriculture and manufacture. The industry is expected to reduce their greenhouse gas emissions as the realization of CSR [8] [9].

This study uses a framework that is based on stakeholder theory and the theory of legitimacy aims to determine the effects of profitability and leverage extensive disclosure of greenhouse gas emissions in Indonesia. This study uses an index developed by [4] under the direction of ISO 14064-1. Selection of the index is done because this research focuses on greenhouse gas emissions. In a previous study [2] states that in December 2009 Indonesia through the National Standardization Agency adopted ISO that associated with greenhouse gases such as ISO 14064 and 14065. *Carbon Emission Disclosures* one example of environmental disclosure that is part of the additional reports have stated in SFAS No. 1 (revised 2009) paragraph twelve.

## 2. Profitability, Leverage and Carbon Emissions Disclosure

Return on Investment (ROI) is used to measure profitability. ROI is more appropriate to describe the level of disclosure of greenhouse gas emissions because of the cost of disclosure is an indirect investment that is expected to enhance the company's net profit. This is consistent with the study [9] which is the company with the ability to better financial performance, the more likely to try to reduce emissions from their corporate activity.

$$ROI = \frac{Net\ Profit\ After\ Tax}{Total\ Assets} \times 100\%$$

Leverage may be implicated in a financial company. This is in line with research [10] in [11] is companies with high leverage may not be able to absorb the adverse financial impact of the disclosure of carbon information. In this study, Debt to Equity Ratio (DER) is used to measure leverage.

$$Debt\ to\ Equity\ Ratio = \frac{Total\ Debt}{Equity} \times 100\%$$

Carbon emissions disclosure captured in the sustainability report that contained in the annual report of the companies and registered consecutively in Sustainability Reporting Award Period 2014-2016. Carbon emissions disclosure item using an index developed by [5] were constructed from the request sheet developed by CDP (Carbon Dislcosure Project). If the company's disclosure in accordance with the specified item will be given a score of 1, whereas if the item is determined not disclosed it will be given a score of 0. Then the score 1 overall summed and divided by the maximum number of items that can be expressed and then multiplied by 100%. There are 18 items identified.

## 3. Theoretical Framework

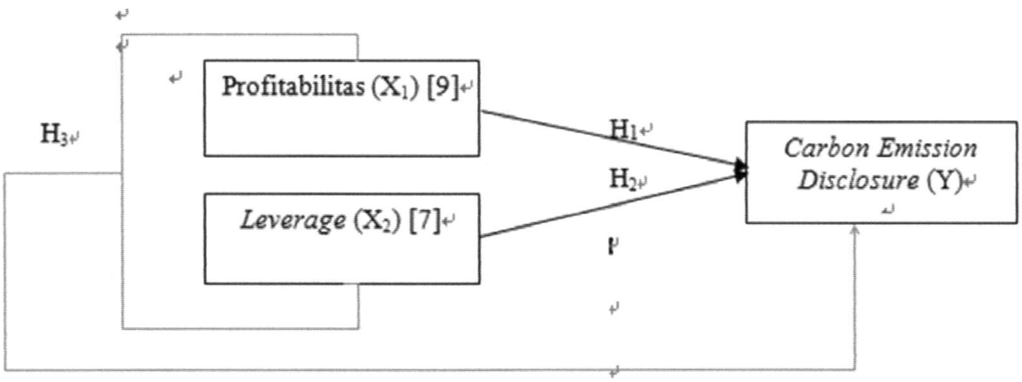


## 4. Hypothesis and Research Method

### *4.1 Hypotheis*
The hypothesis is formulated as below:

H1: Profitability effect on Carbon Emission Disclosure
H2: Leverage effect on Carbon Emission Disclosure
H3: Profitability and Leverage effect to Carbon Emission Disclosure

### *4.2 Research method*
The data collected were analyzed using statistical analysis tools F test, t test, coefficient of determination and multiple regression analysis with a model of the following equation:

$Y = \alpha + \beta 1ROI + \beta 2Leverage + e$

Information :

| | |
|---|---|
| Y | = Carbon Emission Disclosure |
| α | = constant |
| β1- β2 | = Regression Coefficients |
| ROI | = Return on Investment (Measurement of profitability) |
| leverage | = Leverage (Total Debt / Total Assets) |
| e | = Error |

## 5. Results and discussion
Before performing hypothesis testing, performed classical assumption in advance, which includes multicoloniarity test, autocorrelation test, heteroscedasticity test, and normality test.

**Table 1: Coefficient Determination and F test Model Summary[b]**

| Model | R | R Square | Adjusted R Square | Std. Error of the Estimate | Change Statistics R Square Change | F Change | df1 | df2 | Sig. F Change | Durbin-Watson |
|---|---|---|---|---|---|---|---|---|---|---|
| 1 | .675[a] | .455 | .432 | .753 | .455 | 20.050 | 2 | 48 | .000 | 1.502 |

a. Predictors: (Constant), Leverage, Profitabilitas
b. Dependent Variable: CED

**Table 2: Multiple Regression and t test Coefficient**

| Model | Unstandardized Coefficients | | Standardized Coefficients | | | 95.0% Confidence Interval for B | | Correlations | | | Collinearity Statistics | |
|---|---|---|---|---|---|---|---|---|---|---|---|---|
| | B | Std. Error | Beta | t | Sig. | Lower Bound | Upper Bound | Zero-order | Partial | Part | Tolerance | VIF |
| 1 (Constant) | -.311 | .117 | | -2.651 | .011 | -.546 | -.075 | | | | | |
| Profitabilitas | .381 | .094 | .432 | 4.057 | .000 | .192 | .570 | .441 | .505 | .432 | 1.000 | 1.000 |
| Leverage | -.542 | .113 | -.511 | -4.791 | .000 | -.770 | -.315 | -.518 | -.569 | -.510 | 1.000 | 1.000 |

a. Dependent Variable: CED

Based on the table, can be concluded that:

1. Based on F test: Profitability and leverage have simultaneous effect to Carbon Emission Disclosures, this means that there is a relationship between profitability and leverage to the Carbon Emission Disclosure.
2. Based on t test: Profitability effect to Carbon Emission Disclosures, and leverage effect to Carbon Emission Disclosures.
3. Multiple Regression:
   Structural equation: **Y = -0311 + 0.381X1 - 0.542X2**
   ***a*** = -0.311 means that, if the profitability variable (X1) and leverage (X2) is equal to zero or constant then CED has a value of = -0.311 units.
   ***β1*** = 0.381, meaning that is when profitability (X1) increased by one unit, and the other variables are constant or equal to zero then the value of the variable to be predicted CED will increase 0.381 units.
   ***β2*** = -0.54, meaning that is when leverage (X2) increased by one unit, and the other variables are constant or equal to zero then the value of the variable CED would be predicted to be down -0542 units.
4. Coefficient determination:
   Profitability and Leverage has a sufficient relationship and significant to the Carbon Emission Disclosure (CED), as much as 45.52%, and 54.5% is influenced by factors beyond Profitability and Leverage. Partial result can be shown in the table below based on the calculations it can be seen that the profitability variable (X1) gives effect to the CED (Y) of 19.05% and leverage variable (X2) gives effect to the CED (Y) of 26.47%.

**Table 3: Partial results of the calculation coefficient of determination**

| Research variable | Beta coefficient X Zero Order | Result |
|---|---|---|
| (X1) | 0432 x 0441 | 19:05% |
| (X2) | -0511 x -0518 | 26.47% |
| **TOTAL** | | 45.52% |

This shows the profitability (X1) give effect smaller than the leverage (X2) to CED (Y), which occurs because the value of fluctuations in profitability and reduced leverage annually, making it partially or each variable profitability and Leverage effect to CED.

## 6. Conclusion

1) Profitability and Leverage effect to the Carbon Emission Disclosure. This means there is a relationship between profitability and leverage to the Carbon Emission Disclosure.

2) The regression coefficient profitability has a positive sign that means increased profitability will be followed by increasing the Carbon Emission Disclosure in Companies listed consecutively in the Sustainability Reporting Award from the year 2014-2016.
3) Leverage regression coefficient has a negative sign that means increasing leverage then decreasing Carbon Emission Disclosure in Companies listed consecutively in the Sustainability Reporting Award from the years 2014-2016.
4) Based on the conclusions that have been mentioned above, the researchers advise that for the company to report CED fuller, clearer and more transparent. CED will establish and improve the brand image of the company, to investors are expected to be thorough in choosing the company that will make a place to invest. Investors also must pay attention to social responsibility is not only concerned with profit, and academic researchers to further recommended to increase the number of samples in research as well as adding other variables that affect the disclosure of carbon emissions. Several other dependent variables that can be used are the media exposure, the scope of the company's operations, and environmental performance.

**References**

[1] Berthelot, S., & Robert, A.-M. (2011). Climate change disclosures: An examination of Canadian oil and gas firms. Issues in Social & Environmental Accounting, 5(1), 106–123.

[2] Jannah, R., & Muid, D. (2014). Analisis Faktor-Faktor Yang Mempengaruhi Carbon Emission Disclosure Pada Perusahaan Di Indonesia ( Studi Empiris pada Perusahaan yang Terdaftar di Bursa Efek Indonesia Periode 2010-2012 ), 3(2013), 1–11.

[3] Prado-Lorenzo, J.-M., Rodríguez-Domínguez, L., Gallego-Álvarez, I., & García-Sánchez, I.-M. (2009). Factors influencing the disclosure of greenhouse gas emissions in companies world-wide. Management Decision, 47(7), 1133–1157

[4] Rankin, Michaela, Carolyn Windsor, D. W. (2010). An investigation of Voluntary Corporate Greenhouse Gas Emissions Reporting in a Market Governance System, 24(8).

[5] Choi, B. B., Lee, D., & Psaros, J. (2013). An analysis of Australian company carbon emission disclosures. Pacific Accounting Review, 25(1), 58–79.

[6] Luo, L., Tang, Q., & Lan, Y. (2013). Comparison of propensity for carbon disclosure between developing and developed countries. Accounting Research Journal, 26(1), 6–34. http://doi.org/10.1108/ARJ-04-2012-0024

[7] Undang-undang RI No. 17 Tahun 2004 tentang Pengesahan Kyoto Protocol To the United Nations Framework Convention on Climate Change (Protokol Kyoto atas Konvensi Kerangka Kerja Perserikatan Bangsa-Bangsa tentang Peru-bahan Iklim)

[8] Majid, R. A., & Ghozali, I. (2015). Analisis Faktor-Faktor Yang Mempengaruhi Perusahaan Di Indonesia, 4, 1–11.

[9] Pradini, H. S. (2013). The Analysis of Information Content towards Greenhouse Gas Emissions Disclosure in Indonesia Companies, 1–76.

[10] Clarkson, P. M., Li, Y., Richardson, G. D., & Vasvari, F. P. (2008). Revisiting the relation between environmental performance and environmental disclosure: An empirical analysis. Accounting, Organizations and Society, 33(4-5), 303–327. http://doi.org/10.1016/j.aos.2007.05.003

[11] Luo, L., Tang, Q., & Lan, Y. (2013). Comparison of propensity for carbon disclosure between developing and developed countries. Accounting Research Journal, 26(1), 6–34. http://doi.org/10.1108/ARJ-04-2012-0024

# Removal of Pesticides Using Aquatic Plants in Water Resources: A Review

**Premla Devi Chander, Chow Ming Fai, Chai Mee Kin**

College of Engineering, Universiti Tenaga Nasional, Jalan IKRAM-UNITEN, 43000 Kajang, Selangor.
Corresponding email : premla2890@gmail.com

**Abstract.** According to the Food and Agriculture Organization of United Nation (FAO), there is a demand of increasing 70 % of food production to sustain 2.3 billion people by the year of 2050. This shows the need of rapid growth of agriculture applications to improve the food production worldwide. It's an undeniable truth that pesticides induced major role in the economic production. Notwithstanding the beneficial effects of pesticides, the negative impacts on human health, environment quality have well documented worldwide and raise the concerns to the health hazard globally. Pesticides need to be closely regulated for quality control monitoring on fruits, vegetables and water resources for safety purposes. European Regulation 396/2005 and amendments were established on the amount of maximum quantities of pesticide residues permitted in products of animal or vegetable for human or animal consumption. Phytoremediation technology using aquatic plants acts as cost effective alternative over conventional technologies to remove harmful heavy metals, organic and inorganic pollutants from various water bodies. To best of our knowledge, there is little or no data are available on the effectiveness of aquatic plants for the phytoremediation of pesticides. Therefore, in this review paper, the main focus is to study the effectiveness of aquatic plant factors in removal of pesticides from water sources and to provide insight for the future development.

## 1. Introduction

According to the Food and Agriculture Organization of United Nation (FAO), there is a demand of increasing 70 % of food production to sustain 2.3 billion people by the year of 2050[1]. These shows the urge of rapid growth of agriculture applications are the possible option to improve the food production worldwide. It's an undeniable truth that pesticides induced major role in the economic production of wide ranges of vegetable, fruit, cereal, forage, fibre and oil crops which now constitute a large part of successful agricultural industry in many countries [1,2]. Crop loss reduction, help to increases the revenue income of farmers from the additional marketable yield and reduces the cost of crop production per unit output are the important keys that drive farmers to support pesticides usage on their agriculture field [2]. Notwithstanding the beneficial effects of pesticides, the negative impacts on human health, environment quality have well documented worldwide and raise the concerns to the health hazard globally [3]. In addition, European Regulation 396/2005 and amendments were established on the amount of maximum quantities of pesticide residues permitted in products of animal or vegetable for human or animal consumption. As this to ensure the pesticide residues in foodstuffs do not constitute an unacceptable risk for consumer and animal health. Currently, maximum residue limits (MRLs) in force for the following foodstuffs are established in Directives 76/895/EEC (fruit and vegetables), 86/362/EEC (cereals), 86/363/EEC (foodstuffs of animal origin), 90/642/EEC (products of vegetable origin and honey) and amendments [4].

Pesticides are divided into many classes, of which the most important are organochlorine and organophosphorous compounds. Organochlorine pesticides are known to resist biodegradation and therefore they can be concentrated through food chains and produce a significant magnification of the original concentration at the end of the chain [3,4]. The general progression of pesticide development has moved from highly toxic, persistent and bio-accumulating pesticides such as DDT, to pesticides that degrade rapidly in the environment and are less toxic to non-target organisms. The developed countries have banned many of the older pesticides due to potential toxic effect to man and or their impact on ecosystems, in favor of more pesticide formulations [5]. In developing countries, organochlorine pesticides still remain as cheapest to produce and, for some purposes, remain highly effective. Developing countries maintain that they cannot afford, for reason of cost or efficacy, to ban certain older pesticides. The dilemma of cost and efficacy, versus ecological impacts, including long range transport, and access to modern pesticides formulations at low cost remains a contentious global issue [5,6].

Pesticides need to be closely regulated for quality control monitoring on fruits and vegetables for safety purposes [6]. Biological control such as phytoremediation also known as plant assisted clean up exploits the ability of plants to remove, degrade, assimilate, metabolize, or render harmless both inorganic and organic contaminants . It is being promoted as more cost effective over conventional technologies for the remediation of polluted environment. Aquatic plants are well known for the removal of harmful pollutants from water resources .However, little or no data are available on the effectiveness of aquatic plants for the phytoremediation of pesticides. In this review paper, the main focus is to study the effectiveness of aquatic plants factors in removal of pesticides from water sources and to provide insight for the future development.

## 2. Aquatic Phytoremediation Technology Applications

Phytoremediation is a cost-effective and eco-friendly technology that has the ability to remove environment pollutants such as an organic, inorganic compounds and heavy metals by using specific plants or certain plant associated bacteria strains from water resources [7,8]. Also, able to remove many pollutants from shallow depth contaminated water to hydrologic control of ground water within short period of time [9]. Four mechanisms are involved in phytoremediation of organic pollutants: direct uptake and accumulation of contaminants and subsequent metabolism in plant tissues, transpiration of volatile organic hydrocarbons through the leaves, release of exudates that stimulate microbial activity and biochemical transformations around the root system and enhancement of mineralization at the root–soil interface that is attributed to mycorrhizal fungi and microbial concortia associated with the root surface [10].

Aquatic plants such as *Eichhornia crassipes*, *Lemna minor* and *Elodea canadensis* has been used in water treatment due to its photosynthetic activity and growth rate of plants easy to harvest and high absorption rate of pollutants [11].

*Eichhornia crassipes* known as water hyacinth is a free-floating aquatic macrophyte that rapidly grow to high depth (over 60kg/m$^2$) in an open pond, irrigations and other water bodies cause the clogging of water bodies and triggers unfavorable environment effect, human health and affects the economic development as well [11,12]. Despite, *Eichhornia crassipes* shoot covers the surface of water bodies prevent the entry of sunlight impacts the algae growth and also aquatic organisms and this disturbs the balance of ecosystem. The common *Eichhornia crassipes* species has long pendant roots, rhizomes, stolon and it can grow in the range of 40 cm to 100 cm height. It has the ability to float due to the stem and leave contain air-filled tissue that provides substantial buoyancy and also able to grow in drastic condition such as moist sediments for several months [12]. According to *Xia and Ma, 2006* [13], a phytodegradation experiment was conducted using *Eichhornia crassipes* to remove malathion, ethion from aqueous solution and also from polluted water. It was known that 56% of 10 ppm of malathion in 250 mL was degraded by (approx. 11g of weight) *Eichhornia crassipes* . Meanwhile the accumulated ethion in live *Eichhornia crassipes* plant decreased by 55-91% in shoots and 74-81% in roots after the plant growing one week in ethion free culture solutions. This shows the

*Eichhornia crassipes* uptakes on pesticides and its phytodegradation can used as potential, economical and alternative biological method to remove pesticides from the water bodies.

*Lemna minor* known as duck weed aquatic macrophyte that withstands cold weather (1.7°C to 35°C) and grow rapidly within a week under optimum pH 6 [14]. Also, it decontaminates heavy metal and organic pollutants such as pesticides by rhizofiltration [15]. Cost effective, long storage capacity and minimal usage of chemicals are the other potential benefits from this plant [16]. According to Dosnon-Olette et al., 2010 [17], a study was conducted using two types of duckweed species to test the efficiency of removing fungicide dimethomorph from agricultural waste water. From the result obtained, duckweed able to remove dimethomorph as long as the concentration does not become too toxic and inhibit depuration mechanisms. The maximal removal of 41 and 26 μg $g^{-1}$ fresh weight of dimethomorph (at 600 μg $L^{-1}$ of dimethomorph and an initial density of 0.10 g E-$flask^{-1}$) by *Lemna minor* and *Spirodela polyrhiza*, respectively. *Lemna minor* sensitivity towards dimethomorph was correlated with their ability to remove it from water. An increase of removal (in μg $L^{-1}$) was correlated with the increase of growth rate inhibition ($r^2 > 0.86$).

Aquatic plants have great potential to function as on-site biosinks and biofilters of aquatic pollutants because of their abundance and limited mobility [18]. They have been successfully used to sequester selected heavy metals and nutrients through their root systems and by uptake through their plant bodies [19]. *Elodea canadensis* is a free floating, rapid growth and fast cultivate macrophyte aquatic plant. This plant is a native to North America and considered as weed due to its fast growing at drastic conditions. This species has been used commonly in phytoremediation technology to treat organic pollutants such as pesticides. Olette et al., 2008 [19] carried research using *Elodea canadensis* to remove three types of pesticides such as copper sulphate (fungicide), flazasulfuron (herbicide) and dimethomorph (fungicide). It was known that the toxicity of the pesticides were same for this aquatic plant and occurred in descending order such as flazasulfuron > copper sulphate > dimethomorph. The maximal removal rates for copper sulphate in Elodea Canadensis was 16.5 % within three days, dimethomorph was 5.5% within two days and 12% within 4 days for dimethomorph.

Chemical and physical properties of pesticide compounds and interaction of environment factors play vital role for the removal efficiency in aquatic plants. Different absorption, volatilization and translocation rates of pesticides causes different removal rate in *Eichhornia crassipes* , *Lemna minor* and *Elodea canadensis*. High removal rate of pesticides occurs when pesticide mobility proportional to surface adsorption on part of plant [20].

## 3. Future Trends

There are many studies has been conducted widely in removal of heavy metals from various water resources. Yet, there were less study been carried out on removal of pesticides from water sources using aquatic plants and also limited in details of the four mechanisms in phytoremediation using aquatic plants. More researches should focus on the pesticide removal from water bodies using aquatic plants as wide application of pesticide usage by the year 2050 will damage human health and pollute the environment as well.

## 4. Conclusion

Large scale application of pesticides to crops in agricultural field contributes to the presence of toxic contaminants in the water reservoirs, streams and rivers. This triggers negative impacts on human health as well as aquatic organisms. Phytoremediation technology known as plant assisted cleanup exploits the ability of plants to remove, degrade, assimilate, metabolize, or render harmless both inorganic and organic contaminants and prove to be cost effective over conventional technologies for the remediation of polluted environment such as pesticides [21]. Aquatic plants such as *Eichhornia crassipes* , *Lemna minor* and *Elodea canadensis* has been used in water treatment due to its photosynthetic activity and growth rate of plants easy to harvest and high absorption rate of pollutants. The studies of the effectiveness of aquatic plant factors in removal of pesticides from water sources were discussed.

**References**

[1] Farina, Y., Abdullah, M., Bibi, N. and Khalik, W. (2017). Determination of pesticide residues in leafy vegetables at parts per billion levels by a chemometric study using GC-ECD in Cameron Highlands, Malaysia. Food Chemistry, 224, pp.55-61.

[2] Abdulra'uf, L. and Tan, G. (2015). Chemometric approach to the optimization of HS-SPME/GC–MS for the determination of multiclass pesticide residues in fruits and vegetables. Food Chemistry, 177, pp.267-273.

[3] Popp, J., Pető, K. and Nagy, J. (2012). Pesticide productivity and food security. A review. Agronomy for Sustainable Development, 33(1), pp.243-255.

[4] Directive 98/44/EC of the European Parliament and of the Council; New U.K. Domestic Law. (2000). Biotechnology Law Report, 19(5), pp.661-705.

[5] Fernando, C. (2017). Pesticide, Environment and Food Safety: A review. *Food and Energy Security*, 6(2), pp.48-60.

[6] Abdulra'uf, L. and Tan, G. (2015). Chemometric approach to the optimization of HS-SPME/GC–MS for the determination of multiclass pesticide residues in fruits and vegetables. *Food Chemistry*, 177, pp.267-273.

[7] Shalini, T., Radhika, D. and Ankita, P. (2016). Removal Of Pesticide Carbofuran Using Wetland Plants. *International Journal Of Advance Biological Research*, 6(4), pp.482-485.

[8] Singh, T. and Singh, D. (2017). Phytoremediation of organochlorine pesticides: Concept, method, and recent developments. *International Journal of Phytoremediation*, 19(9), pp.834-843.

[9] Romeh, A. (2014). Phytoremediation of cyanophos insecticide by Plantago major L. in water. Journal of Environmental Health Science and Engineering, 12(1),pp.38.

[10] Melania, B., Valer, M. and Simona, E. (2014). Study on the mechanisms of phytoremediation. *Journal of Environmental Research and Protection*, 11(3), pp.67-73.

[11] Abdul Waheed, A., Awang, S. and Sarva, P. (2014). The Comparison of Phytoremediation Abilities of Water Mimosa and Water Hyacinth. *ARPN Journal of Science and Technology*, 4(12), pp.722-731.

[12] Rezania, S., Ponraj, M., Din, M., Songip, A., Sairan, F. and Chelliapan, S. (2015). The diverse applications of water hyacinth with main focus on sustainable energy and production for new era: An overview. *Renewable and Sustainable Energy Reviews*, 41, pp.943-954.

[13] Xia, H. and Ma, X. (2006). Phytoremediation of ethion by water hyacinth (Eichhornia crassipes) from water. *Bioresource Technology*, 97(8), pp.1050-1054.

[14] Prasertsup, P. and Ariyakanon, N. (2011). Removal of Chlorpyrifos by Water Lettuce (Pistia stratiotesL.) and Duckweed (Lemna minor L.). *International Journal of Phytoremediation*, 13(4), pp.383-395.

[15] Sasmaz, M., Obek, E. and Sasmaz, A. (2017). The accumulation of La, Ce and Y by Lemna minor and Lemna gibba in the Keban gallery water, Elazig Turkey. *Water and Environment Journal.*

[16] Li, S., Feng-Ying, Z., Yang, H. and Jian-Cong, N. (2011). Thorough removal of inorganic and organic mercury from aqueous solutions by adsorption on Lemna minor powder. *Journal of Hazardous Materials*, 186(1), pp.423-429.

[17] Dosnon-Olette, R., Couderchet, M., El Arfaoui, A., Sayen, S. and Eullaffroy, P. (2010). Influence of initial pesticide concentrations and plant population density on dimethomorph toxicity and removal by two duckweed species. *Science of The Total Environment*, 408(10), pp.2254-2259.

[18] Gupta, P., Roy, S. and B. Mahindrakar, A. (2012). Treatment of Water Using Water Hyacinth, Water Lettuce and Vetiver Grass - A Review. Resources and Environment, 2(5), pp.202-215.

[19] Olette, R., Couderchet, M., Biagianti, S. and Eullaffroy, P. (2008). Toxicity and removal of pesticides by selected aquatic plants. *Chemosphere*, 70(8), pp.1414-1421.

[20] Dosnon-Olette, R., Couderchet, M., El Arfaoui, A., Sayen, S. and Eullaffroy, P. (2010).

Influence of initial pesticide concentrations and plant population density on dimethomorph toxicity and removal by two duckweed species. *Science of The Total Environment*, 408(10), pp.2254-2259.
[21] Razzaq, R. (2017). Phytoremediation: An Environmental Friendly Technique - A Review. Journal of Environmental Analytical Chemistry, 04(02).

**Acknowledgement**
The authors would like to acknowledge the grants provided by Tenaga Nasional Berhad (Project No.U-TG-RD-17-06 and Universiti Tenaga Nasional (Project No. 10289176/B/9/2017/50) and thank the supports from Innovation and Research Management Centre (iRMC), Universiti Tenaga Nasional.

IOP Conf. Series: Earth and Environmental Science **164** (2018) 012028 doi:10.1088/1755-1315/164/1/012028

# Corrosion Study of Carbon Steel in $CO_2$ Loaded Amine-Amino Acid Solutions-Case of Mixtures of N-Methyldiethanolamine and L-Arginine

**A G Talkhan, A Benamor* and M S Nasser**

Gas Processing Centre, Qatar University, 2713 Doha, Qatar

*E-mail: benamor.abdelbaki@qu.edu.qa

**Abstract.** This research presents an electrochemical corrosion investigation to quantify the corrosiveness of carbon steel in blends of methyldiethanolamine (MDEA) and arginine (Arg) at two molar concentration and different solvent proportions. The selected temperatures range varied from 20 °C to 50 °C and the $CO_2$ partial pressure was in the range of 49-100 kPa. The obtained data showed increased corrosion rate with increased $CO_2$ partial pressure and increased the temperature of the solution. However, when added to MDEA, arginine was found to act as an inhibitor, the higher Arginine proportion in the mixture, the more inhibition effect was observed.

## 1. Introduction

The $CO_2$ absorption process using aqueous amine solutions is still the choice for the removal of $CO_2$ from natural and industrial gas streams. Common industrial applications include natural gas processing, coal gasification, and manufacturing of hydrogen and ammonia. Besides such industrial applications, the $CO_2$ absorption process is considered to be a highly rated potential technique for reducing greenhouse gas emission from flue gas streams [1]. $CO_2$ absorption process using aqueous amine solutions has long encountered corrosion problems. Corrosion seems to take place in several plant locations including the bottom portion of the absorber, the rich-lean heat exchanger, the regenerator, and the reboiler. Both uniform corrosion and localized attacks such as pitting, galvanic, erosion, stress cracking, and intergranular corrosion were commonly found. Methyldiethanolamine is widely used in combination with Piperazine for natural gas treatment, in tail gas treating and hydrogen sulfide enrichment units for selectively removing hydrogen sulfide from gas streams. The corrosion effect of MDEA on plant material had been studied by a number of investigators, they showed that metals corrosion rate in MDEA solutions [4, 5] is lower than that of secondary and primary amines.

Amino acids present potential benefits over amines in $CO_2$ separation. Amino acids typically have significantly lower vapor pressures, resulting in reduced solvent losses. Furthermore, they demonstrated greater resistance to oxidative degradation and lower toxicity than alkanolamines. Amino acids, such as arginine, glycine, alanine, and taurine, have been proposed as alternatives to amines in recent years [6, 7] and have been used commercially in acidic gas treating processes. Amino acids are used as green inhibitors against steel corrosion [8, 9]. In this work, electrochemical techniques were used to investigate the corrosiveness of blends of MDEA and Arginine on carbon steel at a total two molar concentration with different solvent proportions, temperatures, and different $CO_2$ partial pressures. The obtained results indicate that the corrosion rate increased with increased $CO_2$ partial pressure and increased temperature. However, when added to MDEA, arginine was found to act as an inhibitor, the higher arginine proportion in the mixture, the more inhibition effect was observed.

## 2. Experimental

### *2.1 Materials*

Reagent-grade MDEA with a mass purity of 99% from Sigma-Aldrich and arginine 99% purity from Riedel-de Haen were used. All chemicals were used without further purification. $CO_2$ with a purity of 99.999% was purchased from the local market. Distilled water was used for solvent preparation. The working electrode made from carbon steel (CS1018) consisted of a 7.85-mm diameter with an active area of 0.484 $cm^2$.

### *2.2 Experimental setup and procedure*

A schematic diagram of the experimental set up is shown in Figure. 1. A 125 ml corrosion cell was used, a reference electrode made of Ag/AgCl filled with 3 M KCL solution. The counter electrode, a rod from graphite 6 mm diameter. The cell temperature was controlled using a water circulator. Gamry Ref 600 potentiostat was used for electrochemical measurements with General Purpose Electro-chemical Software (GPES version 4.9.007) to control the instrument and record corrosion data. Echem Analyst$^{TM}$ Software was used to analyze the data. Carbonated solutions were prepared by absorbing $CO_2$ in solutions of MDEA and Arg. Three different $CO_2$ concentrations were applied, 50, 70 and 100%. The temperature was varied from 20 to 50 °C. The total solution concentration was kept at 2 M with different arginine proportions. The $CO_2$ loading was determined by gas chromatograph. Once the $CO_2$ absorption has reached equilibrium, corrosion tests were performed. The open circuit potential (OCP) of the cells were measured until they were stable (defined as OCP < 5 mV/10 min). The Tafel polarization curves were obtained over a potential range ± 250 mV versus (OCP) with a scan rate of 1 mV/ s. The EIS (AC impedance) (frequency range: 100 mHz to 100 kHz applied DC potential OCP, the amplitude of applied potential: 10 mV). The linear polarization resistance (LPR) method was applied to investigate the validity of Tafel and EIS results. LPR was at potential range ±20 mV vs. OCP, the scan rate of 0.125mV/s.

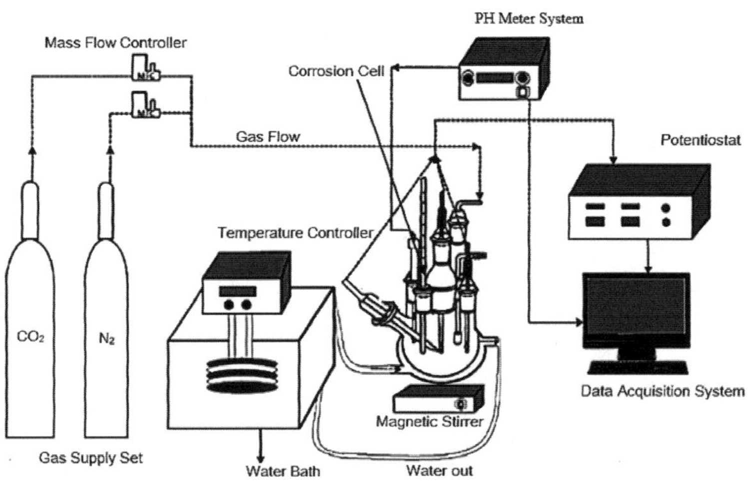


**Figure 1.** Schematic diagram of the experimental setup.

## 3. Results and discussion

### *3.1 Effect of $CO_2$ partial pressure on carbon steel corrosion*

***3.1.1 Polarization behavior of $CO_2$ partial pressure.*** Three typical partial pressure values of $CO_2$ (50, 70, and 100%) were applied at 20 °C and (1.95M / 0.05M) mixture of MDEA/Arg. Figure. 2 shows that the polarization curves (Tafel plot) have shifted to the right with increased $CO_2$ partial pressure. This indicates that the corrosion current density increased with increased $CO_2$ partial pressure [10]. Increasing $CO_2$ partial pressure led to increased $CO_2$ loading in the solution, which in turn led to an increase in the corrosion rate.

***3.1.2 EIS, Nyquist behavior of $CO_2$ partial pressure effect on corrosion.*** Nyquist diagrams (Figure. 3) shows a one-time constant in the EIS curve that is the capacitive loop, which indicates that the sample surface was covered with corrosion scale.

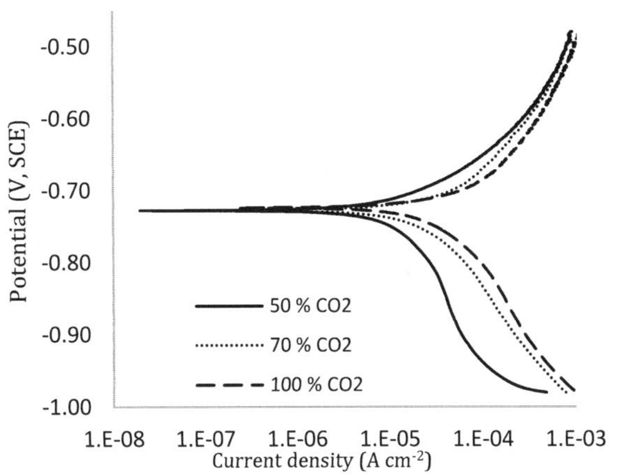


**Figure 2.** Tafel curves showing the effect of $CO_2$ partial pressure on the corrosion of carbon steel.

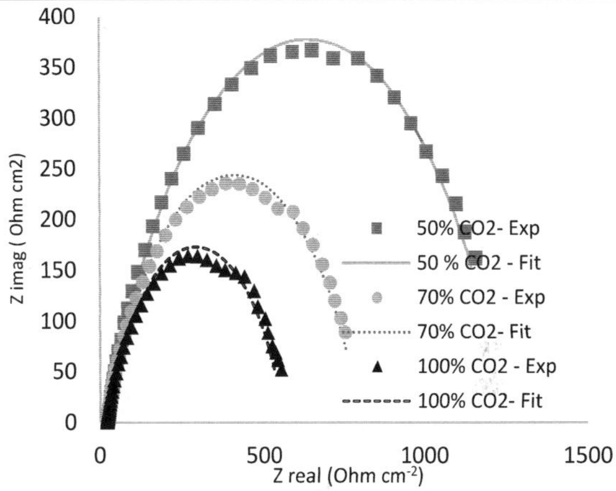


**Figure 3.** EIS Nyquist curves show the effect of $CO_2$ partial pressure on the corrosion.

The EIS data measured was fitted to an electrochemical equivalent circuit (Figure. 4). $R_S$ represents the solution resistance, $R_{ct}$, $R_p$ is the charge transfer resistance, and $Q_{dl}$ is constant phase element (CPE) used as a substitute of the double layer capacitance due to the non–ideal capacitive response of the steel/electrolyte interface. Fitted curves are given in Figure 3. The features of EIS graphs measured are identical while the diameter of EIS curves decreases with increased $CO_2$ partial pressure. The EIS results confirm the results validity of polarization curves.

***3.1.3 Linear polarization resistance (LPR) behavior of $CO_2$ partial pressure.***
In Figure 5, polarization resistance lines for the influence of CO2 partial pressure are shown. The slope of lines decreased with increasing CO2 partial pressure indicating that the transition resistance (LPR) between the electrodes and the electrolyte decreased. This means that the corrosion current density increased, which in turn led to an increase in the corrosion rate of carbon steel.

***3.1.4* Effect of *temperature* on corrosion of carbon steel**
***3.1.5 Polarization behavior of temperature.*** Tafel plots on the effect of temperature on carbon steel corrosion are shown in Figure 6. At 100% $CO_2$ partial pressure, four temperatures (20, 30, 40 and 50 °C) were applied to the blend MDEA/Arg solution. We observe that Tafel curves move right with increased temperature, which indicates that the corrosion current density increases when the temperature goes up [11], increasing the temperature pushes the rate of metal dissolution and hence the oxidation-reduction reaction.

***3.1.6 EIS, Nyquist behavior of temperature effect on corrosion.*** Nyquist plots showing the temperature effect are presented in Figure 7. We observe that only one capacitive loop exists. This indicates that corrosion scale covered the sample surface. The EIS graphs have identical features with decreasing diameter of the curve as the temperature increased. The EIS data was fitted to the same electrochemical equivalent circuit. In general, $R_{ct}$ and $R_s$ decreased with increased temperature indicating that the corrosion rate of the steel increase with temperature.

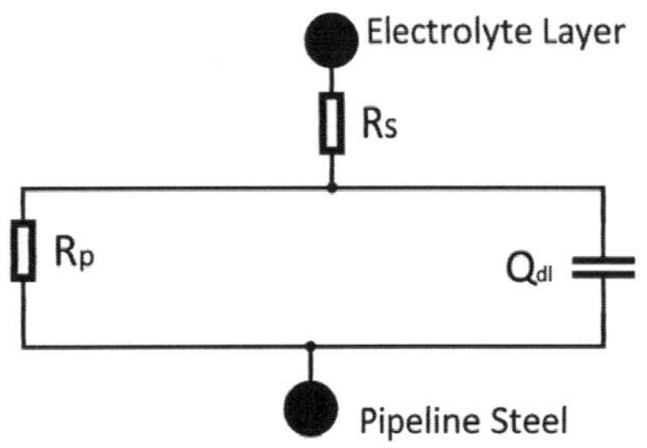


**Figure 4.** The electrochemical equivalent circuit for EIS fitting.

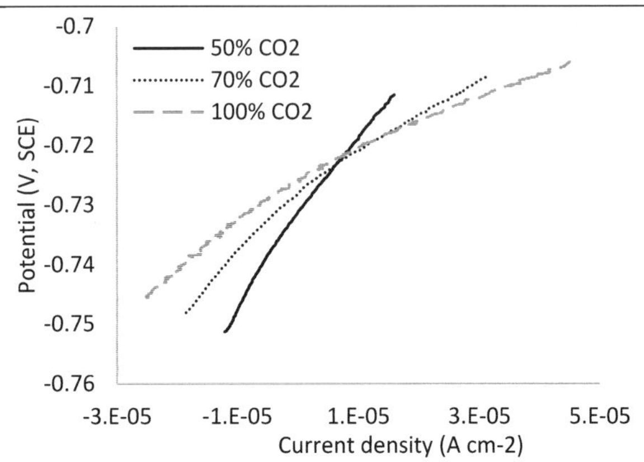


**Figure 5.** LPR curves show the effect of $CO_2$ partial pressure on the corrosion of carbon steel.

***3.1.7 LPR behavior of temperature effect on corrosion.*** Temperature effect on the corrosion of carbon steel was further investigated by linear polarization resistance technique (LPR) as shown in Figure 8. The slopes of the curves decreased with increased temperature indicating that the linear polarization resistance between the electrodes and the electrolyte decreased leading increased corrosion current density, which in turn led to an increase in the corrosion rate of carbon steel in the solution.

***3.1.8*** **Effect of MDEA/Arg mixing ratio on carbon steel corrosion**

Tafel plots, electrochemical impedance spectroscopy, and linear polarization resistance were used to investigate the inhibition effect of Arginine. Arginine concentration was chosen to be 0.03 M, 0.05 M and 0.1 M in a total solution concentration of 2 moles. The temperature was set at 20 °C and $CO_2$ concentration was fixed at 70%.

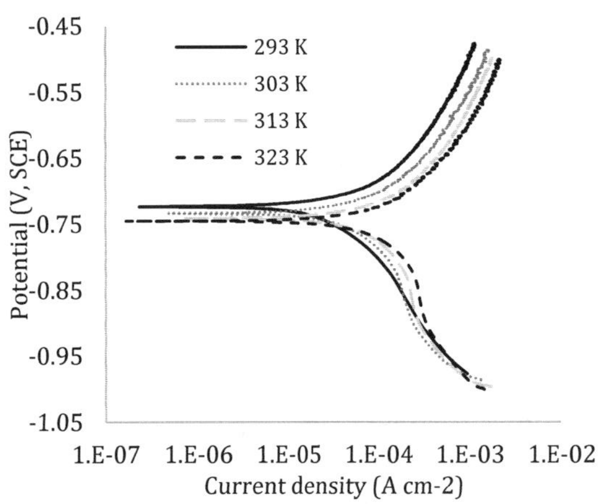


**Figure 6.** Tafel curves show the effect of temperature on the corrosion of carbon steel.

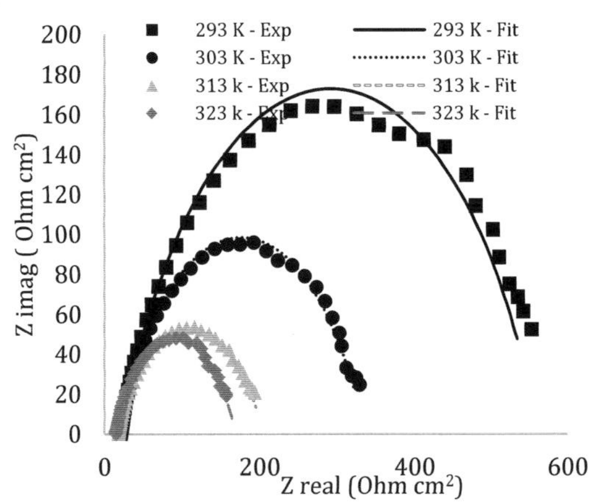


**Figure 7.** EIS Nyquist curves show the effect of temperature on the corrosion.

***3.1.9 Tafel curves behavior of MDEA/Arginine ratio effect.*** From Tafel plot shown in Figure 9, we can see that the curves move left with increased Arginine concentration, which means that corrosion rate decreases. The inhibition effect of different amino acids against steel corrosion had been investigated in different solutions such as HCl [8] and sulfamic acid [9] solution. Inhibition efficiency tends to increase by increasing the inhibitor concentration [16].

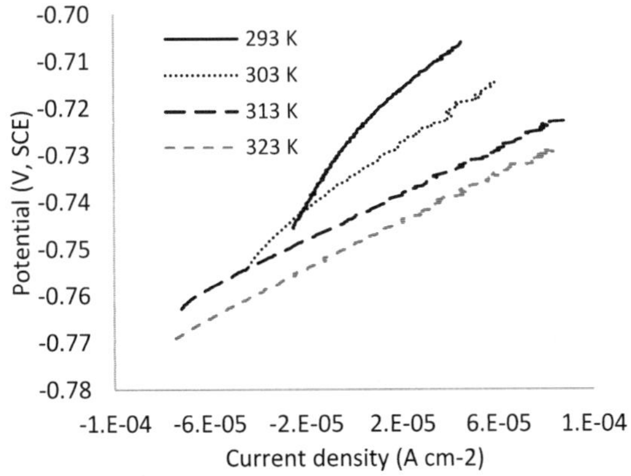


**Figure 8.** Temperature effect on corrosion investigated by LPR method.

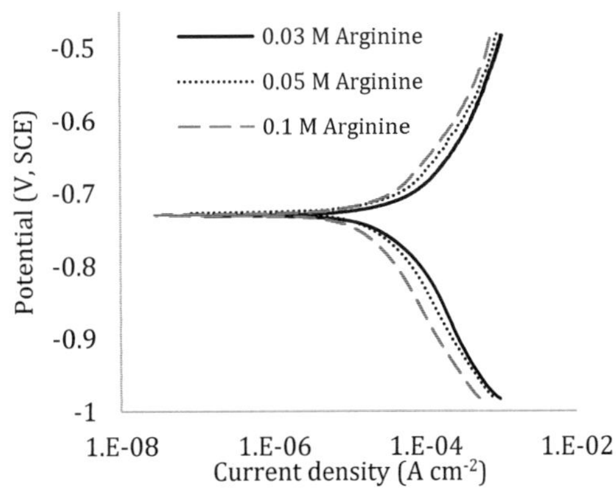


**Figure 9.** Tafel curves showing the effect of MDEA/Arginine mixing ratio on the corrosion.

***3.1.10 EIS, Nyquist behavior of arginine concentration effect on corrosion.*** Nyquist plots showing Arginine effect are presented in Figure 10, only one-time constant exists in the EIS curves that is a capacitive loop which indicates that the sample surface was covered with corrosion scale. Using the same electrochemical equivalent circuit to fit the collected EIS data, fitted curves are given in Figure 10 where the diameter of EIS curve increased as the Arginine concentration increased.

***3.1.11 LPR behavior of arginine concentration effect on corrosion.*** Linear polarization resistance as shown in Figure 11 investigated the effect of Arginine (Arg) concentration on the corrosion of carbon steel. The slope of lines increases with increasing Arginine concentration, which means that the linear polarization resistance (LPR) between the electrodes and the electrolyte also increases. This means that the corrosion current density decreased which causes a decrease in the corrosion rate of carbon steel in the solution.

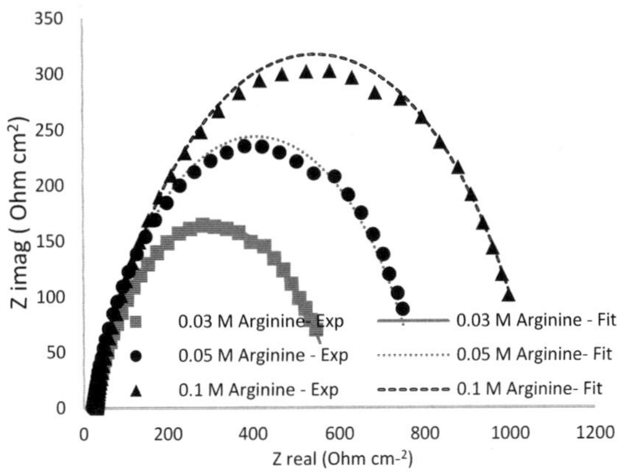


**Figure 10.** EIS Nyquist curves show the effect of MDEA/ARG ratio on the corrosion.

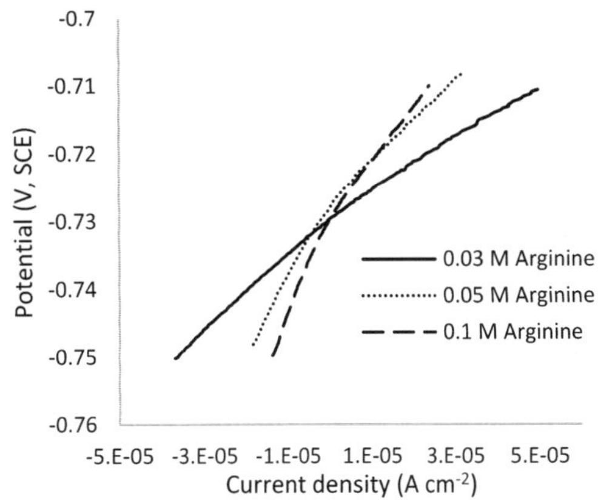


**Figure 11.** The corrosion effect of MDEA/ARG mixing ratio.

## 4. Conclusion

The corrosion of carbon steel (1018) in aqueous mixtures of MDEA and Arginine was investigated using different electrochemical techniques. Effects of $CO_2$ partial pressure, temperature, and

Arginine/MDEA proportions were considered. Increasing $CO_2$ partial pressure in the solution caused an increase in the corrosion rate of carbon steel. The carbon steel Corrosion was found to increase with increasing temperature due to the increased metal dissolution rate and hence the oxidation-reduction reaction. Arginine was found to play inhibitor role, the corrosion rate decreased with increased Arginine concentration.

## 5. References

[1] A. Veawab, P. Tontiwachwuthikul, and A. Chakma, "Corrosion behavior of carbon steel in the $CO_2$ absorption process using aqueous amine solutions," *Industrial & engineering chemistry research,* vol. 38, no. 10, pp. 3917-3924, 1999.

[2] F. Tamajón, E. Álvarez, F. Cerdeira, and D. Gómez-Díaz, "$CO_2$ absorption into N-methyldiethanolamine aqueous-organic solvents," *Chemical Engineering Journal,* vol. 283, pp. 1069-1080, 2016.

[3] N. J. Penders-van Elk, S. M. Oversteegen, and G. F. Versteeg, "Combined Effect of Temperature and p K a on the Kinetics of Absorption of Carbon Dioxide in Aqueous Alkanolamine and Carbonate Solutions with Carbonic Anhydrase," *Industrial & Engineering Chemistry Research,* vol. 55, no. 38, pp. 10044-10054, 2016.

[4] A. L. Kohl and R. Nielsen, *Gas purification*. Gulf Professional Publishing, 1997.

[5] F.-Y. Jou, A. E. Mather, and K. A. Schmidt, "Solubility of Nitrous Oxide in Aqueous Methyldiethanolamine Solutions," *Journal of Chemical & Engineering Data,* 2017.

[6] S. Paul and K. Thomsen, "Kinetics of absorption of carbon dioxide into aqueous potassium salt of proline," *International Journal of Greenhouse Gas Control,* vol. 8, pp. 169-179, 2012.

[7] D. Guo *et al.*, "Amino acids as carbon capture solvents: chemical kinetics and mechanism of the glycine+ $CO_2$ reaction," *Energy & Fuels,* vol. 27, no. 7, pp. 3898-3904, 2013.

[8] C. Kamal and M. G. Sethuraman, "Caulerpin. A bis-indole alkaloid as a green inhibitor for the corrosion of mild steel in 1 M HCl solution from the marine alga Caulerpa racemosa," *Industrial & Engineering Chemistry Research,* vol. 51, no. 31, pp. 10399-10407, 2012.

[9] C. Kamal and M. Sethuraman, "Spirulina platensis–A novel green inhibitor for acid corrosion of mild steel," *Arabian Journal of Chemistry,* vol. 5, no. 2, pp. 155-161, 2012.

[10] Y. Xiang, M. Xu, and Y.-S. Choi, "State-of-the-art overview of pipeline steel corrosion in impure dense $CO_2$ for CCS transportation: mechanisms and models," *Corrosion Engineering, Science and Technology,* pp. 1-25, 2017.

[11] M. Desimone, G. Gordillo, and S. N. Simison, "The effect of temperature and concentration on the corrosion inhibition mechanism of an amphiphilic amido-amine in $CO_2$ saturated solution," *Corrosion Science,* vol. 53, no. 12, pp. 4033-4043, 2011.

[12] Y. Maocheng, X. Jin, Y. Libao, W. Tangqing, S. Cheng, and K. Wei, "EIS analysis on stress corrosion initiation of pipeline steel under disbonded coating in near-neutral pH simulated soil electrolyte," *Corrosion Science,* vol. 110, pp. 23-34, 2016.

[13] A. A. Gürten, H. Keleş, E. Bayol, and F. Kandemirli, "The effect of temperature and concentration on the inhibition of acid corrosion of carbon steel by newly synthesized Schiff base," *Journal of Industrial and Engineering Chemistry,* vol. 27, pp. 68-78, 2015.

[14] M. Harris, "Effects of Corrosion Inhibitor on the Corrosion Behavior of Low Carbon Steel in $CO_2$ Environment," 2017.

[15] W. Emori *et al.*, "Corrosion behavior of carbon steel in amine-based $CO_2$ capture system: effect of sodium sulfate and sodium sulfite contaminants," *Materials and Corrosion,* vol. 68, no. 6, pp. 674-682, 2017.

[16] K. Khaled and N. Abdel-Shafi, "Chemical and electrochemical investigations of l-arginine as corrosion inhibitor for steel in hydrochloric acid solutions," *Int. J. Electrochem. Sci,* vol. 8, pp. 1409-1421, 2013.

**Acknowledgement**

This paper was made possible by an NPRP Grant # 7-1154-2-433 from the Qatar National Research Fund (a member of Qatar Foundation). The statements made herein are solely the responsibility of the authors.

IOP Conf. Series: Earth and Environmental Science **164** (2018) 012029 doi:10.1088/1755-1315/164/1/012029

# Effects of slopes on the stormwater attenuation performance in extensive green roof

**M F Chow[1*], M F Abu Bakar[2], M H Mohd Razali[2]**

[1]Center for Sustainable Technology and Environment (CSTEN), Universiti Tenaga Nasional, 43000 Kajang, Selangor, Malaysia.

[2]Department of Civil Engineering, Universiti Tenaga Nasional, 43000 Kajang, Selangor, Malaysia.

*Corresponding author email: mingfaichow12345@gmail.com

**Abstract.** This study is aimed to determine the effects of slope on stormwater retention and peak flow attenuation performances in extensive green roof in Malaysia. One green roof test bed with dimensions of 2.0 m (length) x 1 m (width) x 0.2 m (height) was constructed in Universiti Tenaga Nasional. The substrate layer was fixed as 130 mm depth and placed on top of the filter fabric and drainage layers. The green roof test bed was planted with Portulaca Grandiflora (sedum species) as the native plant in Malaysia. Different slopes (0 °, 2 °, 5 ° and 7 °) of green roof test bed were tested in this study for its stormwater retention and peak flow attenuation performances. The results show that stormwater retention percentages are ranged from 56.9% to 52.3%. The stormwater retention percentages are decreased as slope increased in this study. Flat slope showed the longest observed time of peak flow while the 7 ° slope has the shortest observed time of peak flow. This finding suggests that slope has a significant impact to the stormwater retention and peak flow attenuation performances in extensive green roof.

## 1. Introduction

The urbanization process is taking place rapidly in the recent decades and the urban city population growth is expected to reach 85% by 2030 [1]. The impervious surfaces such as street, building and roof have changed the hydrological processes in the natural catchment which cause the increased of runoff volume, decreased evapotranspiration and ground water recharges, as well as enhancing river channel erosion [2,3,4,5]. As the impervious surfaces are increasing, more frequent flash floods will be happening in urban city. Rooftop and road systems commonly account for 80% of all the impervious surfaces in the urban area [6]. Rooftop runoff poses a greater threat to water quantity in urban catchment than rural catchment as the runoff can enter the receiving water bodies more rapidly due to the greater connectivity of roofs to gutters and drainages.

Green roof have been introduced as a sustainable stormwater management technology to control the urban stormwater runoff. Green roofs have been widely implemented in many temperate countries like European countries, USA and Canada. Numerous studies had proved that green roofs can effectively reduce the peak flow and extend the delay of stormwater runoff [7,8,9,10,11,12,13,14,15,16]. The runoff can be delayed between 10 min [13,17,18] to 4 hours [19] from the green roofs if compared to conventional roofs of which the runoff is nearly instantaneous. Carter and Rasmussen [20] observed 57% of peak flow in a vegetated roof was delayed up to 10 minutes as compared with that from a conventional roof. Liu [7] found that runoff from the green roof was reduced to 0.5 mm/h when the initial rainfall was 2.8 mm/h. Meanwhile, Moran et al. [9] in North Carolina, USA found 57% to 87% reduction in flow rates from a green roof. Fassman-Beck et al. [21] monitored four green roofs (depth 5-15 cm) in New Zealand and observed 73% to 89% of peak flow reduction compared to control roofs. Alfredo et al. [22] also found 22% to 70% reduction in peak discharges from green roofs compared to the control roof.

Though green roofs is widely implemented in many developed countries, climatic variations between the temperate and tropical regions have hindered the direct transferal of roof designs, substrate specifications, and plant species. Therefore, it is important to carry out local study on the green roof system with regards to its suitable plant species, substrate composition and roof designs [23]. Choosing native plants in green roof systems will provide many benefits such as better adaption to local condition, require less watering, low maintenance, increase biodiversity as well as more aesthetically pleasing than non-native plant. The roof design is also very important for tropical countries like Malaysia as it subjected to short and frequent high intensity storm events. The research related to the green roof systems is still lacking in Malaysia. Therefore, this study is carried out to determine the effects of slope on stormwater retention and peak flow attenuation performances in extensive green roof in Malaysia.

## 2. Methodology

### 2.1. *Setup of experimental green roof*

This experiment was conducted at the civil engineering laboratory in Universiti Tenaga Nasional (UNITEN). The dimension of green roof test bed is 2.0 m length x 1 m width x 0.2 m height, and installed at a height of 1m above the ground level. Figure 1 shows the experimental green roof test beds that established in this study.

**Fig. 1**. Experimental green roof test beds that established in this study.

The typical potting soil was chosen in this study for investigating its stormwater retention performance and attenuation of peak flow. The choice of this substrate type was based on its availability and popularity as common horticultural substrate used by the local landscaping industry. The substrate layer was fixed as 130 mm depth and placed on top of the filter fabric and drainage layers (Fig. 1). The green roof test bed was planted with Portulaca Grandiflora (sedum) as shown in Fig. 1. All surface areas inside the green roof test bed were ensured that 100% covered by the respective plant. Different slopes (0 °, 2 °, 5 ° and 7 °) of green roof test bed were tested in this study for its stormwater retention and peak flow attenuation performances.

*2.2. Data collection and analysis*

An artificial rainfall simulator was installed at 0.5 m height above the green roof test bed. The simulated rainfall was set to run for 5 minutes duration and time was recorded after the rainfall initiated. A total of 400 liter of simulated rainfall volume was poured onto the green roof test bed during the experiment. Once the first runoff is observed, time was recorded for each experiment. The stormwater runoff that discharged from drainage layer within the green roof test beds was drained out by using a PVC pipe. A runoff harvesting tank was placed under the green roof test beds and connected to the PVC pipe in order to collect the runoff from green roofs. The water volume in the runoff harvesting tank was measured and time of each measurement was collected simultaneously during the simulated rainfall period. The measured runoff volume and time were converted to runoff flow rate in order to plot the hydrograph for each experiment.

**Fig. 2.** Simulated rainfall (a) and measurement of flow rate (b) during the experiment.

The stormwater retention percentage of green roof is generally calculated by dividing the difference between rainfall and runoff depths with the total rainfall depth as shown in Equation 1.

$$Retention\ (\%) = \frac{Rainfall\ (mm) - Runoff\ (mm)}{Rainfall\ (mm)} x\ 100\%$$ Equation 1

The curve number (CN) is calculated for every green roof test bed by using Equation (2):

$$CN = \frac{25400}{254 + S}$$ Equation 2

Where, S = rainwater retention in soil (mm)

The hydrograph for each experiment with respect to different slope values (0 °, 2 °, 5 ° and 7 °) was plotted in order to determine the peak flow rate and time to peak flow.

## 3. Results and discussions

The effects of slope on the stormwater retention performance and delay of peak flow have been investigated during a 3 month study in Universiti Tenaga Nasional. The stormwater retention percentage is calculated for every storm event monitored in each green roof configuration. The stormwater retention percentages are ranged from 56.9% to 52.3%. The highest stormwater retention percentage is observed at the 2 ° slope (56.9%) while the least at the 7 ° slope (52.3%). It is observed that stormwater retention percentages are decreased as slope increased in this study. The cumulative runoff depths for green roof test beds with different slope values are presented in Fig. 3. The curve number was also calculated for each green roof slope in the study. The curve numbers for 0 °, 2 °, 5 ° and 7 ° slopes are 81.7, 81.8, 82.0 and 82.9, respectively. The curve number started to show significant different at the 7 ° slope. All of these curve numbers are lower than a conventional roof curve number of 98. Using these curve numbers in the equation CN= 25,400/(254 + S) and solving for potential retention (S) we find that S ranges from 52.3 mm to 56.9 mm. These findings are greater than that reported by Carter and Rasmussen [20] who found a curve number of 86 (S=40.5 mm) for a green roof with <2% slope and 7.62 cm of substrate. This may be due to difference in substrate depth, antecedent substrate moisture status and rainfall pattern [24]. The hydrographs for every green roof configurations are shown in Fig. 4. Great variations were observed for the time of peak flow in green roof with respect to different slopes. Flat slope (0 °) showed the longest observed time of peak flow. Meanwhile, the 7 ° slope exhibited 45% faster than the observed time of peak flow in flat slope. The magnitude of peak flow for 7 ° slope is 1.75 time greater that the peak flow in flat slope. This finding suggests that slope has a significant impact to the attenuation of peak flow in extensive green roof.

**Table 1.** Water retention percentage, curve number and observed times of runoff

| Slope (°) | Water retention percentage (%) | Curve number | Observed time | |
|---|---|---|---|---|
| | | | First runoff | Peak flow |
| 0 | 56.9 | 81.7 | 11 min 5 s | 23 min 5 s |
| 2 | 56.4 | 81.8 | 9 min 23 s | 19 min 23 s |
| 5 | 55.9 | 82.0 | 7 min 46 s | 15 min 46 s |
| 7 | 52.3 | 82.9 | 4 min 37s | 12 min 37 s |

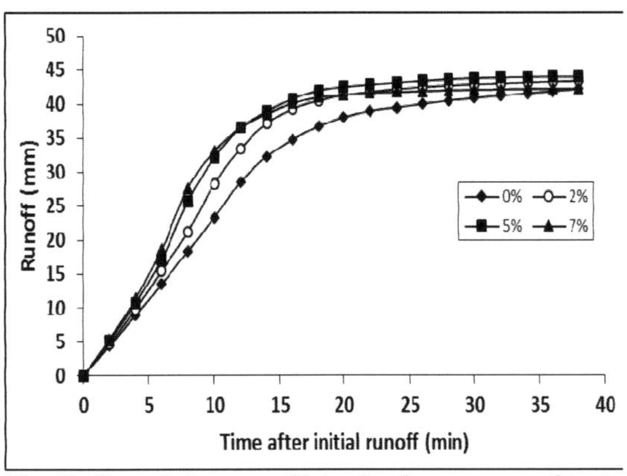


**Fig. 3.** Cumulative runoff depths for green roof test beds with different slope values

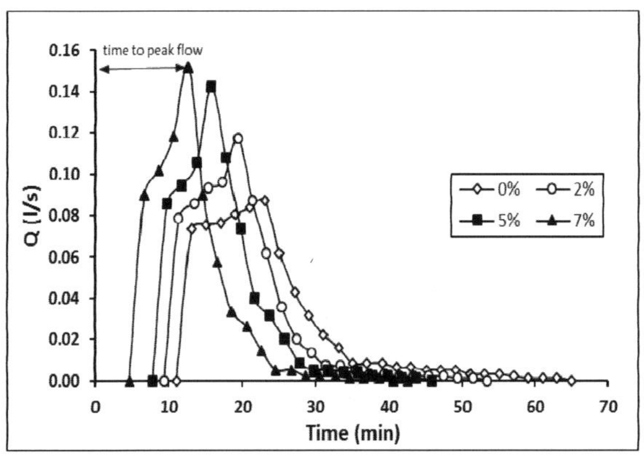


**Fig. 4.** Hydrographs for green roof test beds with different slope values

## 4. Conclusions

This study has investigated the effects of slope on the stormwater retention performance and delay time of peak flow in extensive green roofs. The results showed that green roofs with 0 ° and 7 ° slopes exhibited the highest and lowest stormwater retention percentages which recorded 56.9% and 52.3%, respectively. Flat slope showed the longest observed time of peak flow while the 7 ° slope has the shortest observed time of peak flow. The magnitude of peak flow for 7 ° slope is 1.75 time greater than the peak flow in flat slope. This finding suggests that slope has a significant impact to the stormwater retention and peak flow attenuation performances in extensive green roof.

**References**

[1] United Nations 2002. World Urbanization Prospects: The 2001 Revision. United Nations, New York.

[2] Stone Jr., B., 2004, Paving over paradise: how land use regulations promote residential imperviousness, *Landscape and Urban Planning* **69**:101–113.

[3] Roy, A. H., M. C. Freeman, B. J. Freeman, S. J. Wenger, W. E. Ensign, and J. L. Meyer, 2005. Investigating hydrological alteration as a mechanism of fish assemblage shifts in urbanizing streams. *Journal of the North American Benthological Society*. **24**:656–678.

[4] Arnold, C.L., Gibbons, C.J. 1996. Impervious Surface Coverage: The Emergence of a Key environmental Indicator. Journal of the American Planning Association. 62(2): 243-258.

[5] Walsh, C. J., T. D. Fletcher, and A. R. Ladson, 2005, Stream restoration in urban catchments through redesigning stormwater systems: Looking to the catchment to save the stream. *Journal of the North American Benthological Society* **24**:690–705.

[6] Slonecker, E. T., and Tilley, J. S., 2004. An evaluation of the individual components and accuracies associated with the determination of impervious area. *GIScience and Remote Sensing*. **41**(2): 165-184.

[7] Liu, K., 2003. *Engineering performance of rooftop garden through field evaluation*. Proc. Of the 18$^{th}$ International Convention of the Roof Consultant Institute. 93-103.

[8] DeNardo, J.C., A.R. Jarrett, H.B. Manbeck, J. Beattie, and R. D. Berghage, 2004. Stormwater Mitigation and Surface Temperature Reduction by Green Roofs. *Transactions of ASAE (In Press)*. **48**(4): 1491-1496.

[9] Moran, A., Hunt, B., Smith, J., 2005. Hydrological and water quality performance from green roofs in Goldsboro and Raleigh, North Carolina. In: Green Roofs for Healthy Cities Conference, Washington, DC,

[10] Bengtsson, L., 2005. Peak flows from thin sedum-moss roof, *Nordic Hydrology*, **36**:269-280.

[11] Bengtsson, L., Grahn, L., Olsson, J., 2005. Hydrological function of a thin extensive green roof in southern Sweden. *Nord. Hydrol.*, **36**(3):259–268.

[12] Villarreal, E. L., and L. Bengtsson, 2005. Response of a sedum green-roof to individual rain events. *Ecological Engineering* **25**(1):1-7.

[13] VanWoert, N. D., Rowe, D. B., Andresen, J. A., Rugh, C. L., Fernandez, R. T., Xia, L., 2005. Green roof stormwater retention: Effects of roof surface, slope, and media depth. *Journal Environment Quality* **34**:1036-1044.

[14] Chow, M. F., M., Abu Bakar, M. F., Roslan, M. A. A., Fadzailah, F. A. Idrus, M. F. Z.,Ismail, N. F., Sidek, L. M., Basri, H., 2015 Hydrological Performance of Native Plant Species within Extensive Green Roof System in Malaysia. *ARPN Journal of Engineering and Applied Sciences*. **10**(15): 1819-6608.

[15] Chow, M. F., Abu Bakar, M.F. 2016. A review of the development and challenges of green roof systems in Malaysia. International journal of Architecture and Environmental Engineering. 10(1): 16-20.

[16] Kok, K.H.; Sidek, L.M.; Chow, M.F.; Zainal Abidin, M.R.; Basri, H.; Hayder, G. 2016. Evaluation of green roof performances for urban stormwater quantity and quality controls. Int. J. River Basin Manag. 14, 1–7.

[17] Simmons, M.T., Gardiner, B., Windhager, S., Tinsley, J., 2008. Green roofs are not created equal: the hydrologic and thermal performance of six different extensive green roofs and reflective and non-reflective roofs in a sub-tropical climate, *Urban Ecosystems*. **11**:339–348.

[18] Chow, M.F., Abu Bakar, M.F., Sidek, L.M., Basri, H 2017 Journal of Engineering and Applied Sciences 12(21) 5379-5383.

[19] Moran, A., Hunt, B., and Jennings, G., 2004, A North Carolina field study to evaluate green roof quantity, runoff quality, and plant growth. *2nd Greening Rooftops for Sustainable Communities Conf.*, Portland, 2-4 June, pp. 446-460.

[20] Carter, T.L. and Rasmussen, T.C., 2006. Hydrologic behavior of vegetated roofs. *Journal of the American Water Resources Association*, **42**(5):1261-1274.

[21] Fassman-Beck, E., Voyde, E., Simcock, R., and Hong, Y. S., 2013. 4 living roofs in 3 locations: does configuration affect runoff mitigation? *J. Hydrol.* **490**:11–20.

[22] Alfredo, K., Montalto, F., and Goldstein, A., 2010. Observed and modelled performances of prototype green roof test plots subjected to simulated low- and high-intensity precipitations in a laboratory experiment. *J. Hydrol. Eng.* **15**(6):444–457.

[23] Williams, N. S. G., Raynor, K. J., 2010. Green roofs for wide brown land: opportunities and barriers for rooftop greening in Australia. *Urban forest. Urban Green.* **9**(3): 245-251.

[24] Getter, *K.L., Rowe, D.B., Andresen, J.A. (2007). Quantifying the effect of slope on extensive green roof stormwater retention. Ecological Engineering. 31: 225-231.*

## Acknowledgments

The authors would like to thank Ministry of Higher Education, Malaysia (MOHE) for providing the research grant (Vot no: 20140130FRGS). We would like to acknowledge the BOLD Research Grant provided by Universiti Tenaga Nasional (Project No. 10289176/B/9/2017/50) and TNB Seeding Grant provided by Tenaga Nasional Berhad (Project No: U-TG-RD-17-06).

# Numerical Groundwater Modeling of an Eogenetic Karst Catchment Using Analytical Head-Guided Zonation Method: A New Analytical Approach

**D K S Y Klaas[1,2,3], M A Imteaz[1], A Arulrajah[1], I Sudiayem[3], E M E Klaas[3] and E C M Klaas[3]**

[1] Department of Civil & Construction Engineering, Swinburne University of Technology, Melbourne, VIC, Australia. dklaas@swin.edu.au
[2] Department of Civil Engineering, Politeknik Negeri Kupang, Indonesia.
[3] Department of Health, Science and Technology, Klinik Garuda, Kupang, Indonesia

**Abstract.** In parameterisation step, implementing suitable calibration methods that accurately represents the actual system is crucial in developing a dependable groundwater model in data-scarce areas. In this study, the Head-Guided Zonation (HGZ) method was applied to accommodate the limitation of subsurface data, such as the lack of comprehensive spatial distribution of hydraulic conductivity and specific yield values. This study demonstrated that the spatial distribution of groundwater heads can be used to assume the spatial distribution of sub-surface parameters, hence, hydraulic conductivity zones in the model domain can be established using the distribution of hydraulic gradient. We conclude that the HGZ method is practical and applicable for development of a physically-based groundwater model.

## 1. Introduction

Employing appropriate calibration methods that achieves the modelling objective (good fit between simulated and observed values) and properly represents the corresponding system is an important task in developing a robust groundwater model in data-limited areas [1]. In this study, the HGZ method was chosen to accommodate the limitation of subsurface data, e.g. the absence of spatial distribution of hydraulic conductivity and specific yield values. Therefore, the main objectives of this paper are:

1. to develop a physically-based groundwater model for the recharge-bounded area of Oemau Spring in Rote Island, Indonesia using the Head-Guided Zonation (HGZ) method.
2. to calibrate and validate the model using statistical measures.

## 2. Materials and methodology

### *2.1 Study area*

The experimental area is Oemau recharge area (ORA) enclosing an area of 20.11 km$^2$. Geographically located between latitudes 10°46’42.17”S ~ 10 ° 43’36.91”S and longitudes 123 ° 3’14.84”E ~ 123 ° 9’17.64”E, ORA’s terrain is characterised by highly undulated topography with surface elevation ranges from around 98 m to 340 m above sea level. Areas of low and medium slopes (< 10 degree) are mainly found in the middle and lower end of the catchment where rice paddies and settlements are located, whereas steeper slopes largely dominate the upper areas of the catchment [2]. Geologically, the subsurface is distinctively dominated by two main carbonate formations, i.e. permeable Holocene coralline limestones, Bobonaro formations, featured as karst landscape [3-10]. Governed by a monsoonal climate with two distinct seasons: dry (May-November) and wet (December-April), ORA has mean annual precipitation of 1000 - 2300 mm and humidity between 75 and 92% [8]. The long-term daily precipitation records are available from Lekunik station located around 3 km upstream of the spring (Figure 1).

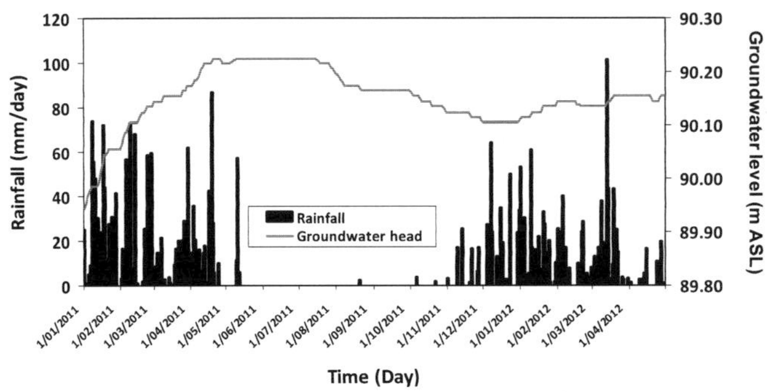


**Figure 1**. Time series of rainfall at Lekunik Station and groundwater level at Oemau Spring

*2.2 HGZ method and zone delineation*

HGZ method assumes that each zone has the same hydogeologic characteristics represented by similar hydraulic gradient of groundwater flow. In regions with parallel hydraulic conductivity, the hydraulic gradients are equally similar. A low hydraulic gradient is expected in areas with high conductivity while a high gradient suggests a slow groundwater flow in a low conductivity area. Therefore, hydraulic conductivity zones in the model domain can be established using hydraulic gradient distribution. In this study head slope calculation was incorporated to delineate the zones. Using dimensionless hydraulic gradient ($i$) between two groundwater heads, the locations of zones are decided empirically by:

$$\frac{dh}{dL} = i \tag{2}$$

where, $dh$ is the difference between two groundwater heads (m), and $dL$ is the length of flow line between the two heads (m). The flow line is drawn perpendicular to groundwater contour.

*2.3 Groundwater model*

To perform groundwater flow simulations under both steady-state and transient conditions, this study employed MODFLOW [11] which solves by means of the following groundwater flow equation:

$$\frac{\partial}{\partial x}\left(K_x h \frac{\partial h}{\partial x}\right) + \frac{\partial}{\partial y}\left(K_y h \frac{\partial h}{\partial y}\right) + \frac{\partial}{\partial z}\left(K_z h \frac{\partial h}{\partial z}\right) + W = S_y \frac{\partial h}{\partial t} \tag{1}$$

where, $K_{x,y,z}$ is the hydraulic conductivity along the $x$, $y$, and $z$ coordinates in "m/day", $W$ is the volumetric flux that represents sources or sinks of water (1/s), $h$ is the hydraulic head (m), $S_y$ is the specific yield, and $x$, $y$, and $z$ are coordinate directions, and $t$ is time (day).

The conceptual model was set up as one horizontal layer and discretised into 50 x 50 m grid cells using Layer Property Flow, LPF package [12]. Assuming a continuum representation of unconfined eogenetic carbonate aquifer in the study area, the model is comprised of 210 columns and 106 rows resulting in 8,061 active cells. Using Drain package, DRN [13] the distribution of surface drainage

(Figure 4) was input into model. Drain conductance values ranging from 2,700 to 8,125 m/d represents drain dimension and hydraulic conductivity values derived from local pumping test analysis. A constant porosity value of 0.3 was assigned into the model to embody dominating carbonate formation of the study area [14]. Two model boundaries (Figure 2), were assigned: (1) no-flow boundary; and (2) specified head boundary (Dirichlet) using Time-Variant Specified-Head, CHD.

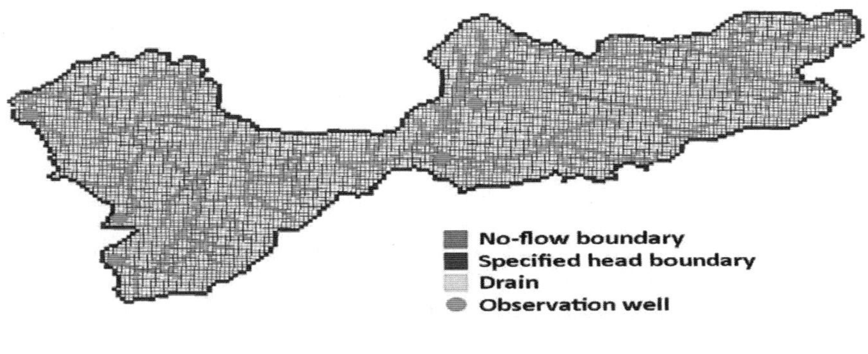


**Figure 2**. Model discretisation and boundary conditions.

## 3. Results and discussion

### *3.1 Zone delineation*

The groundwater level contours were interpolated and drawn [15-16] using available groundwater level data inside and outside ORA (Figure 3a). The result is distribution of groundwater contour (Figure 3b). Hydraulic gradients between groundwater heads were then calculated to delineate the zones. The next step is to divide the model domain into three zones based on the hydraulic gradient values. The zone division resulted in six parameters to be calibrated and validated (each zone represents two parameters; hydraulics conductivity and specific yield).

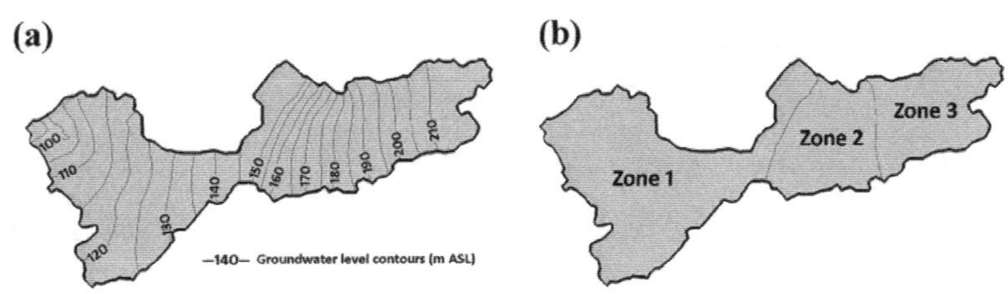


**Figure 3**. (a) Groundwater contour distribution; and (b) Zones obtained from hydraulic gradient calculation.

### *3.2 Model calibration and validation*

In steady-state calibration, the maximal rainfall of 86.5 mm/day on 19 April 2011 and its corresponding groundwater levels at seven observation wells were initially used to represent the response of the highest groundwater heads to the maximal daily recharge. Head differences between the modelled and observed values are insignificant, characterised by small $ME_h$,. The negative bias values of $ME_h$ of -0.57 and -0.09 m in both steady-state and transient calibrations respectively suggest relatively minor underestimations of simulated heads by the model. The total residuals between observed and simulated heads demonstrated consistent results represented with $RMSE_h$. The model performances in both transient calibration and validation were considered acceptable with $RMSE_h$ values of 0.36 and 0.48 respectively, indicating an overall well calibrated and validated model.

*3.3 Sensitivity analysis*

The analysis shows that, for $K_h$ and *Sy* respectively, zone 1 has the highest *css* values of 0.040 and 0.008. Parameters in zone 2 and 3 are considerably less sensitive than those in zone 1. This indicates that among all calibrated parameters $K_h$ and *Sy* in zone 1 provide the greatest constraints to the model.

## 4. Conclusions

In this study, a new analytical approach of HGZ method was applied to accommodate the limitation of subsurface data, such as the lack of comprehensive spatial distribution of hydraulic conductivity and specific yield values. This study showed that hydraulic conductivity zones in the model domain can be established using the distribution of hydraulic gradient. The calibrated and validated model shows a well fit between simulated and observed values. Therefore, it is concluded that the HGZ method is practical and applicable for development of a physically-based groundwater model.

## References

[1] Klaas D K S Y and Imteaz M A 2017 Investigating the impact of the properties of pilot points on calibration of groundwater models: case study of a karst catchment in Rote Island, Indonesia *Hydrogeol J* **25**(6) 1703–1719 doi: 10.1007/s10040-017-1590-4

[2] Klaas D K S Y, Imteaz M A and Arulrajah A 2016 Evaluating the impact of grid cell properties in spatial discretization of groundwater model for a tropical karst catchment in Rote Island, Indonesia *Hydrol Res* doi: 10.2166/nh.2016.250

[3] Klaas D K S Y, Imteaz M A, Sudiayem I, Klaas E M E and Klaas E C M 2017 On the development of a new methodology in sub-surface parameterisation on the calibration of groundwater models *Proc. 1st Int Conf on Adv Environ Eng (AEE 2017). IOP Conf. Series: Earth and Environmental Science* **92** 1-5

[4] Klaas D K S Y and Imteaz M A 2017 Development of a groundwater model for a tropical eogenetic karst aquifer in Rote Island, Indonesia *Proc. 37th Hydrol and Water Res Symp Queenstown, New Zealand* 228-235

[5] Klaas D K S Y, Imteaz M A and Arulrajah A 2017 Development of groundwater vulnerability zones in a data-scarce eogenetic karst area using Head-Guided Zonation and particle-tracking simulation methods *Water Res* **122**, 17-26 doi: 10.1016/j.watres.2017.05.056

[6] Klaas D K S Y, Imteaz M A, Sudiayem I, Klaas E M E and Klaas E C M 2017 Novel approaches in sub-surface parameterisation to calibrate groundwater models *Proc. 3rd Int Conf on Water Res and Environ (WRE 2017). IOP Conf. Series: Earth and Environmental Science* **82** 1-8 doi: 10.1088/1755-1315/82/1/012014

[7] Klaas D K S Y, Imteaz M A, Sudiayem I, Klaas E M E and Klaas E C M 2017 On the development of a new methodology in subsurface parameterisation on the calibration of groundwater models. *Proc. 1st Int Conf on Adv Environ Eng (AEE 2017). IOP Conf. Series: Earth and Environmental Science* **92** 012026 doi: 10.1088/1755-1315/92/1/012026

[8] Klaas D K S Y, Imteaz M A, Sudiayem I, Klaas E M E and Klaas E C M 2017 Assessing the impact of grid cell properties in spatial discretization of groundwater model. *Proc. 1st Int Conf Adv on Eco Eng Dev (ICEED 2017). IOP Conf. Series: Earth and Environmental Science* **109** 012037 doi: 10.1088/1755-1315/109/1/012037

[9] Klaas D K S Y, Imteaz M A and Arulrajah A 2017 Application of head-guided zonation method to delineate groundwater vulnerability zones in a data-scarce catchment. *Europ Water* **57**:395-398

[10] Klaas D K S Y 2008. *Indigenous Water Management: Sustainable Water Conservation Strategies in a Karstic Dominated Area in Rote Island, NTT Province, Indonesia* (MEngSc Thesis). Monash University, Melbourne, Australia.

[11] McDonald M and Harbaugh A (1988 *A modular three-dimensional finite-difference groundwater flow model*, U.S.G.S., Book 6, Chapter A, 1258p.

[12] Harbaugh A W 2005 *MODFLOW-2005*, the U.S. Geological Survey Modular Ground-Water Model, the Ground-Water Flow Process (USGS Techniques and Methods 6-A16, USGS)
[13] Harbaugh A W, Hill M C and McDonald M G 2000 *MODFLOW-2000*, the U.S. Geological Survey Modular Ground-Water Model: User Guide to Modularization Concepts and the Ground-Water Flow Process (USGS Open-File Rep 00-92, USGS)
[14] Johnson A I 1967. *Specific Yield – Compilation of Specific Yields for Various Materials*, pp. 1-74
[15] Fetter C W 2001 *Applied hydrogeology*, 4th edn. (Upper Saddle River: Prentice-Hall)
[16] Hudak P F (2010) *Principles of hydrogeology*, 2nd edn. (Boca Raton: CRC)

**Acknowledgments**
The authors wish to acknowledge the Australian government for supporting the research through the Australia Awards Scholarship (AAS) and Allison Sudrajat Prize Award in the Swinburne University of Technology, Australia.

IOP Conf. Series: Earth and Environmental Science **164** (2018) 012031 doi:10.1088/1755-1315/164/1/012031

# The Influence of shales characteristics on $CO_2$ adsorption behaviour under sub-critical conditions

Ahmed Mohammed Al-Mutarreb[1] , Shiferaw Regassa Jufar[1*], Hesham Abdulelah [1] Eswaran Padmanabhan [1] , Ghareb Mostafa[1] and Maqsood Ahmed[1]

[1]*Shale Gas Research Group (SGRG), Institute of Hydrocarbon Recovery, Faculty of Geoscience & Petroleum Engineering, Universiti Teknologi PETRONAS, Malaysia*
*Email: ahmed.mutarreb@gmail.com,* shiferaw.jufar@utp.edu.my

**Abstract.** Shale gas has been an important source of natural gas and is expected to contribute 30% towards the global output by 2040. Shale formations also have $CO_2$ sequestration potential, which could be used to mitigate the greenhouse gas emissions. A quarter of Malaysian's sedimentary comprises of black shale formations. In this article, we evaluated the $CO_2$ sequestration potential of some of Malaysian's formation under sub-critical conditions, and we correlate shale's characteristics to $CO_2$ storage behaviour under sub-critical conditions. The evaluation is based on total organic carbon (TOC wt. %) contents, mineralogical compositions, particle size distribution PSD of mesopores and micropores. The results show that total organic carbon measured by TOC analyser ranges between 0.5wt. % to 12.1wt%. Bulk mineralogical composition was semi-quantified using X-ray Diffraction (XRD). Non-clay composition varies between ~35 wt.% to ~75 wt. %, which is dominated by quartz. All samples also contain clay minerals such as kaolinite and illite at different range. The $CO_2$ adsorption isotherm results show that KH sample adsorb the most while in contrast the KL samples adsorb the least of $CO_2$. Positive correlation between organic content and $CO_2$ adsorption was observed in most samples, indicating that formations with high organic content has higher $CO_2$ sequestration potential. Mineralogical composition also influence the CO2 adsorption, for instance, illite has lower surface area and pore volume compare with some other clays, subsequently, its presence reduced the CO2 storage potential on KL sample. However, the application of this observation may not fit to other formations due to the high degree of heterogeneity of shale formations.

## 1. Introduction

Shale is a common fine grained sedimentary rock laid below as consolidated mud and can be split easily into fragile slabs. It is usually comprises of organic matter and minerals such as quartz, and clays; illite, smectite and kaolinite and other minerals[1]. Shales occupy a large percentage of the world land. For example, black Palaeozoic shale rocks exist in Peninsular Malaysia, where they alone consistent 25% of the land area [2]. In addition to natural gas interest on shale formations, several studies have proven that shale formations can be used as $CO_2$ geological sinks. $CO_2$ sequestration in shales is identified by mass-transport paths and mechanisms simulation that micro-pores in organic matter acts as molecular sieves which allow only linear molecules for instance $CO_2$ to access their pore-spaces[3]. Furthermore, despite the ultra-low permeability shale have the capacity to store $CO_2$ approximately five times more than $CH_4$ storage capacity. Several authors have contributed to the literature by investigating experimentally the influencing factors on the adsorption behavior in shales. The factors are and not limited to; total organic carbon wt.%, micro and mesopores volume system, mineral compositions[4]. Only few papers have published the characteristics of Malaysian's black shale formations regarding the aforementioned factor. In addition, to our knowledge, no publication has evaluated the $CO_2$ adsorption quantity for Malaysian's black shales and correlate shale's characteristics to $CO_2$ storage potential. In this paper, we characterized several shale samples collected from five locations of Malaysian's formations. TOC analyzer, X-ray diffraction (XRD), and $N_2$ low pressure adsorption were utilized in order to characterize the samples. The characterization is based on total organic carbon (TOC wt.%) contents mineralogical compositions and (pore size distribution PSD of mesopores and micropores). We also correlate the $CO_2$ volume adsorption capacity at sub-critical conditions to these characteristics. The evaluation may provide useful reference prospect in designing $CO_2$ Sequestration process in the future.

## 2. Sampling and Experiments

Eleven (11) shale samples were collected from five malaysian's shale formations locations namely; Batu Gaja (BG), Kenny Hill (KL), Dohol at Johor (J), Setap (S) at Sarawak, and Pending Kroh (KH). Longitude and latitude are described at table 1. The samples were collected from outcrops. Outcrops might have been affected by some degree of weathering. However, it is reported that collecting a newly outcrops from excavated engineering sites is recommended [5]. Therefore, this could reduce significantly the difference on the experimental results between drilling-cuttings and outcrops in the same formations. Around 30 grams of the each collected sample was crushed and sieved. All experimental tests, in this report were performed in crushed powder samples size between 200μm -300 μm.

### *2.1. Total Organic Carbon wt.%*

TOC wt.% was determined based on the GB/T 19145-2003 standard [6]. 1.0 gram of each powdered samples were measured and place in beakers. The samples were acidified with HCL to remove inorganic carbon friction from samples. They were then left in the fume chamber combustion over 650°C before using rinsed with reverse osmosis water for 3 times and dried in the oven at 60 °C for 24 hours, after drying, 60mg of each samples were being weighted and placed in ceramic boats. Percentage of organic carbon is measured using Multi n/c 3100. The measurements were ran in duplicate and then results to be averaged, the precision reaches 0.6 ppm

### *2.2. Mineralogical Compositions*

The X Ray Diffraction method allows mineralogical analysis of fine-grained shale samples with an absolute accuracy of 0.01_ of 2h. This process involves an analysis of the diffraction patterns that are produced when X-rays are passed through the sample to reveal the structural properties of crystalline solids.

### *2.3. The Microspores and Mesopores System*

$N_2$ low pressure adsorption/desorption were conducted using Micromeritics ASAP 2020 at -196 °C on ~ 1 gram crushed sample. The pretreatment was carried out for 12 hours at 383.15 K to remove any moisture and volatile matters. The samples were also purged and degassed under vacuumed to eliminate any further residual moisture[7].

### *2.4. $CO_2$ Adsorption Measurement*

Experimental studies of the $CO_2$ adsorption were conducted at subcritical conditions at relative pressure of 1bar and room temperature 293 K using BELSORP-mini II instrument (BEL Japan, Inc.). Before commencing the measurement, the samples were pre-treated and degassed at 383.15 K for at least 12 hours to remove the moisture and any volatile materials. Employing the volumetric technique, starting with measuring the dead volume of the reference cells and injecting $CO_2$. A target pressure is gradually increasing up to 1 bar, and at each pressure points, the amount of $CO_2$ is measured at each pressure segment.

IOP Conf. Series: Earth and Environmental Science **164** (2018) 012031 doi:10.1088/1755-1315/164/1/012031

## 3. Results and Discussion

### *3.1. Total Organic Carbon*

The total organic carbon content results of the collected samples showed a wide range of variety between 0.52wt. % to12.13 wt. % table 1. The volume of stored gas is generally correlated with total organic carbon of the sample. The gases have high affinity to be stored in the organic matter small pores. Shale gas plays usually exhibits TOC higher than 2 wt.% [1]. However, TOC is not the only indicator of hydrocarbon storage potentiality, and rock eval pyrolysis analysis is needed to identify the more important details of the rock formations such as kerogen types, and thermal maturity [8].

**Table 1.** Location and coordinates of collected samples and toc contents (wt. %).

| Formation | State | Sample ID | Longitude | Latitude | TOC% |
|---|---|---|---|---|---|
| **Batu Caja** | Perak | BG3 | 101°0'7.32" | 4°47'35.25" | 1.72 |
| | | BG4 | 101°0'7.32" | 4°47'35.36" | 2.02 |
| | | BG7 | 101°0'7.32" | 4°47'36.06" | 1.84 |
| | | BG9 | 101°0'7.32" | 4°47'36.27" | 2.06 |
| | | BG15 | 101°0'7.15" | 4°47'55.28" | 0.52 |
| **Kenny Hill** | Kuala Lumpur | KL1 | 3.1745° | 101.6798° | 1.13 |
| | | KL5 | 3.1765° | 101.6798° | 1.71 |
| | | KL7 | 3.1765° | 101.6795° | 1.53 |
| **Dohol** | Johor | JB | 103 11.288' | 01 55. 097' | 1.81 |
| **Setap** | Sarawak | S | 114°00′30.7″ | 03°59′22.1″ | 2.61 |
| **Kroh** | Kedah | KH | 05° 36' 57" | 101°01' 49" | 12.10 |

TOC: Total organic carbon

### *3.2. Bulk Mineralogical Compositions*

Five samples are analyzed using X-ray diffraction XRD that provides the possible mineralogical composition. We applied Bragg equation to determine the D-spacing. Semi quantitative mineralogical composition for the samples were estimated using USGS program RockJock to a degree less than 1 of the fitting between the measured and calculated pattern. Table 2 describes the mineralogical compositions approximation weight percent for the samples. KH bedrock contains the highest of quarts and Kspar compare to other samples. Similar trend is shown in S setap shale but with ~ 10 wt.% obsidian, and no Kspar content. While BG9 and JB have lower quartz compositions (~ 29 wt. % and ~ 32 wt. % respectively), but they contain the highest clays such as illite, chlorite and muscovite.

IOP Conf. Series: Earth and Environmental Science **164** (2018) 012031 doi:10.1088/1755-1315/164/1/012031

**Table 2.** Mineralogical Compositions for Shales Samples.

| Possible Minerals | Weight % & Sample ID | | | | |
|---|---|---|---|---|---|
| | **KH** | **BG9** | **S** | **JB** | **KL5** |
| **Quartz** | 65.9 | 29.3 | 48.0 | 32.5 | 46.6 |
| **Kspar** | 7.2 | 12.8 | - | 1.0 | 1.6 |
| **Plag** | 0.7 | - | 8.5 | 1.5 | 2 |
| **Calcite** | 0.8 | 1.3 | 1.5 | - | 1.3 |
| **Obsidian** | - | - | 10.6 | 0.5 | - |
| **Total non-clays** | **74.5** | **43.4** | **68.6** | **35.5** | **51.5** |
| **CLAYS** | | | | | |
| **Kaolinite** | 2.8 | 6.3 | 3.5 | 4.5 | 1.4 |
| **Smectite** | 9.7 | 12.7 | 5.1 | 6.7 | - |
| **Illite** | 9.7 | 21.1 | 9.1 | 9.0 | 41.5 |
| **Chlorite** | 1.9 | 5.0 | - | 7.8 | 1.6 |
| **Muscovite** | - | 9.5 | 7.3 | 30.7 | - |
| **Dickite** | 0.5 | - | 1 | - | - |
| **Others** | 1 | 2 | 5.5 | 5.8 | 4 |
| **Total clays** | **25.5** | **56.6** | **31.5** | **64.5** | **48.5** |

Clay minerals have the capacity to adsorb gases to their internal Pore structure, the amount is a clay type dependents. For example, Ross and Bustin [9] stated that the small nano-pore structure of 1-5 nm radius, which exists at our samples based on pore system analysis next section, between crystal layers of clay in such orders smectitie, kaolinite, illite provides adsorption sites for gases[10]. This makes the mineralogical composition important for gases adsorption capacity, subsequently the amount of $CO_2$ that might be sequestered in the formations.

### *3.3. Microspore and Mesoporous Systems*

According to International Union of Pure and Applied Chemistry (IUPAC) classifications, Shale pores are classified into; micropores <2nm diameter, mesopore 2 nm<diameter<50 nm, and macropores that includes diamater >50 nm [4]. Figure.1 describes the incremental pore volume for five samples. micropore and mesopore volumes is determined using BJH model which can be found in the literature [7]. The smallest micropore size among the sample is 1.09nm in Setap shale S which also has the smallest average pore size 9.33nm comparing to all other samples. The smallest pores provide the sample with the largest surface area and total micro and meso pores volume. There is a noticeable positive correlation between TOC wt.% pore volume for samples. This suggests that the mesopores and micropore of these formations are created during the migration of organic matter or probably during the hydrocarbon generation[11]. The reason maybe that, KL is illite rich, and illite has the lowest surface area and pore volume comparing to other clays[5, 12]. KL exhibits the lowest pore volume among all samples using nitrogen adsorption. Another reason maybe due to the limitation of nitrogen molecules to access nanopores of specific type of organic matters. In the contrary of CO2 linear molecules structure which have higher potential to penetrate into all organic nanopores volumes. Consequently, it's recommended to investigate the reactive surface area with the adsorbate gas, and to [3]conduct pore system analysis using $CO_2$ adsorption-desorption [13]. Overall, micro and meso pores less than 5 nm is the major contributor to the bulk micropore and mesopores. volume which indicate, these pores would contribute significantly in providing volume for adsorption potential [7].

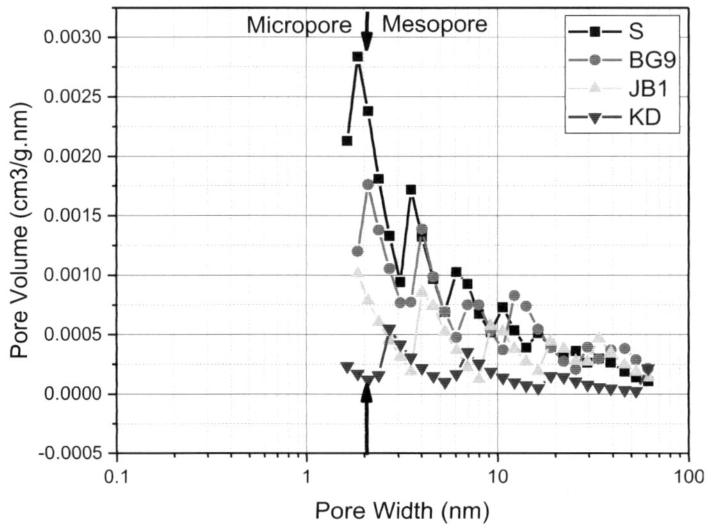


**Figure 1.** Pore-size distribution, pore volume using N2 adsorption analysis. Pore diameter 1.09nm and 100nm

### *3.4. Carbon Dioxide Storage Potential of Shales*

The amount of adsorbed $CO_2$ on all shale samples at constant temperature 298K increased with pressure. The adsorbed measured quantity is shown in figure.2. From the shape of the curves, we could mention the samples show different adsorption models. The highest adsorbent three samples KH JB and BG exhibits mono-layer adsorption property (Langmuir shape). The S and KL1 sample is showing a multi-layer adsorption (BET model)[14]. It is clearly that the amount of $CO_2$ adsorbed is highest at Kedah KH adsorbs ~ 0.8 cc/g where it has the highest TOC content. Organic rich shales has $CO_2$ wettability, while low organic shale has water-wet wettability[15] which is the reason KH has the highest $CO_2$ uptake. Moreover, KH contains higher micro and mesopore volume compare with other samples. Lowest $CO_2$ volume adsorbed for the both KL samples ~0.1 cc/g to ~ 0.22 cc/g at 1 bar. KL5 samples do not show any significant increase when increasing pressure. BG9 and JB samples adsorb almost similar amount of $CO_2$ over the course of pressure values. Both samples have also similar values of specific area and pore volume discussed in the previous section, which indicate the $CO_2$ adsorption characteristics is directly related to the pore system. The smaller pore size provides more volume and surface area for $CO_2$ molecules to be adsorbed. However, this trend seems to disagree with S that exhibits a relatively low $CO_2$ adsorption. This could be due to the $CO_2$ affinity to the mineral or the kerogen type comprised the bedrock[3].

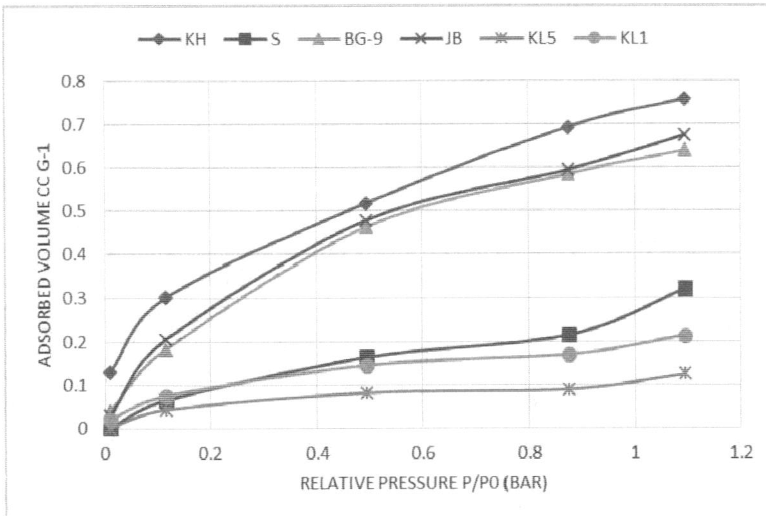


**Figure 2.** $CO_2$ adsorbed on Malaysian Shales at 298 K and at 1 bar.

## 4. Conclusion

TOC analyser, X-ray diffraction, $N_2$ experimental studies were utilized in order to characterize collected samples from different location of Malaysian's black shale formations and to correlate the shale's characteristic's to CO2 adsorption behavior. The characterizations include total organic carbon content, mineralogical composition, micro and meso pore distribution and $CO_2$ adsorption capacity under sub-critical conditions.

The sample's total organic carbon content is between 0.5 to 12.10 wt.%. Generally, the samples can be considered quartz rich shales. Clay minerals also present in all the samples mainly illite, kaolinite, and smectite. Micropores and mesopores size range between 1.1nm to pore size to more than 50nm. The specific surface area SSA as measured with BET is from ~ 2 $m^2/g$ to ~ 12 $m^2/g$. All samples exhibit high volume contribution from pores within a dimeter of less than 5nm. KH sample adsorb the highest ~ 0.8 cc/g, and JB sample comes second ~ 0.7cc/g. KL5 and KL1 adsorb the lowest at ~0.1 cc/g to ~ 0.22 cc/g respectively.

There is a positive correlation between TOC content and pore volume for samples. This suggests that the mesopores and micropores of these formations may have been created during the migration of organic matter or probably during the hydrocarbon generation.

KH exhibits the highest organic content and pore volume, therefore it has the highest $CO_2$ adsorption capacity. There is a positive correlation between $CO_2$ adsorption capacity and total organic carbon content and pore volume for the samples (i.e. KH, JB, and BG-9). It is recommended for future work to evaluate $CO_2$ storage capacity under geological sub-surface conditions high pressure high temperature.

## Acknowledgment.

The authors would like to thank Universiti Teknologi PETRONAS for utilizing the facility, and Shale Gas project grant for financial support.

## References

[1] K. S. Lee and T. H. Kim, *Integrative understanding of shale gas reservoirs*. Springer, 2016.
[2] F. K. YEE, "The palaeozoic sedimentary rocks of Peninsular Malaysia—stratigraphy and correlation," in *Workshop on stratigraphic correlation of Thailand and Malaysia*, 1983, pp. 1-19.

[3] S. M. Kang, E. Fathi, R. J. Ambrose, I. Y. Akkutlu, and R. F. Sigal, "Carbon dioxide storage capacity of organic-rich shales," *Spe Journal,* vol. 16, no. 04, pp. 842-855, 2011.

[4] R. Rezaee, *Fundamentals of gas shale reservoirs*. John Wiley & Sons, 2015.

[5] S. Chen, Y. Han, C. Fu, Y. Zhu, and Z. Zuo, "Micro and nano-size pores of clay minerals in shale reservoirs: Implication for the accumulation of shale gas," *Sedimentary Geology,* vol. 342, pp. 180-190, 2016.

[6] G. Xu, Y. Gao, S. Dong, and D. Wang, "GB/T 19145-2003, Determination of Total Organic Carbon in Sedimentary Rock," ed: Standards Press of China, Beijing, 2003.

[7] G. R. Chalmers, R. M. Bustin, and I. M. Power, "Characterization of gas shale pore systems by porosimetry, pycnometry, surface area, and field emission scanning electron microscopy/transmission electron microscopy image analyses: Examples from the Barnett, Woodford, Haynesville, Marcellus, and Doig units," *AAPG bulletin,* vol. 96, no. 6, pp. 1099-1119, 2012.

[8] K. E. Peters and M. R. Cassa, "Applied Source Rock Geochemistry: Chapter 5: Part II. Essential Elements," 1994.

[9] D. J. Ross and R. M. Bustin, "The importance of shale composition and pore structure upon gas storage potential of shale gas reservoirs," *Marine and Petroleum Geology,* vol. 26, no. 6, pp. 916-927, 2009.

[10] R. Heller and M. Zoback, "Adsorption of methane and carbon dioxide on gas shale and pure mineral samples," *Journal of Unconventional Oil and Gas Resources,* vol. 8, pp. 14-24, 2014.

[11] T. Li, H. Tian, J. Chen, and L. Cheng, "Application of low pressure gas adsorption to the characterization of pore size distribution of shales: An example from Southeastern Chongqing area, China," *Journal of Natural Gas Geoscience,* vol. 1, no. 3, pp. 221-230, 2016.

[12] U. Kuila and M. Prasad, "Specific surface area and pore‐size distribution in clays and shales," *Geophysical Prospecting,* vol. 61, no. 2, pp. 341-362, 2013.

[13] E. Fosu-Duah, E. Padmanabhan, and J. G. Vintaned, "Variations in Specific Surface Area of Some Oligocene-Miocene Shales," 2016.

[14] K. S. Sing, "Reporting physisorption data for gas/solid systems with special reference to the determination of surface area and porosity (Recommendations 1984)," *Pure and applied chemistry,* vol. 57, no. 4, pp. 603-619, 1985.

[15] M. Arif, M. Lebedev, A. Barifcani, and S. Iglauer, "Influence of shale‐total organic content on CO2 geo‐storage potential," *Geophysical Research Letters,* 2017.

# The Environmental Load Characteristic Analysis of LCA-based IPC Girder Bridge

**G C Choi, J S Kim, S Sackey and B S Kim**

Department of Civil Engineering, Kyungpook National University, 80 daehakro Bukgu, Daegu, Korea

Email:choigyeongchan@naver.com;kimjoonsoo@knu.ac.kr;s.sackey123@knu.ac.kr;bskim65@knu.ac.kr

**Abstract.** The carbon emission amount of the Republic of Korea ranks 8th in the world including China and the US and accordingly, the government announced its emission reduction plan by 37% compared to BAU (8.51 million tons) by 2030. As a result of the emission trading scheme that has been implemented since 2015, efforts are being made to reduce environmental load. Accordingly, this study analyzed the material-specific environmental load distribution and environmental impact category for 60 IPC girder bridges of national road constructions in Korea by utilizing the life cycle assessment (LCA) method. The analysis result was found to be in the order of ready-mixed concrete (RMC) (64.00%), rebar (9.1%), timber (7.8%) and plywood (5.6%). As for environmental impact category-specific result, it was found to be in the order of global warming (50.60%), abiotic resource depletion (32.20%), photochemical oxidant creation (POC) (10.10%) and human toxicity (HT) (7.10%), thereby considered to be helpful when selecting the form of bridge while considering environmental load during the planning and design phases.

## 1. Introduction

### *1.1. Background*

At the 21st United Nations Climate Change Conference (COP21) held in Paris in December 2015, the Paris Agreement, a new climate system containing temperature elevation goal, reduction implementation review and climate management fund for developing countries, was signed. Contrary to 1997 Kyoto Protocol that demanded greenhouse gas reduction obligation only for developed countries, it is the first international climate agreement that must be observed by 195 countries, and it substitutes the Kyoto Protocol system that expires in 2020 [1].

The Paris Agreement includes the effort to limit global mean temperature even lower than before industrialization by 2100 [2]. Forecasting greenhouse gas emission to reach the peak by 2030, it presented 'carbon neutral' goal for reaching the equilibrium of greenhouse gas emission and absorption by reducing greenhouse gas through forest greenification, carbon capture & storage and energy technology.

At present time where environmental issue is a global concern, various efforts are being demanded in Korea's construction sector that occupies 40% of the country's total energy consumption [3]. Ultimately, construction industry requires eco-friendly approaches such as eco-friendly material use &

construction method and maximization of heavy equipment energy consumption to solve such issue [3, 4]. Accordingly, Ministry of Environment has been measuring the amount of carbon being discharged by companies by performing product-specific LCA through "carbon labeling" system and in construction industry, environmental impact assessment studies in construction area have been conducted based on LCA method since the late '90s [1].
However, this study analyzed material & construction-specific environmental load characteristics by analyzing 60 IPC (Incrementally Prestressed Concrete) girder bridges.

### *1.2. Scope & Method*
In this study, detailed breakdown information on IPC girder bridges of national road construction in Korea was collected first before computing the materials used for each construction based on breakdown system analysis. Upon quantifying the amount of emission and resources used throughout the product life cycle, the environmental load was computed by using LCA analysis, an environmental impact assessment method for comprehensively assessing their impact on the environment. In this study, the environmental load is composed of eight impacts categorize, namely, abiotic resource depletion (ARD), acidification (AD), Eutrophication (EU), global warming (GW), ozone depletion (OD), photochemical oxidant creation (POC), terrestrial eco-toxicity (TET) and human toxicity (HT), albeit their measurement unit varies. Since there is the need to compare them at equivalent level, environmental impact assessment indexes were assessed through eco-point unit that has been converted through the process of classification, characterization, normalization and weighting [1, 5]. In addition, representative construction types were computed to explain the total environmental load of IPC girder bridges upon identifying major material-specific environmental load to deduce the environmental load characteristics of representative construction types through the eco-point distribution. Fig.1 represents the method and process of the research. The research process starts with the collection of information for further analysis of IPC girder bridge environmental load, analysis of material-specific environmental and through to the deduction of environmental load characteristics.

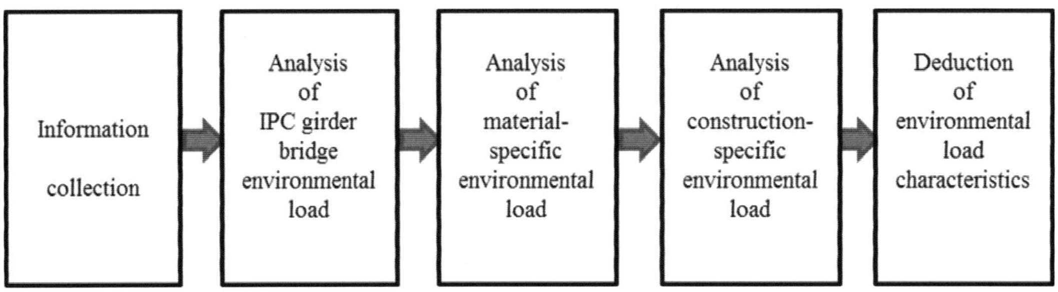


**Figure 1.**Method and process of the Research.

In this study, environmental load was computed from the energy consumption occurring while equipment operation and materials used during the construction phase. Basically, environmental load occurring during the construction phase can be computed if it is connected with LCI DB that can be described as a quantification of environmental load occurring when extracting, transporting and processing raw materials based on the BOQ (Bill of Quantity) prepared through breakdown information. In this study a method was applied by first calculating breakdown cost based on detailed statement before converting resources and energy used into environmental load. In addition, some of the items that have been applied as actual construction cost and items applied as project-specific actual construction cost are very different. Accordingly, National Construction Project Design Guideline and Standard of Estimate were used to apply items of same construction type to calculate energy used while operating equipment and resources used for the same construction type.

## 2. Analysis of IPC girder bridge environmental load characteristics

### *2.1. IPC girder bridge*

IPC (Incrementally Prestressed Concrete Girder) bridge is a multi-step tension type prestressed concrete girder bridge that allows minimum girder height and maximum span up to 60m, while having the advantage of being able to lower the height of existing girder or reduce cross-section area by allowing incremental implementation of prestressing by considering weight increase during construction step. IPC Girder Bridge is cheaper in construction cost as compared to that of conventional structural steel girder because of its lower girder height & weight and long-span construction, and makes repair & reinforcement easy when needed by using unbounded prestressing strand inside PC girder, thereby allowing maintenance & repair cost-savings [6].

### *2.2. Characteristics of data*

To analyze the environmental load characteristics of IPC Girder Bridge, LCA analysis was performed for 60 IPC beam bridge cases. The number of bridges analyzed in this study is 60, and the distribution of bridge span length was 30-45m. In addition, they were divided into 25m 6.67%, 30m 30.00%, 35m 10.00%, 40m 38.33%, 45m 15.00%. In terms of the cut-off level indicating the LCI DB amount percentage of Girder Bridge; it was found to be within 70.26-96.33%, 85.01% mean and standard deviation. For analyzing environmental load upon reviewing various cases of IPC girder bridge, “Facility-specific Carbon Emission Estimation Guideline” presented by Ministry of Land, Transport and Maritime Affaires was applied in this study. This criteria states that “While ensuring over 90% of cut-off level based on material cost when estimating greenhouse gas emission at construction step, assessor may adjust and decide considering the characteristics of project.”, and that “In the case where cut-off level 90% is not achieved when applying both domestic & foreign DB, achievement of at least 80% is required to officially proclaim that carbon emission has been estimated, in which case analysis rate must be clearly presented.” [7].

### *2.3. Distribution of environmental load*

As for the distribution of construction-specific environmental load, it can be said that ‘material’, ‘IPC GIRDER’ and ‘non-shrinkage concrete’ occupy about 90% of the total environmental load. Mould had the least percentage weight of 3.97%. Accordingly, it is thought that the key point of reducing environmental load will be how to control the amount of materials during designing since the majority of environmental load occurs due to RMC and metal products used while constructing IPC Girder Bridge. Fig. 2 depicts the distribution of construction-specific and material-specific environmental load. The left side figure represents the distribution of construction environmental load while the right side figure depicts the material-specific environmental load.

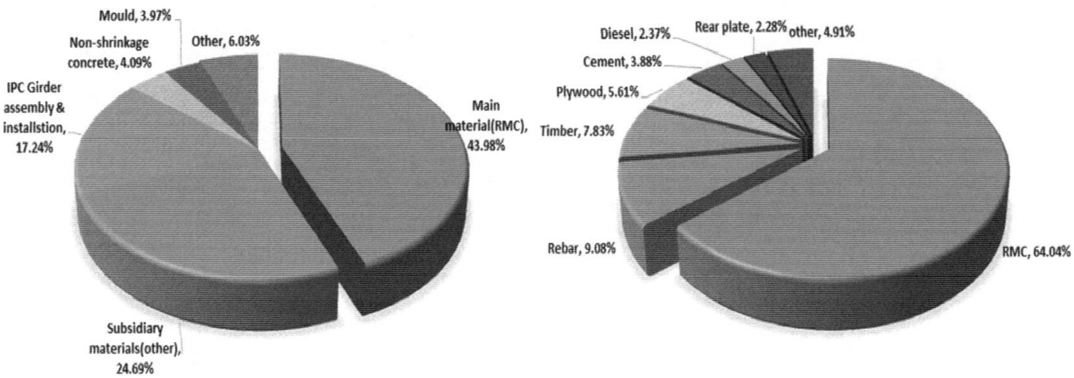


**Figure 2.**Distribution of construction- & material-specific environmental load.

As for the distribution of material-specific environmental load of IPC girder bridge, RMC occupied the highest percentage with 64.04%, followed by rebar with 9.08%, timber with 7.83%, plywood with 5.61%, cement with 3.88%, diesel with 2.37%, rear plate with 2.28% and other with 4.91%.

### *2.4. Distribution of 8 major impact categories*

As for the distribution of 8 major impact category-specific environmental load, it was found to be in the order of GW with 50.60%, ARD with 32.0%, POC with 10.1%, HT with 7.1%, TET with 2.8%, EU with 1.8%, AD with 1.6% and OD with 0.5%. Fig. 3 is the graphical representation of the distribution of 8 major impact categories. From fig.3, the eco-point distribution of GW and ARD represents 82.60% among the 8 major impact categories. Ultimately, it can be known that RMC, metal products and equipment fuel used while constructing IPC girder bridge discharge much environmental pollutants.

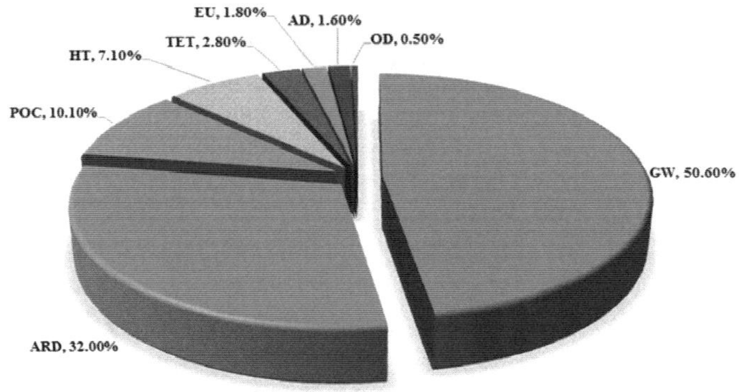


**Figure 3.**Distribution of 8 major impact categories.

In regards to the representative materials with much environmental loads, namely, RMC (64.04%), rebar (9.08%) and timber (7.83%), their major pollutants were researched. As shown in Table 1, the result showed GW (47.41%), ARD (28.74%) and POC (12.82%) to be the major pollutants of RMC. In the case of rebar, GW (51.73%), ARD (39.02%) and POC (4.53%) were deduced, while GW (75.73%), ARD (18.56%) and AD (2.50%) for timber.

**Table 1.** Eco-point of Material-specific Major Pollutant.

| Material | Weighting | | | | | | | | |
|---|---|---|---|---|---|---|---|---|---|
| | ARD | AD | EU | GW | OD | POC | TET | HT | Sum |
| | Eco-point | Eco-point | Eco-point | Eco-point | Eco-point | Eco-point | Eco-point | Eco-point | Eco-point |
| RMC | 3.969E+01 | 1.869E+00 | 6.960E-01 | 6.547E+01 | 8.038E-01 | 1.771E+01 | 6.084E-01 | 1.125E+01 | 1.381E+02 |
| | 28.74% | 1.35% | 0.50% | 47.41% | 0.58% | 12.82% | 0.44% | 8.15% | 100.00% |
| Rebar | 8.108E+00 | 1.894E-01 | 1.473E-03 | 1.075E+01 | 2.935E-02 | 9.413E-01 | 1.863E-01 | 5.737E-01 | 2.078E+01 |
| | 39.02% | 0.91% | 0.01% | 51.73% | 0.14% | 4.53% | 0.90% | 2.76% | 100.00% |
| Timber | 3.646E+00 | 4.910E-01 | 1.038E-01 | 1.488E+01 | 1.537E-02 | 6.914E-02 | 4.313E-02 | 3.997E-01 | 1.965E+01 |
| | 18.56% | 2.50% | 0.53% | 75.73% | 0.08% | 0.35% | 0.22% | 2.03% | 100.00% |

## 3. Conclusion

This study was conducted as an advanced study for establishing a model considering environmental load during the planning and design phases by analyzing the environmental load characteristics of IPC Girder Bridge. The result showed that material-specific environmental load of IPC Girder Bridge was in the order of RMC 64.00%, rebar 9.10%, timber 7.80%, plywood 5.6%, with cement as 3.9% and diesel 2.4%. As for the construction types that can describe environmental load that occurs while constructing IPC girder bridge, 'materials', 'IPC GIRDER' and 'non-shrinkage concrete' occupy about 90% of total environmental load, while GW and ARD were shown as environmental load on earth. In addition, it was not possible to obtain 16.84% of LCA among total construction materials since there are many materials used in construction industry that are not commonly used in other industries. It is thought that conducting a follow-up study after sufficiently obtain LCI DB with supplemented information of construction industry materials will be helpful in establishing a model that assists in making rational decisions considering environmental load during the planning and design phases based on high quality IPC girder bridge environmental load characteristics.

## 4. References

[1] Cho, N. H., Yun, W. G., Lee, W. R., Kim, K. J., 2016, An Analysis of the Characteristics of Environmental Impact for PSC Beam Bridges using Life Cycle Assessment, *Journal of Korean Society Of Civil Engineers*, vol.36, no.2, pp.297-305

[2] Hur, S. S., 2016, Climate Change and Concerted Actions by Mankind *Journal of Korean Social Trend and Perspective,* vol.2016, no.2, pp.214-220

[3] Chun, J. K., Kim, B. S., 2012, The Measurement Method of $CO_2$ Emission for Construction Equipment using ZigBee Sensor *Journal of Korean Society Of Civil Engineers,* vol.32, no.2D, pp.167-174

[4] Kim, B. S., Jang, W. S., 2013, A Study on Comparing the $CO_2$ Emission Estimating Result for Construction Equipment *Journal of Korean Society Of Civil Engineers,* vol.33, no.4, pp.1675-1682

[5] Cha, G. W., 2011, A Study on the Calculation of CO2 Emissions and Environmental Cost according to Construction Waste Types on the basis of LCA, *Master Thesis*, Kyungpook National University

[6] Hwang, H. W., 2011, A study on the work crew-combination for IPC Girder Bridge, *Master Thesis*, Kyungpook National University

[7] Choi, D. S., Lee, M. E., Cho, G. H., 2012, A Study on Environmental Impact Assessment in Domestic Construction Industry Using Life Cycle Assessment *Journal of Korean Institute of Architectural Sustainable Environment and Building Systems*, vol.6, no.1, pp.46-56

**Acknowledgments**
This research was supported by a grant (Grant No. 16SCIP-C085707-03) from Construction Technology Research Program funded by the Ministry of Land, Infrastructure and Transportation of the Korean Government.

# Analysis of Construction-specific Environmental Load Characteristics in the Road Paving Work Maintenance & Management Phase

**J G Im[1], D H Kim[1], J N Liu[1], J Y Park[2] and B S Kim[1]**

[1] Department of Civil Engineering, Kyungpook National University, 80 daehakro Bukgu, Daegu, Korea
[2] Department of Civil Engineering, Gyeongbuk Provincial College, 114 Doripdaehak-gil, Yecheon-eup, Yecheon-gun, Gyeongsangbuk-do, Korea

zakoim88@gmail.com

**Abstract.** In accordance with the Paris Agreement that was signed to cope with recent climate change, the Republic of Korea has presented its greenhouse gas reduction goal as 37% compared to BAU. Accordingly, construction industry that occupies 40% of total industrial energy consumption also needs to make efforts to reduce greenhouse gas. In particular, studies are needed on the effort to reduce emission amount of environmental pollutants that occur during road constructions. In the case of road construction project, which is the subject of this study, large amounts of environmental pollutants are discharged during not only construction process but also operation and maintenance & management process. For analyzing the characteristics of environmental load occurring during the road maintenance & management phase, this study conducted life cycle assessment (LCA) for 10 cases of maintenance & repair projects for local roads that have been constructed in Korea to analyze environmental load characteristics according to construction types. As for the distribution of weighted values, TET and GW showed significant results with respectively 42.45% and 27.65%, followed by ARD (13.67%) and HT (11.00%).

## 1. Introduction

### *1.1. Background*

According to the Paris Agreement signed on recent climate change, Korea has presented greenhouse gas reduction goal as 37% compared to BAU by 2030. The energy consumption of construction industry occupied 40% of total industrial energy consumption, and greenhouse gas emission in construction industry is 2.67 million tons as of 2010 [1]. In addition, it mainly occurs from construction site and construction equipment fuel combustion and production of materials used during construction phase. As the scale of public works has become large, many areas are being performed through mechanization, thereby resulting in large amounts of environmental pollutant emission during construction phase.

In the case of road construction project, in particular, large amounts of environmental pollutants are being discharged not only during construction process but also operation and maintenance & management process [2]. Accordingly, it is necessary to actively respond to estimate and alleviate environmental load occurring not only during construction phase but also during maintenance and repair phase to reduce environmental load [3]. If the characteristics of construction-specific environmental load emission can be identified, environmental load reduction plan can be established based on the information and resource implementation plan can be accordingly established and implemented for materials and equipment [4].

Accordingly, this study was conducted to provide basic information in establishing environmental load reduction plan by analyzing the characteristics of maintenance & repair project-specific environmental load upon estimating environmental load occurring during the road maintenance & repair phase by conducting life cycle assessment (LCA).

### *1.2. Scope & Method*

To analyze the characteristics of environmental load emission during the road maintenance & repair phase, 10 local road construction project cases performed domestically were used. In this study, road paving project of road maintenance & repair phase was exclusively targeted. Based on the analysis of design documents (quantity statement, volume estimation, unit cost estimation and breakdown cost) of each project, materials, and equipment used for each project and amount of energy usage were identified to identify the amount of resources used. For calculating environmental load, LCA method was applied. Fig.1 represents the method and process of the research. The figure illustrates the procedure for computing carbon emission which is divided into three phases of case project selection, construction type analysis, and environmental load computation. These three were combined for further analysis by classifying them into construction-specific environmental load characteristic analysis and after which correlation analysis of environmental load and design information & utilized resources was performed. As for the related database, data that has been established by Ministry of Trade, Industry, and Energy and Ministry of Environment, and international DB were referred to a study conducted by [5].

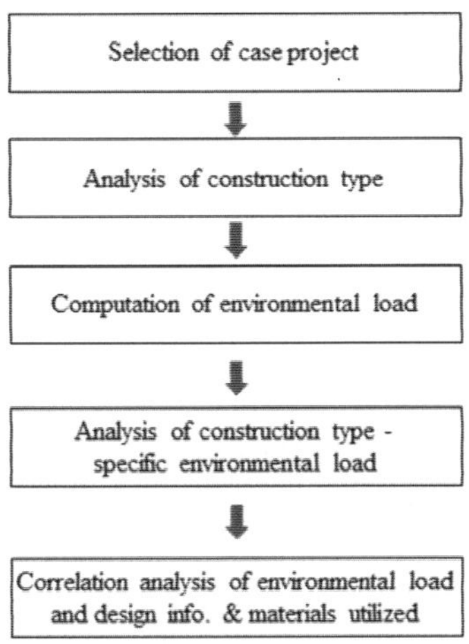


Figure 1. Method and Process of the Research

## 2. Project collection & construction type analysis

### *2.1. Project collection*

As for the cases collected to analyze maintenance & repair cases, 10 cases of projects implemented for road maintenance construction (except for bridge and tunnel repair) in Milyang, Gyeongsangnam-do during 2015-2016, and the overview was classified into construction period, repair width, repair length, repair area and repair construction type.

### *2.2. Construction type analysis & application scope*

The construction type in the road paving maintenance & repair phase can be classified into asphalt concrete pavement overlaying, asphalt concrete pavement cutting and overlaying, asphalt pavement overlaying and cutting and overlaying mixture.

As for its scope of application, overlaying will be performed in the service index (PSI) 2 – 1.1 section according to the PSI-specific construction type application scope specified in AASHTO Road Pavement Structure Design Guideline, and in the service index (PSI) below 1.0 section for cutting and overlaying [6].

## 3. Environmental load estimation & analysis

### *3.1. Environmental load estimation & analysis*

Environmental load was estimated by linking each resource used that has been aggregated with LCI DB (Life Cycle Inventory Data Base). Table 1 provides the total environmental load estimation of maintenance and repair cases. The average percentage values are provided as weighted values.

**Table 1.** Total Environmental Load Estimation of Maintenance & Repair Cases (Weighted).

| Case | ARD | AD | EU | GW | OD | POD | TET | HT | Sum |
|---|---|---|---|---|---|---|---|---|---|
| | Eco-Point | Eco-Point | Eco-Point | Eco-Point | Eco-Point | Eco-Point | Eco-Point | Eco-Point | Eco-Point |
| Sum | 1.09E+01 | 8.15E-01 | 1.02E+00 | 2.23E+01 | 2.29E-01 | 1.82E+00 | 3.81E+01 | 9.46E+00 | 8.46E+01 |
| Average (%) | 13.67% | 0.99% | 1.18% | 27.65% | 0.29% | 2.77% | 42.45% | 11.00% | 100% |

Environmental load occurring during pavement repair work was divided into 8 major impact categories to analyse the characteristics, and the result revealed that TET showed the higher distribution of weighted value with 42.45%, followed by GW with 27.65%, ARD with 13.67%, HT with 11.00%, POC with 2.77%, EU with 1.18% and AD with 0.99%.

Fig. 2 is the graphical representation of environmental load according to 8 major categories and environmental characteristics distribution according to materials and construction type. The right side figure represents the environmental load according to 8 major categories while the left figure represents the environmental characteristics distribution according to materials and construction type. Meanwhile, in regards to the environmental load distribution according to the 8 major impact categories of road construction projects, ARD and GW show the highest environmental load distributions with respectively 37.15% and 31.34% [2].

As for its reason, it is thought that environmental load of GW and ARD was significantly assessed from fossil fuel consumption since environmental load occurring in each construction type consisting road construction project such as earthwork, slope stabilization, drainage works, pavement work, traffic safety facility work and subsidiary work is mainly determined by the consumption amount of oil used for construction equipment.

In regards to the characteristics of environmental load caused by major processes and materials that consist of pavement repair work, on the other hand, 80.82% of environmental load occurring from entire pavement repair work is caused by asphalt concrete material, while 11.26% and 6.40% were caused respectively by RMC material and paint for lane repainting.

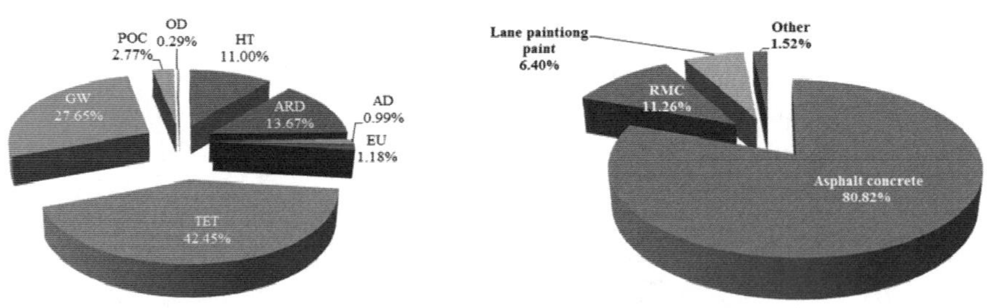


Figure 2. Environmental load distribution according to 8 major impact categories; environmental load characteristics distribution according to materials & construction types.

In regards to the percentage of environmental load occurring from asphalt concrete, RMC and paint materials, it was 98.48% of total environmental load occurring from pavement repair work. In addition, the amount of environmental load of pavement repair work that occurs from construction work was found to be extremely insignificant and it was also found to be occurring from main materials used.

As shown in Table 2 that displays the sum of eco-point of material-specific pollutant for the degree of material-specific pollutant, the result showed TET 54.70% and GW 23.81% for ASCON, GW 47.41% and ARD 28.74% for RMC and ARD 49.55% and GW 41.35% for lane painting paint, thereby showing significant percentages of TET, GW and ARD among the 8 major impact categories.

Since measurement unit of environmental load measured through each impact category varies, environmental impact assessment index was assessed in this study through eco-point unit that converted characteristic value into weighted value.

**Table 2.** Sum of Material-specific Pollutant Eco-Point (Weighted).

| Material | ARD | AD | EU | GW | OD | POC | TET | HT | Sum |
|---|---|---|---|---|---|---|---|---|---|
| | Eco-Point | Eco-Point | Eco-Point | Eco-Point | Eco-Point | Eco-Point | Eco-Point | Eco-Point | Eco-Point |
| ASCON | 4.07E+00 | 5.53E-01 | 8.64E-01 | 1.64E+01 | 1.68E-01 | 5.71E-01 | 3.78E+01 | 8.62E+00 | 6.91E+01 |
| | 5.89% | 0.80% | 1.25% | 23.81% | 0.24% | 0.83% | 54.70% | 12.48% | 100.00% |
| RMC | 2.59E+00 | 1.22E-01 | 4.54E-02 | 4.27E+00 | 5.24E-02 | 1.15E+00 | 3.97E-02 | 7.34E-01 | 9.00E+00 |
| | 28.74% | 1.35% | 0.50% | 47.41% | 0.58% | 12.83% | 0.44% | 8.15% | 100.00% |
| Lane painting paint | 1.31E+00 | 1.20E-01 | 4.96E-02 | 1.10E+00 | 1.10E-03 | 2.44E-02 | 1.21E-03 | 4.45E-02 | 2.65E+00 |
| | 49.55% | 4.54% | 1.87% | 41.35% | 0.04% | 0.92% | 0.05% | 1.68% | 100.00% |

*3.2. Environmental load design info. & resources utilized correlation analysis*

Table 3 shows the correlation analysis result between environmental load and pavement repair work construction period, repair width, repair length, repair area, repair material, repair construction type.

**Table 3.** Correlation Analysis.

| Division | Coefficient of determination ($R^2$) | Pearson correlation coefficient |
|---|---|---|
| Pavement repair length | 0.794 | 0.891 |
| Pavement repair area | 0.976 | 0.976 |
| Asphalt concrete amount | 0.924 | 0.961 |

The correlation between pavement repair length and environmental load showed 0.794 coefficient of determination ($R^2$) and 0.891 Pearson correlation coefficient.

Next, correlation between pavement repair area and environmental load showed 0.952 coefficient of determination ($R^2$), which is an improvement by 20% compared to pavement length, and 0.976 Pearson correlation coefficient, which is an improvement by 10% compared to pavement length. Accordingly, increase in the pavement repair area was found to have significantly contributed in linearly increasing environmental load of pavement repair work.

Lastly, correlation between asphalt concrete amount and environmental load showed 0.924 coefficient of determination ($R^2$) and 0.961 Pearson correlation coefficient.

Upon collecting cases on overlaying, cutting and overlaying and mixed constructions of pavement repair construction works, environmental load was estimated and correlation of case-specific characteristics & main attribution information and environmental load was analyzed. The result showed that environmental load occurring as a result of asphalt concrete material quantity was dominant for pavement repair work, along with high correlation between repair length and repair area.

## 4. Conclusion

This study examined the construction type-specific environmental load characteristics of road maintenance & repair for 10 cases of domestic local roads, and the result of environmental load

distribution according to 8 major environmental impact categories showed a concentrated result of TET and GW with respectively 42.45% and 27.65%.
In regards to the characteristics of environmental load caused by main processes and materials consisting of pavement repair work, 80.82% of environmental load occurring in the entire pavement repair work was from asphalt concrete material, while 11.26% and 6.40% were respectively from RMC and lane painting paint.
Compared to existing road construction, it is shown that environmental load occurs in the majority of material items in road maintenance & repair work, along with an extremely insignificant amount of environmental load caused by construction work. In addition, it was found that there is a close correlation between environmental load and pavement repair length, pavement repair area and asphalt concrete amount.

## 5. References

[1] Park, J. Y., 2017, An Environmental Load Estimation Model for Road Earthwork in the Planning Phase of National Highway Project Based on LCA, Ph.D. thesis, Kyungpook Nat. University, Korea
[2] Park, J. Y., Kim. B. S., 2016, Development of Environmental Load-Estimating Model for Earthwork in the Road Planning Phase, *Proceeding of Journal of Korean Society of Civil Engineering,* pp.98~99
[3] Wi, S. D., 2016, Maintenance of package, Han Young, Korean
[4] Cha, G. W., 2014, A Study on the Calculation of CO2 Emissions and Environmental Cost according to Construction Waste Types on the basis of LCA, Master Thesis, Kyungpook National University, Korea
[5] Kim, S. R., 2017 Analysis of Environment Emission Characteristics Each Construction Type for Road Field *Proceeding of Journal of Korean Society of Civil Engineering*, vol.37, no. 143-151
[6] Southgate, H. F., 1991, Sensitivity study of 1986 AASHTO Guide for Design of Pavement Structures

**Acknowledgment**
This research was supported by a grant (Grant No. 16SCIP-C085707-03) from Construction Technology Research Program funded by the Ministry of Land, Infrastructure, and Transportation of the Korean Government.

# The Impact of Taiwan's Urban-renewal Policies, Laws and Decrees on Urban Redevelopment and Urban Disaster Prevention

**Jen-Chih Chao**[1, a], **Kuo-Wei Hsu**[2, b]

[1]PhD program in Architecture and Urban Design, Department of Architecture, Chaoyang University of Technology, Taiwan
[2]Department of Landscape and Urban Design, Chaoyang University of Technology, Taiwan

E-mail: [a]jenchihchao@gmail.com; [b]kwhsu@cyut.edu.tw

**Abstract.** Highly developed cities dating from 1970s of Taiwan have recently encountered a set of problems commonly known as urban decay, resulting in the relocation of urban business districts and high levels of out-migration. This study examines the impact of the various urban-renewal policies, laws and decrees that have been enacted to remedy this situation, as well as the effects of these measures on urban disaster prevention. Based on historical data, textual analysis and field research, it concludes that these governmental strategies, while important to the goal of urban renewal, have also been the cause of new urban-development problems such as traffic congestion, extreme population density, and deficiencies in the amount open space. This is traced to a lack of integration of large-scale development goals and strategies with specific urban-renewal areas' plans Part of the problem appears to be that the Taiwanese government usually adopts only a floor area incentive strategy to promote public participation. The authors recommend that it deploy a more diverse array of strategies, including but not limited to tax incentives, accelerated renewal processes, and regional rewards. The authors also highlight how real-estate development enterprises' design strategies and quality-control processes for residential construction can serve the goals of urban disaster prevention as well as urban renewal.

## 1. Introduction

### *1.1 Background and motivation*

In the late 1990s, many in Taiwan began advocating urban-renewal policies as a means to achieving the goal of urban redevelopment, and issues surrounding such policies have since been widely discussed. Their supporters have cited benefits including the regeneration of blighted urban areas, improvements to urban quality of life, the promotion of sustainable development, and urban disaster prevention. In addition to passing various measures with these aims in mind, the government has provided a range of incentives to promote the planning and implementation of related projects including green buildings and new public spaces.

Urban development in Taiwan has been threatened by global-warming-related disasters, particularly flooding, as well as by urban heat-island effects and earthquakes, among other problems. Nevertheless, return on investment (ROI) and other economic benefits have consistently been seen as key to urban-renewal decision-making there, and this has generally led planners to ignore disaster-prevention, quality-of-life and sustainability considerations.

### *1.2 Purposes of the Study*
Urban-renewal strategies have considerable potential to aid disaster prevention and mitigation, urban sustainable development, quality-of-life, public interest in people and the development of a green economy. To date, however, most studies relating to urban-renewal issues have focused on rights transfers, financial risks, details of policy, and the people involved in urban-renewal initiatives. Few have examined urban-renewal policies, laws and decrees from the perspective of the urban disaster prevention. Thus, this study will be guided by the following two research questions:

1) *The influence of urban-renewal policy, laws and decrees on Urban-renewal.*
2) *The influence of urban-renewal policy, laws and decrees on urban disaster prevention.*

## 2. LITERATURE REVIEW

### *2.1 History of urban renewal in Taiwan*
Taiwan's Urban Renewal Act has been in force m and its genesis is divided into several distinct phases by the Construction and Planning Agency (CPA) of the Ministry of the Interior (CPA website) [1]. These are discussed in turn in the following six sections.

1) *The legislation: urban renewal policy, 1997-1999*

A 1974 amendment to the Urban Planning Act titled "The urban renewal of the old city districts" is recognized as the first milestone of urban-renewal legislation in Taiwan. However, both the modes of implementation that it specified – compulsory collection and sectional expropriation – were insufficient to their purposes, due to lack of funds and personnel. Therefore, a system of encouraging private investment initiatives was launched, under the general rubric Construction Agreement between Property Owners and Developers. Still, problems in acquiring land and other property remained an immense obstacle to effective urban renewal.

2) *Reconstruction following the 921 earthquake, 1999-2003*

A violent earthquake of magnitude 7.3, later named the 921 earthquake, struck Taiwan on September 21, 1999. In its northern and central cities in particular, many people became homeless and numerous buildings were in desperate need of reconstruction. Incidentally, however, the catastrophe provided an opportunity to test the Urban Renewal Act and other elements of the existing legal and economic framework for regeneration projects The experience of radical physical reconstruction in the disaster-hit areas led to the Act and its implementation system being modified considerably, with the objective of creating safer as well as higher quality living environments.

3) *Urban renewal acceleration scheme, 2004-2008*

To enhance Taiwan's domestic prosperity, environmental protection and urban functionality, the government launched the 2005-2008 Urban Renewal Demonstration Project in 2005 and the Urban Renewal Acceleration Scheme in 2006. Love-Taiwan Infrastructure Project, 2009-2014

In 2009, the Ministry of the Interior launched the 12-part Love-Taiwan Infrastructure Project, targeting six key areas of the five largest cities in Taiwan, and announced the aim of completing it within eight years. Originally proposed by the CPA in 2009, this project's objectives were to promote comprehensive urban renewal along three pathways: 1) from "base redevelopment" to "area redevelopment" to "urban regeneration"; 2) from "reconstruction" to "rehabilitation and conservation"; and 3) from "investment" to "community self-help".

The CPA's urban-renewal projects reflect the Taiwanese government's enthusiasm for integrated approaches to spurring economic and industrial development, improving employment opportunities, eliminating obstacles to private-sector participation in urban-renewal projects and urban sustainability. However, conflicts of interest between public-sector and private-sector participants resulted in a range of challenges, to the point that the question of how to build a mutually beneficial cooperation mechanism for these two sectors emerged as an intractable problem.

4) *Ideas and strategies for disaster prevention as an aspect of to urban renewal, 2013*

The urban development of older areas of Taiwan has been noticeably impacted by climate change and natural disasters such as earthquakes. The CPA's Urban Renewal Division, created in 2013, proposed that local governments propose ideas for disaster-prevention/urban-renewal pilot projects and build the relevant operational mechanisms and models. Then, government would evaluate disaster-prevention-type areas relocation and renewal, and oversee changes in urban special projects, urban design regulations, disaster prevention strategies, and so forth. For low-level and medium-level disaster areas, the reconstruction, rehabilitation conservation, and shock-resistance improvement of old buildings would be implemented. The CPA would also draft subsidy notices and raise funds for improving the shock resistance of pre-1970 buildings, and assist local governments to set up dedicated disaster prevention agencies.

*5) CPA urban-renewal projects, 2015-2018*

The urban-renewal projects that have been proposed by the CPA from 2015 onward are based on the National Development Projects (2013-2016), which aimed at "living justice": defined as improving the living environments of Taiwanese people, enhancing the attractiveness and competitiveness of towns and cities, improving the speed of responses to seismic disasters, and maintaining the momentum of urban renewal. The CPA also defined the five key objectives of sustainable urban renewal, as: 1) carefully monitoring relevant urban-renewal laws and regulations; 2) ensuring that urban-renewal pilot projects are well-integrated with one another' urban-renewal pilot projects; 3) promotion of urban renewal, with the wider aim of spurring regional redevelopment; 4) encouragement of non-governmental implementation of renewal projects 5) ensuring that information about urban-renewal projects is transparent, and that popular participation in them is as wide as reasonably possible.

We can divide strategies for the evaluation of levels of disaster risk confronting different urban areas into three categories: prevention and control, preparation and response, and land development and adaptation. Calculating floor area ratio (FAR) and assessing land-use strength are two fairly simple ways of prioritizing areas' environmental status, which in turn can accelerate the whole process of urban renewal. However, it should be noted that there is some conceptual overlap between disaster-prevention urban renewal and green infrastructure. Disaster-prevention construction strategies, meanwhile, include not only building techniques *per se* – e.g., shock-resistance measures, green roofs, and energy-saving designs and materials – but also the layouts of public facilities and schemes of reorganization for regional ensembles. In other words, sound implementation of urban renewal cannot be executed on a building-by-building basis.

The urban-renewal development approach currently espoused by the CPA has been improved by past experience that revealed the problems of existing urban-renewal processes: in particular, low levels of public participation, lagging private-sector investment, and lack of transparency. These were successfully addressed through adjustments in policy orientation and loosening of regulations in a manner that satisfied the needs of the public sector, private sector, and general public. The focus of urban renewal in Taiwan shifted to climate change, urban disaster prevention, and sustainable development that balances the needs of society, the environment and the economy.

### 2.2 Urban-renewal Literature

Despite an increasing number of urban-renewal policies having been based on the perspective of urban disaster prevention in recent years, studies of such policies remain rare.

Urban-renewal research, including studies focused on environmental quality and urban disaster prevention, is very diverse: involving aspects of culture, society, economics, ecology, and policy studies. Lien (2015), based on an overlay analysis, found that Taichung City's Central District had the highest potential for renewal as well as the highest disaster risk, making it a suitable model for urban renewal aimed at disaster prevention in Taichung City as a whole [2]. Chiu (2016), meanwhile, found that urban-renewal processes aimed at disaster prevention tended to follow the basic requirements of law; that the public had a weak grasp of urban disaster prevention; and that the Urban Renewal Advisory Committee was not adequately focused on the disaster-prevention aspects of its role [3]. Wu (2014) reported that residents of an area tend to think that environmental quality is improved as a result of urban renewal projects, but nevertheless assign low scores to their general feelings about such projects as well as to the quality of life in their own homes [4].

In short, while urban renewal is an important strategy for promoting local development, research has indicated that its successful implementation must acknowledge local people's cultural, social, economic and ecological orientations, along with the targeted area's disaster-prevention and sustainability needs, due to the increasing frequency and magnitude of natural disasters. Thus, urban renewal has taken on new connotations, as redevelopment and disaster prevention become ever more important to it, whether at the level of single buildings, neighborhoods, or entire regions.

## 3. RESEARCH METHODS

This study focusing on general urban-renewal business issues and renewal projects' impacts on the environment, society, and economy. Our preliminary research indicated that the problems of urban renewal are more complicated that such un-systematized approaches allow. This study therefore applies content analysis and field research as well as literature analysis to its examination of the influence of Taiwan's urban-renewal policies, laws, decrees, and actions on urban disaster prevention there.

### *3.1 Content analysis*

Content analysis is utilized to explain the current status of a phenomenon and its development over a particular time period, through data obtained from the relevant literature. It is a combination of quantitative and qualitative analysis, insofar as it involves deducing qualitative change from quantitative change; and it stresses objectivity, systematization and quantification.

## 4. RESULTS

Recent urban-renewal studies have mostly adopted qualitative research methods, especially field research and literature review. Some have examined development performance, literature on financial assessment and quantitative methods, while others have mainly focused on qualitative methods.

### *4.1 The Roles of urban-renewal policies, laws, and decrees in urban development*

*1) The influence of urban renewal on land development in pre-1970' urban areas*

In older districts of cities, urban renewal generally takes place after the end of a period of prosperity that has left behind a complex cultural legacy, and land-development strategies for such areas can therefore be challenging. Problems that emerge during these processes can involve property rights, heritage matters, and the strongly held views of residents. In addition, the sheer length of the development process has always tended to reduce the enthusiasm of potential investors and business enterprises for such schemes. Therefore, Taiwan's government has provided various types of floor area incentive (e.g., green build and open spaces) when facing m and dispersed the construction of important public facilities across different areas to encourage public use of them.

*2) The government's emphasis on policies, laws and decrees as means to achieving the goal of urban renewal*

To accelerate the process of urban renewal in older urban areas, Taiwan's government has provided floor area incentives and floor areas transfers, set up area-specific urban-renewal strategies, and promoted bulk-reward projects themed around urban renewal for sustainable development. However, such reward policies have been difficult to implement, and have failed to attract the anticipated levels of participation from businesses and other investors.

*3) Urban renewal's emphasis on cooperation between the public and private sectors, and its financial approach to attracting public participation*

By its nature, the process of urban renewal is lengthy, and this inevitably influences land-investment and development businesses' costs and ROI. This, in turn, has a profound impact on the willingness of such enterprises to participate in it. On the other hand, because mutual trust between the public and investment/development enterprises is unstable, information about finance is not always correct. Diligent government inspection of each enterprise's implementation process tends to increase such trust as well as, somewhat counterintuitively, the willingness of such enterprises to cooperate with the government. The government has also exhibited a willingness to provide a variety of flexible policies for different urban-renewal environments. In 2009, for example, it proposed urban renewal in 30 areas, and allowed the public to participate in the implementation phase; this decreased the financial pressure on the government while at the same time improving the working relationship between the public and private sectors.

*4) Urban renewal's stress on internalization of external benefits and acceleration of older-area land development*

Taiwanese public-sector bodies tasked with promoting urban renewal have tended to stress the external benefits of urban disaster prevention: by measuring its performance after urban renewal is accomplished, the influences and development for each areas, and the environmental improvement of decayed areas. From the perspective of environmental, economic, cultural and social influence on the promotion of urban disaster prevention and urban sustainable development, the essential benefits are not the buildings that are constructed during the urban-renewal period, but the long-term impact of such projects after they have been accomplished.

### *4.2 The roles of urban-renewal policies, laws and decrees in urban disaster prevention*

The urban disaster prevention aspect of urban-renewal implementation and promotion in Taiwan is principally based on the Urban Renewal Act, which has been gradually amended and expanded to include legal interpretations regarding urban-renewal business operations. An Urban Renewal Operations Manual has also been published, aimed at promoting urban renewal program [5]. In addition, local government bodies have begun publishing local urban-renewal-related self-governance articles, ostensibly to reflect their unique social, economic and cultural situations, though in fact we found that the differences between such sets of articles were few. Accordingly, we have focused on the rules and regulations related to disaster prevention represented by the Urban Renewal Act and the Regulations for the Periodical Overall Review of Urban Planning, and will analyze them using the designation method and disaster factors. After this analysis, we will sum up the key issues and explain their respective relationships to urban disaster prevention.

*5) Designation analysis: the urban renewal act and urban disaster prevention*

The Urban Renewal Act divides urban-renewal areas into two types: prioritized and direct-designated. In prioritized urban-renewal areas, the safety and quality of construction in the face of danger factors such as fire, earthquake, and flooding are enhanced from a single building for area disaster prevention. Though this focuses more on post-disaster measures, it has been found effective in the prevention of serious disasters, as well as the cooperation of important buildings in direct-designated areas. (According to the Ministry of the Interior Constructionand Planning Agency designated priority urban renewal areas.)

*6) Designation analysis: regulations for the periodical overall review of urban planning and urban disaster prevention*

In addition to setting forth general principles of urban design and the development of land for public facilities, this document contains strategies and principles for the periodic review of urban disasters, the designation of prioritized urban-renewal areas, and the establishment of urban disaster prevention and measurement projects.

More specifically, it sets forth basic modes of investigation of the history and characteristics of renewal areas at risk from disasters and the estimation of potential future disaster scenarios. Through such analyses, it is possible to identify the development issues and problems confronting particular projects and propose strategies to solving them. This, in turn, facilitates the prioritization of urban-renewal areas. When processing detailed regulations, the principles of urban ecology, water and green designs are crucial elements in decreasing the likelihood of urban disasters such as fires and floods: for example, through the planning of firebreaks and measures to prevent rainwater infiltration. When processing urban design, urban-renewal projects in older areas are expected to adhere in detail to urban-design principles such as configurations of space for disaster prevention and disaster relief, and the inclusion of facilities that will decrease the impact of disasters on the urban environment, such as greening, water retention, environment-friendly and recyclable facilities, and so forth. When processing detailed regulations for overall review, the urban-renewal fields will be delimited by CPA, and factors such as public safety and shabby buildings investigated and analyzed as a whole. Delimiting urban-renewal areas, renewing basic policies and bring in regulations of plans are the executions for the prioritized areas. Specifically, the surface area of sites for public facilities within old urban-renewal areas should be attain to more than 1 hectare; and taking development into consideration comprehensively, the total surface area of public open space, public facilities, and urban disaster prevention space shall not be less than 10% of the surface area.

In short, a close analysis of the Urban Renewal Act and Regulations for the Periodical Overall Review of Urban Planning suggest that Taiwan's systems of planning urban development and urban disaster prevention do decrease the impact of different disaster factors on areas with dilapidated buildings and/or prior histories of disasters occurring. In the face of the environmental impacts of climate change and natural disasters, the foci of urban disaster prevention range from single buildings to regional-scale response plans for fires, earthquakes and floods. However, having looked at the range, designs, promotion and execution of Taiwan's urban-renewal projects and overall development strategies, it is clear that the single building is the subject of renewal in most cases, and that real-estate development enterprises make efforts to obtain the best floor area incentives that they can under the existing the laws and decrees, in part because this enables them to attract more residents. The inadequacy of an area's urban-renewal strategy will have knock-on effects on the renewal's completion as well as on its urban disaster prevention measures. For example, in the case of insufficient amounts of land being allocated to emergency shelters and open space, many other problems emerge, including traffic congestion and urban heat-island effects. Thus, however hard we try to improve the safety and quality of individual buildings, serious systemic problems are likely to remain.

## 5. CONCLUSION

### *5.1 The influence of urban-renewal policies, laws and decrees on urban development*

Through content analysis of the relevant literature and field research, we can conclude that there are four main characteristics of the relationship between Taiwan's urban development and its urban-renewal policies, laws and decrees. These are:

1) *The strong influence of urban-renewal policies, laws, and decrees on land development in older urban areas.*

2) *The government's emphasis on policies, laws and decrees as the means to achieving the goal of urban renewal.*

3) *Emphasis on cooperation between the public and private sectors, coupled with a broadly financial approach to attracting the public to participate in urban renewal.*

4) *Urban renewal's stress on the internalization of external benefits and the acceleration of land development in older areas.*

Therefore, to solve the problems of urban disaster prevention, urban-renewal policies should do more to promote public participation; and the increasing negative environmental impact on cities of urban-renewal policies should be regulated. Moreover, the participation of the public should not be promoted solely through floor area incentive strategies; in addition to or instead of such rewards, the government could provide tax incentives, shorten renewal times, offer region-specific rewards, and provide more public facilities. Lastly, more should be done to encourage real-estate developers to control the design quality, build quality, and volume of their projects.

### *5.2 The influence of urban-renewal policies, laws and decrees on urban disaster prevention*

Urban-renewal policies, laws and decrees should incorporate urban sustainable development as the core of the Taiwanese government's comprehensive development strategy and its implementation, albeit with an awareness that profit maximization on the part of real-estate development enterprises will always make this difficult, since it tends to result in urban over-crowding. It is largely for this reason that urban-renewal initiatives aimed at mitigating urban disaster risk have become the sources of further disaster-related vulnerabilities. Therefore, greater harmony among urban-renewal policies, laws and decrees should be sought, and ideally amount to a clear statement of an overarching strategy aimed at the planning of people-oriented environments, public facilities and public space.

## 6. References

[1] Construction and Planning Agency, Ministry of the Interior website: http://twur.cpami.gov.tw/reports/dl.aspx?MP=1&DLTYPE=3

[2] Lien, Hung-I (2013). Identifying disaster preventive type urban renewal: A case study in Taichung City. MA thesis, Feng Chia University Department of Urban Planning and Spatial Information.

[3] Chiu, Chia-Wen (2016). The relationship between the planning process of urban renewal and urban disaster prevention: The case of Wanhua District, Daan District, Zhongzheng District in Taipei City. MA thesis, Ming Chuan University Department of Urban Planning and Disaster Management.

[4] Wu, I-An (2014). A study on the impacts of living quality of urban renewal projects on the surrounding area from local residents' perspective. MA thesis, National Cheng Kung University Department of Urban Planning.

[5] Construction and Planning Agency, Ministry of the Interior: Urban Renewal Operations Manual website: http://uract.cpami.gov.tw/module.php?i=rule&type=A31&rid=A3101097115000000

IOP Conf. Series: Earth and Environmental Science **164** (2018) 012035 doi:10.1088/1755-1315/164/1/012035

# Design and Implementation of Earthquake Emergency Response System Based on Grid Data

**J Zhao[1], Z H Ma[1]**

[1] Shanxi Earthquake Administration, 69 Jinci Road, Taiyuan, Shanxi 030021, China

jiezhao2009@qq.com

**Abstract.** Generally speaking, the estimate of casualty, economic loss and damage of building or structures after earthquakes is modelled through the evaluating seismic influence field and the data which is administrative units for census and statistics. Thus, the evaluating results have good application in large scale, but are not accurate enough in the case of post-earthquake disaster assessments. The data based on kilometre grid can better embody the spatial heterogeneity of elements. And then it would improve the accuracy of the assessments. The paper introduces the design and implementation of earthquake emergency response system based on the kilometre grid data in Shanxi province of population, economy, and housing from several facets, such as system architecture, technology implementation, database design, functional modules and algorithms, system workflows and product details. Taking the 2010 M4.8 Hejin earthquake as an example, the evaluation is fast, precise, and spatial heterogeneity of the disaster pre-assessment is obvious..

## 1. Introduction

The earthquake is one of the natural disasters with the greatest impact on human activities. Within a short period of time after the earthquake, details of the disaster will be not too clear. In addition to relying on limited disaster information, the government also needed to rely on the earthquake emergency auxiliary decision results issued by earthquake emergency command centre to launch the emergency response level and formulate the emergency rescue strategy. During the period of the Tenth Five-Year Plan, China seismological bureau has established a nationwide earthquake relief headquarters technology system, which played an important role in the past earthquake emergency events. But the system used the areal weighting to assess the building's collapse and casualties in disaster areas, i.e., the statistics of the population, economy and housing in the administrative unit was treated uniformly according to the area of the unit. Due to the obvious spatial aggregation and directionality of population, economy, housing and earthquake influence, by this method sometimes there would be low accuracy and large errors in the results of disaster assessment. Taking the administrative region as the unit, the statistics in the unit was distributed into the grid according to certain mathematical model [1], the law and the spatial distribution of the population, economy and the building will be easily expressed through the statistics of grid gradiometer data. Using the grid statistical data for earthquake emergency and rapid assessment can improve the estimation precision.

This paper designs a set of earthquake emergency response system based on kilometre grid data, the system uses C/S architecture, integrates the earthquake emergency basic database, professional evaluation model library and Office software into geographic information system (GIS) to assessment the earthquake casualties and distribution of the disaster area, economic losses, building's collapse and so on rapidly, meanwhile outputs standardized reports and thematic maps. This paper expounds the design and implementation of the system from overall system design, database design and function module design.

## 2. Overall design of system

### *2.1 System architecture*

The system adopts object-oriented software design, integrates GIS seamlessly with office software and the professional model. The application scenarios of this system have high security confidentiality, rapid deployment, high portability and other requirements [2]. The system uses the mature C/S three-tier architecture which is shown in figure 1. According to the general working process and characteristics of the rapid assessment of earthquake emergency, the system design is divided into data layer, functional service layer and application layer. Data layer mainly includes regional earthquake relief headquarters earthquake emergency database [3], which contains 42 small class and 9 big categories data, kilometer grid population, economic and housing data, e.g., basic geographic information, social and economic statistics data, engineering seismic data in earthquake, the background of disaster influence, factors related to disaster, disaster relief, emergency contact and earthquake emergency plans and so on. Functional service layer mainly includes the disaster area and impact assessment model, seismic intensity evaluation model, population casualties estimates model, rescue strength evaluation model, the economic loss assessment model, earthquake emergency auxiliary decision document generation services and so on. The application layer mainly includes personnel casualties and distribution assessment, housing damage assessment, rescue force assessment, economic loss assessment, earthquake emergency thematic map and other auxiliary functions.

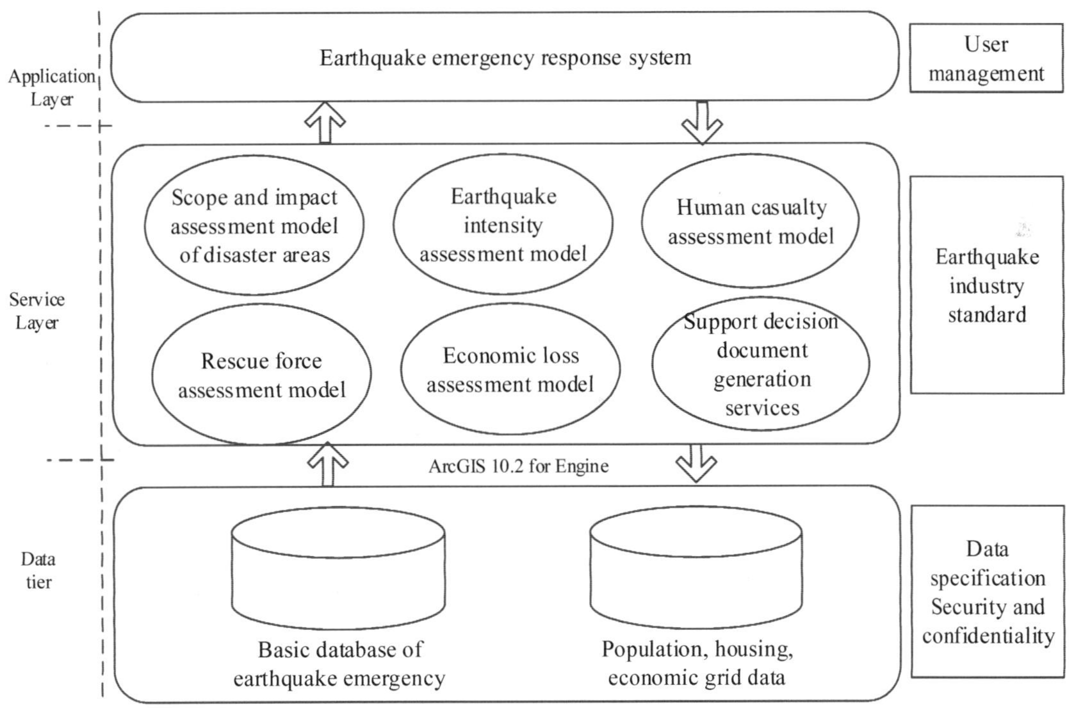


**Figure 1.** System architecture

### *2.2 Technology implementation*

Earthquake emergency response system with high complexity integrates a variety of professional evaluation models and partial functionality of GIS and office. In practical application scenario, it is required to response quickly, have high availability, high compatibility and good portability. It adopts mature, general and stable technical route in system development. The underlying support technology of the system is GIS, and the ArcGIS Engine Developer Kit 10.2 is used as the base platform for the second development of GIS. ArcGIS Engine is a desktop GIS library product for developers and uses Microsoft's COM specification. It provides some map controls and Command button for many common functions that can be used to implement a map-centric application system [4]. ArcGIS Engine is a very mature GIS secondary development product, has been widely applied in the field of mapping, meteorology, emergency, and other industries construction in the domestic and overseas . During the period of the tenth five-year, the secondary development platform used in national earthquake emergency command system was ArcGIS Engine. The daily work environment is Windows.

In order to install and deploy expediently, the.net platform was used as the base development environment, and DevExpress was used as the UI control to make the system application interface beautiful and concise. The technological implementation methods are shown in table 1.

**Table 1.** Technical methods of system implementation

| technique formal | technical method | description |
|---|---|---|
| development model | C/S | system overall architecture |
| development platform | .NET，C# | development framework an d programming language |
| development tools | Visual Studio 2013 | software for coding and testing |
| GIS platform | ArcGIS 10.2 for Engine，ArcMap 10.2 | GIS secondary development platform |
| operational environmen t | Windows XP and above， Microsoft Office 2003 and above，ArcGIS Engine Runtime 10.2 | client operational environm ent |

## 3. Database Design

There are many data types involved in earthquake emergency response system. According to its data format and usage scenarios, the database of this system mainly includes:

### *3.1 Earthquake emergency basic database*

The data contains the administrative division data of provincial, city and county, township and village five levels, the traffic data of high speed, state roads, provincial roads, county roads and other, historical earthquake data, as well as population, economy, geologic disaster danger point, hazards, scenic spots and historical sites, the rescue force data and so on. These data are stored in the Geodatabase based on vector format. Geodatabase is the primary data structure of ArcGIS, which is the main data format for editing and data management, including file geographic database, personal geographic database, and ArcSDE geographical database. File geographic database is suitable for the processing of GIS projection data, and used for personal and small groups with very high performance. It can be extended to store large amounts of data without having to use DBMS, also can be ported across multiple operating systems. The earthquake emergency basic database is stored in the file geographic database format.

### *3.2 Kilometer grid database*

It includes the population of kilometers grid, GDP, housing data, DEM data of 30m resolution, high-resolution historical remote sensing image data, etc. Considering the usage scenario of raster data,

data volume size, and the portability of the system, raster data is stored as the file in the file system [5].

## 4 System function module

The function of earthquake emergency response system is to evaluate the damage loss rapidly after a destroyed earthquake. In the unclear case, it just needs users input earthquake three elements (time, location, magnitude) to trigger rapid assessment function [6] ,then the system will obtain information about earthquake coverage and disaster area, casualties and injured population, damage degree and distribution of buildings in a short term with the support of mathematical model, expert knowledge base and basic geographic information, and generate the basic situation of the earthquake zone, disaster briefing and earthquake relief work three main reports. In addition, the system will automatically produce the provincial emergency command center earthquake emergency thematic map, which will be used by the unit and the field task force. The functional modules of the system are shown in figure 2.

### *4.1 Earthquake triggering*

In the "setting up earthquake" interface, the user can input the earthquake's longitude and latitude, the earthquake time, focal depth, and the magnitude of the earthquake to start the rapid assessment process, at the same time, it can be used to modify the earthquake fault diagnosis parameters and obtain rupture length. After setting up parameter, the system can produce the disaster report, the basic situation of the earthquake zone and the auxiliary decision-making document in minutes, and the emergency special plan for the earthquake is started in the background. In addition, the system supports the dynamic modification of the size and direction of the affected fields based on the sequence information of the aftershock and the disaster, and supports to import external impact field for assessing the disaster loss, then exports vector format of impact field documents for use by other working groups.

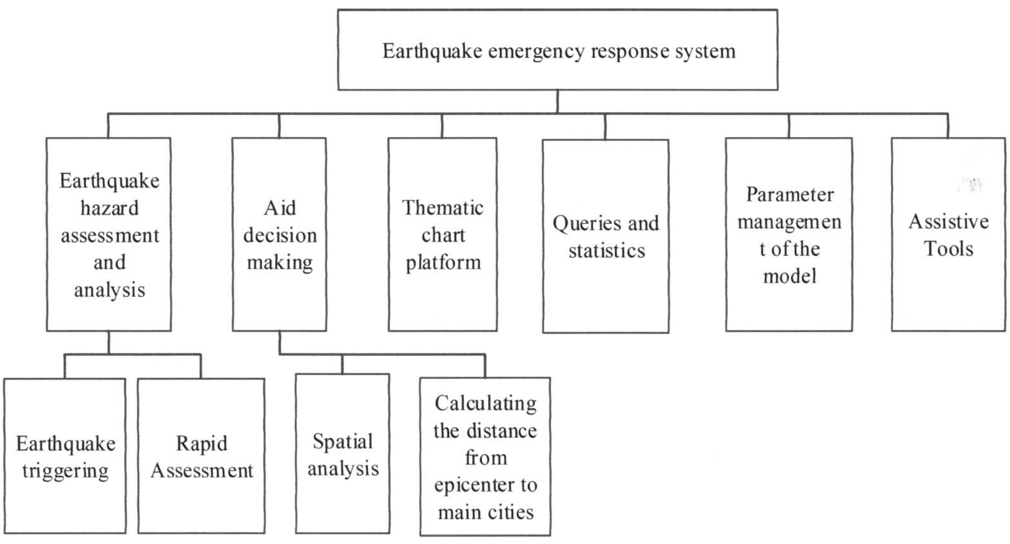


**Figure 2.** System function modules

The assessment of building earthquake damage is the basis of rapid assessment of earthquake damage. The damage of all kinds of houses in the grid cells can be obtained by combining the data of the grid and the vulnerability matrix of the houses in the area. The types of houses are divided into four categories, including the houses built of wood, soil, stone, brick and brick houses, single-story or multi-story masonry houses, reinforced concrete houses, and other classes. The damage grade of the house is basically intact, slight damage, moderate damage, severe damage and destruction of five grades. The death toll is estimated based on the housing damage situation [7]. The empirical formula adopted is as follows:

$$N_D = RP \quad (1)$$

ND is the death toll; P is the total population of grid units; R is the mortality. The relationship between mortality and house collapse rate is shown by the following formula:

$$\lg R = 12.479C^{0.1} - 13.3 \quad (2)$$

Among them, C is house collapse rate.

The number of injured were estimated based historical damage experience to be between three and five times the number of people killed.

*4.2 Aid decision making*

Path analysis includes Shortest path analysis, optimal path analysis, proximity facility point analysis, etc., which can provide auxiliary decision-making suggestions for personnel evacuation path development and distribution of relief supplies. The toxic gas spreading simulation function can provide help for the gas leak disposal of hazardous sources such as chemical factories after the earthquake. In addition, the system integrates the buffer analysis and distance calculation from the epicenter to the cities, counties and towns.

*4.3 Thematic map production*

The user can modify thematic diagram category and order through thematic drawing management function, and design the layout and layer elements of thematic diagram according to the actual demand of earthquake emergency.

Through thematic map output Settings, the user can decide the category of the output thematic map category whether to add the earthquake magnitude, and whether the thematic map is scaled to the influence scope of the earthquake or the resolution of outputting thematic map, etc. After outputting thematic diagram automatically, users can output other thematic drawings on their own through the feature graph output function.

*4.4 Queries and statistics*

Users can query the attribute information of the target layer by drawing points, rectangles, polygons, and round, or get some elements of a layer by fuzzy query.

*4.5 Parameter management of the model*

With the increase of earthquake cases, the continuous improvement of the database, the improvement of research and analysis methods, and the development of localization research, the model and parameters of earthquake emergency assessment need to be adjusted dynamically. With model parameter adjusted, the model coefficients of the intensity attenuation coefficient, building earthquake damage matrix, building reconstruction parameters, building loss ratio, economic loss ratio and human casualty assessment model are adjusted by users.

*4.6 Assistive tools*

This function module provides the functions of map operation functions such as zooming and bookmark management in GIS, and also provides some measure functions such as positioning, length measurement, area measurement and interface adjustment, etc.

## 5 Application case

There was a 4.8-magnitude earthquake in the Hejin City of Shanxi Province at 10:36 a.m. on January 24, 2010. The epicenter was located in urban district of Hejin City with a depth of 12km. According to the emergency plan, the Earthquake Administration of Shanxi Province immediately initiated the emergency response of the magnitude 4 earthquake, the relevant departments immediately activated the system to carry out emergency assessment of earthquake disasters,

Authorities immediately launched this system for earthquake disaster emergency assessment. The results were produced within two minutes. It concluded with three documents: earthquake disaster briefing, earthquake relief work recommendations and basic situation of earthquake zone.

According to the earthquake disaster briefing, the earthquake intensity of the magistoseismic region was Ⅵ degrees, involving one county with a population of about 77,711 people. This GDP of zone was about 85 million yuan. It was estimated that 0 people were killed and 3 people were injured because of damage to collapsed houses. The earthquake may have caused the destruction of 30 buildings, and varying degrees of damage of 4,400 buildings.

According to the earthquake relief work recommendations, the commander provided the following work suggestions: according to Shanxi earthquake emergency plan, this earthquake relief work was led by Yuncheng city. Suggestions were as follows. Firstly, the provincial seismological bureau assisted the earthquake relief headquarters in the disaster area to carry out earthquake emergency response and was responsible for earthquake site emergency work. Secondly, Shanxi earthquake emergency rescue team should always be prepared. Thirdly, other departments assist disaster area to do a good job in earthquake relief.

The basic situation of the earthquake zone mainly contained the following items: the population quantity, total economy and area of the districts involved, the distance from the main administrative residence to the epicenter, earthquake zone's history earthquake catalogue, traffic conditions, geological disaster risk points, nearby rescue forces list, schools, major hazard installations, and the summary table of material requirements including emergency resettlement supplies, rescue team demands, medical personnel and facilities requirements, clothing and food needs, relief and housing costs, etc.

In addition to the above detailed assessment results, thematic maps were outputted synchronously. They were the epicenter location map, seismic intensity effect field initial evaluation distribution map, the population of the earthquake zone map, activity faults and historical earthquakes distribution map, the distribution maps of schools hospitals, major hazard sources, rescue forces and reservoirs, etc.

## 6. Conclusion

The rapid assessment of post-earthquake disaster is the focus of earthquake emergency and rescue. The application of kilometer grid data in disaster assessment is the research direction after the establishment of the national earthquake emergency command system during the tenth five year plan. There are generally three major types of date utilized by earthquake emergency, consisting of population, GDP and buildings. Grid transformation is to divide administrative units data into certain spatial scaling grid basing on mathematical model. GRID-structured data can show the spatial distribution and differentiation law of geographical elements. Accurate data are the basis for post-earthquake rapid assessment. This paper expounds the design and implementation of earthquake emergency response system based on the kilometer grid data from the aspects of system architecture, technical route, and database design and system function module. The software system uses C/S architecture, which meets the requirements of rapid deployment, portability and safety of the system. It integrates the work flow of post-earthquake emergency command center into the system and can quickly produce the evaluation results and thematic drawings to provide helpful suggestions for the government's earthquake relief work. Taking Hejin M4.8 earthquake in 2010 as an example, the software system evaluated the earthquake rapidly, the users obtained a high accuracy simulation results with expert advice and preliminary disaster and had a fundamental grasp of personnel casualties, building damage situation and spatial distribution pattern in the early period of the earthquake, which provided support for earthquake emergency rescue operation. Compared with the conventional method, the method described by this paper is capable to get better results that is accurate and reflects the spatial distribution of casualties, financial loss, population buried below the pressure and broken houses. This is also more conducive to the deployment of rescue operations.

However, the accuracy of current grid transformation models is limited. The grid data is always static. Improving the accuracy and dynamicity of the models need further research. Social media registration data can reflect dynamic changes in the activities of the crowd. The map POI (points of interest) data can embody the Urban Function Partition. To build a grid transformation model considering these above two parameter will improve data quality. The spatial scale of grid data is another influence on the post-earthquake disaster assessment. To choose the appropriate space scale is a valuable research issue.

**Acknowledgments**

This project was supported by 2017 The Spark Program of Earthquake Technology of CEA (XH17009Y).

**References**

[1] H Y Fu, M C Li, J Zhao. A review of the research progress of population data grid model [J]. HUMAN GEOGRAPHY,2006,21(3):115-119,114.

[2] Y Xu. Comparison and analysis of network architecture system based on B/S and C/S [J]. Computer Knowledge and Technology(Academic Exchange),2005,(4):28-31.

[3] J Liu , B J Chen, L Zhang, etc. Base database construction and update experience of Guizhou earthquake emergency command system based on ARCGIS [J]. Urban construction theory research: electronic edition, 2011(25):1-6.

[4] Y L Chen, S H Li, B Shi, etc. *ArcGIS development authority guide* [M]. Beijing: Publishing House of Electronics Industry, 2015:31.

[5] Esri. ArcGIS Help 10.2- the type of Geodatabase [EB/OL]. (2014-05-10)[2016-08-22]. http://resources.arcgis.com/zh-cn/help/main/10.2/index.html#//003n00000007000000.

[6] H Y Xiao, Z Q Du, Q Zhu, etc. Design and implementation of the distributed disaster emergency collaborative research system [J]. Geospatial information, 2014(4):7-10.

[7] J Q Lin , C Q Miao, Y S Li, etc. GB/T 30352-2013《Earthquake disaster emergency assessment》[S]. Chinese Specification Press: China Earthquake Administration, 2014:11.

# Research Direction in Realizing Sustainable IoT Based Smart City Ecosystem

**Iqbal Yulizar Mukti[1], Yudha Prambudia[2]**

[1] Enterprise Resource Planning (ERP) Center, Universitas Widyatama, Jl. Cikutra No. 204A, Bandung
[2] Center for Sustainable Systems (CSS), Universitas Widyatama, Jl. Cikutra No. 204A, Bandung

iqbal.yulizar@widyatama.ac.id

**Abstract.** In order to achieve sustainable goals within urban context, IoT based smart city is a popular concept to endeavor. The utilization of IoT technology is the backbone of this concept. This paper explores the current development of IoT based smart city ecosystem and suggests some issues to be followed up. In this ecosystem, stakeholders are basically categorized into Users, Devices Producer, Middleware Developer and Apps Developer. It is found that the future challenges of IoT based smart city are circling around coordination activity among these stakeholders.

**Keywords**. smart city ecosystem, sustainability, internet of things.

## 1. Introduction

Nowadays, most of the people on this planet live in urban areas. According to the United Nations, 54% of world's population was living in urban areas in 2014, and it is projected that 66% of world's population will live in the urban area by 2050 [1]. This urbanization trend has a direct impact on the growth of the cities and creates significant challenges for the cities' services and infrastructure, including scarcity of resources, inadequate infrastructures, energy shortage, human health concerns, and demand for better economic opportunities [2]. It is approximated that the cities consume 75% of the world's energy and 60% of all water allocated for human use. Moreover, it also produces 80% of greenhouse emissions [3]. This rapid expansion of the cities has leaded the urgency to manage the cities in sustainable manner, a harmony in the aspects of profit, people, and planet.

To manage cities in sustainable manner, government needs to orchestrate multiple systems at once in concert. These systems are including several city governance domains like city administration, transportation, education, health care, public safety, housing, infrastructure, and utilities [4]. Since the systems are essential for urban living, those systems should perform uninterruptible while maintaining its services quality. To this end, tremendous management effort is required to monitor, report, and intervene the systems [3].

In general, management of the systems mentioned above is assigned to the relevant private or public-sector entities [3]. In such condition, coordination between the entities is often difficult and inefficient because of various obstacles in technical, administrative and political domain that may leads to performance issue of the city systems [3]. To address this coordination issue, a new way in governing the city that can integrate the entities in managing city's systems is need to be developed.

## 2. Literature review

### *2.1. Smart City*

A new way in governing the cities is by utilizing intelligent devices and broadband technologies in the smart city concept [3,5]. There are range conceptual definitions of smart city. It is differentiating due to different context and circumstances in each case of smart city implementation [4]. However, [6] has laid the foundation of smart city definitions. They defined smart city as an instrumented, interconnected, and intelligent city. *Instrumented* means that the city has set of instrumentation devices enables the capture and integration of live real-world data. *Interconnected* means the city has capacity to integrate the information gathered from the instrumentation devices into an enterprise-computing platform and communicate the data into various city services. *Intelligent* refer to the ability in analyzing the interconnected information into insights that drive decisions and actions that improve process outcomes or system, organization, and industry value chains.

Further, [6] construct 5 layers of smart city structure as shown in **Figure 1**. In this structure, it is shown that the primary information comes from a bunch of sensors that capture live real-world data to be further analyze to in the upper layers to guide or automate decisions or actions to improve a certain city function.

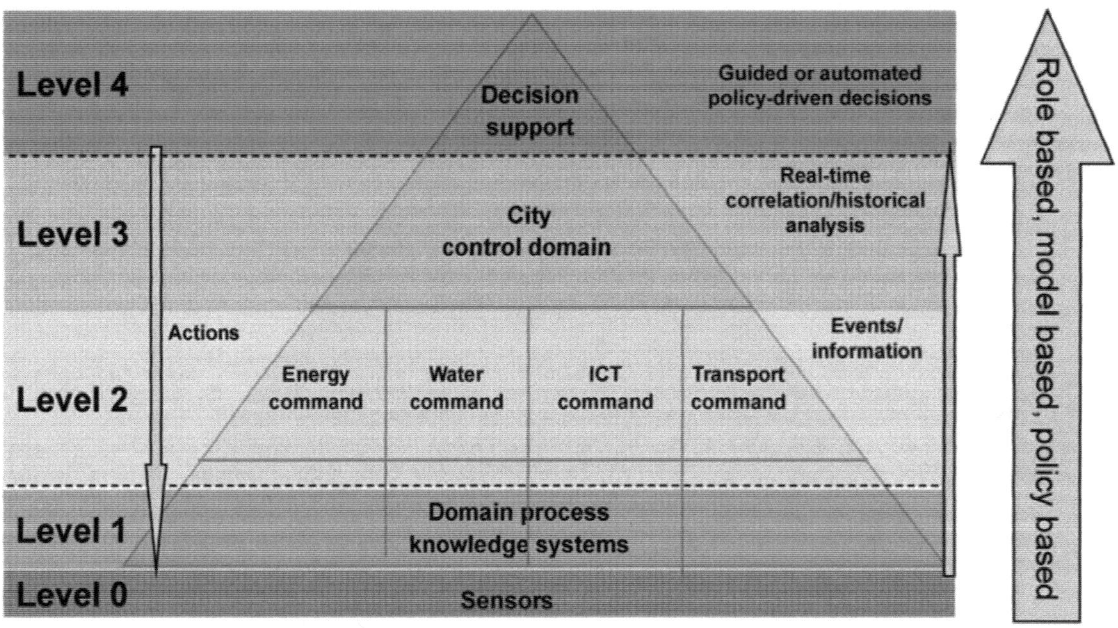


**Figure 1.** Smart city structure [6]

From this structure, interconnected sensors that generally referred the internet of things (IoT) is inevitably required as the inherent feature of the smart city, since it has ability to automatically share data, interact and combine services whenever and wherever is required [3].

### *2.2. Internet of Things*

IoT is an evolution of the current internet into a network of interconnected objects that harvesting data from the physical environment that become source of information of analytical tools, applications and services in certain domain [7]. There is a definition for IoT that do not restrict in any standard of communication protocol such that allow smart technologies to be developed and deployed using available state-of-the art protocol at any given point in time [7]. The definition is,

> "Interconnection of sensing and actuating devices providing the ability to share information across platforms through a unified framework, developing a common operating picture for enabling innovative applications. This is achieved by seamless large scale sensing, data analytics and information representation using cutting edge ubiquitous sensing and cloud computing".

Based on the above the definition, [7] summarize three components of IoT, namely: a) Hardware, in which consist of sensors, actuators and embedded communication hardware b) Middleware, in which consist of on demand storage and computing tools for data analytics and c) Presentation, in which consist of visualization and interpretation tools which can be widely accessed on different platforms and can be designed for different applications.

IoT devices applied in city infrastructures supply enormous amount of data, generally defined as big data, to be further utilizes in various applications supporting city operations. To be able to utilize these big data efficiently and secure, managing these big data flow requires orchestration from dedicated parties known as IoT operator [8]. Particularly in relation with big data flows, [8] suggest four layers for IoT ecosystem in smart city context as shown in **Figure 2**.

As shown in **Figure 2**, at the first layer, there are city infrastructure and citizens as the objects and users that become the enablers of smart city ecosystem. A city infrastructure is where cities' utilities are operates (e.g., energy facilities, water source, street light, parking area, road, etc) and citizens are the users of various smart city services that enhance their life quality. From this first layer, data harnessed by IoT devices implemented in the second layer, and then be collected by middleware solution in the form of standardized APIs that enable interaction with data generated by the IoT devices in open data model at the third layer. Based on the open data delivered by standardized APIs, at the fourth layer, public private partnership (PPP) entity is formed as the IoT service operator to govern the data delivery to third-party application developer. By using apps-store like model, IoT service operator grants API access for third party developer to develop their applications, analytical tools, and services. The developed apps then will be distributed to users by apps store like Apple apps store or Google play store. Regulated access by the IoT service operator can beneficial to citizens since it will generate services that give positive impact on citizens' life quality. On the other hand, utilities provider will also get benefit by improved control over their infrastructure [8].

Many cities around the world have rolled out smart city initiatives based on IoT technologies. This initiatives are leveraging smart technologies in improving city's services, including transport infrastructures, energy distribution networks, and natural resources. For example, City of Melbourne, Australia rolled out smart city initiative in three focus groups of potential IoT applications, which are citizens (health care, emergency services, defence, and crowd monitoring), transport (traffic management and infrastructure monitoring) and services (water, building management, and environment) [4]. Another example, in Santander, Spain, more than 5000 IoT devices already deployed both at static locations (streetlights, facades, bus stops) as well as on mobile vehicles (buses, taxis) [9]. This IoT project, SmartSantander is a unique, city scales IoT experimentation project in developing applications and services for a smart city [3,9]. Several services are already deployed on SmartSantander platform are outdoor parking management, environment monitoring, participatory environmental sensing, irrigation and garden monitoring, and point of interest augmented reality, and energy smart metering [9].

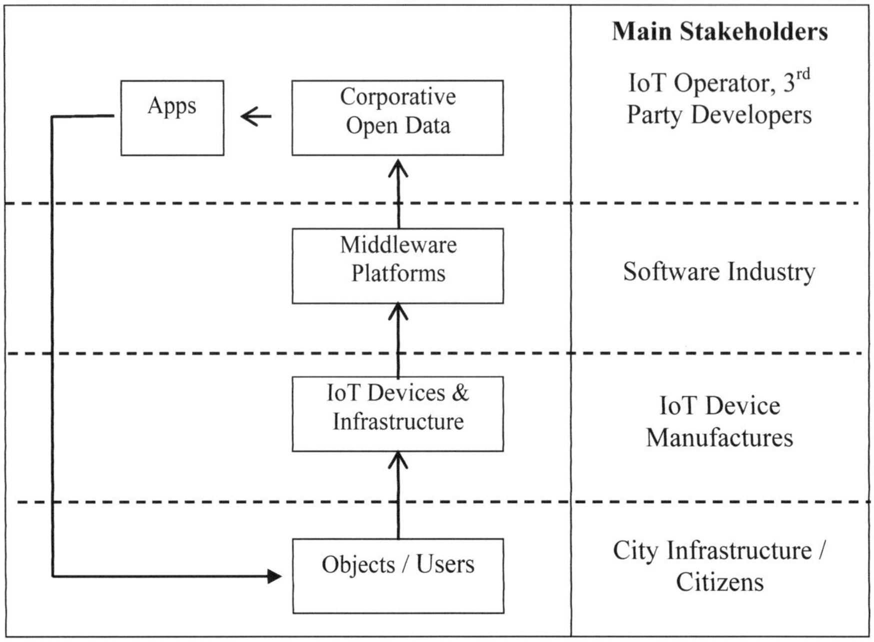


**Figure 2**. IoT ecosystem in smart city context [8]

### 3. Discussion

Based on literature study to the selected articles, there are several challenges in smart city digital ecosystem need to be address such that smart city's services can be sustain and beneficial for all involved stakeholders. Adopted from [8], this paper proposes to categorize these challenges in three categories, namely governance, financial, and technological. Challenges related with interaction of stakeholders within the ecosystem are grouped in governance category, while related with financial and business aspect of the ecosystem are grouped in financial category, and on the other hand challenges related with technical and technological aspect of the ecosystem are grouped in technological category. **Table 1** lists the challenges in each category.

**Table 1.** Smart city challenges

| Category | Challenges | Author |
|---|---|---|
| Governance | Cooperation between involved stakeholders | [8,10,11] |
| | Users or organizational diversity | [11] |
| | Resistance to change | [11,12] |
| | Data and privacy policy | [8,12,13] |
| Financial | No clear business model | [8,10,13,14,15] |
| | High investment and continuation of funding | [8,12] |
| | Lack of appropriate return on investment methodologies and metrics | [8,10] |
| Technological | Big data handling | [8,16,17,18,19] |
| | Interoperability & integration | [8,10,11] |
| | IoT devices standard | [12,13] |
| | Technology infrastructure limitation | [12] |
| | Security & privacy threats | [11,12,13,15] |

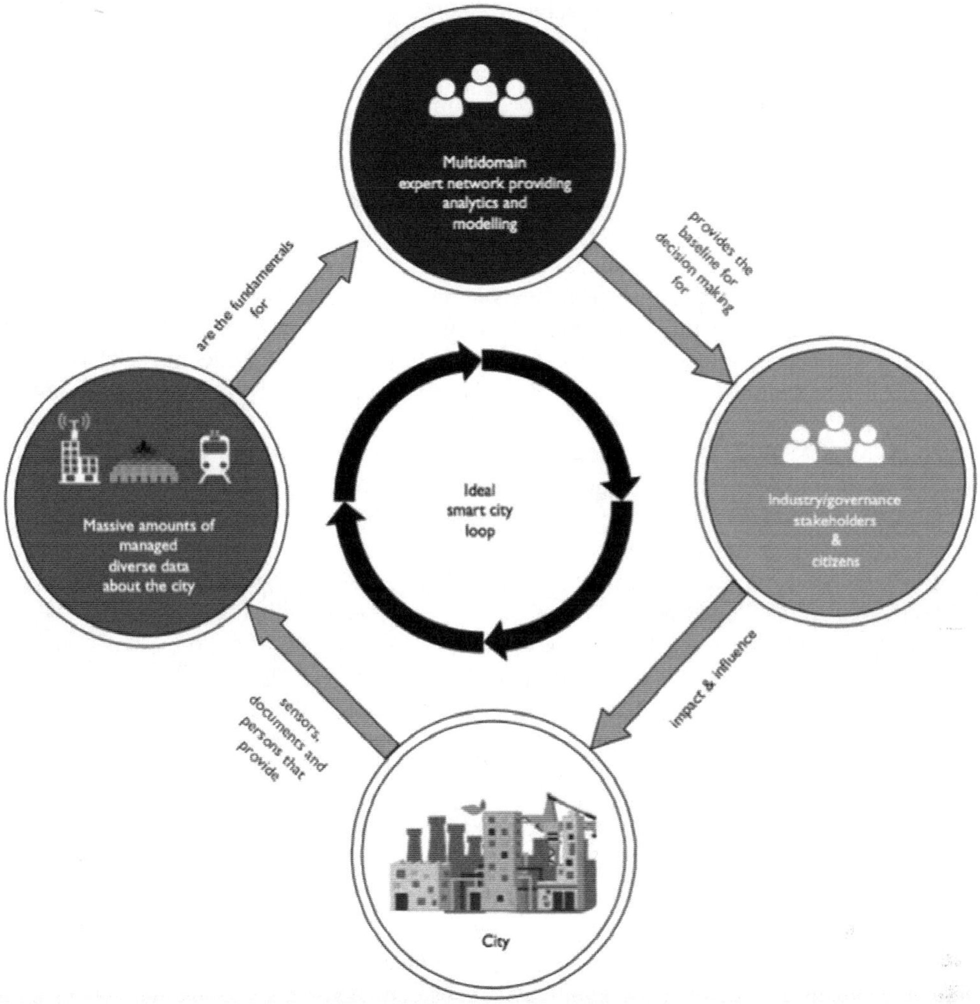


**Figure 3**. Smart city loop [21]

Particularly in governance category, most of the challenges are related with coordination of the stakeholders within the ecosystem and its derivatives, including challenges related to user behavior and data policy. These are important challenges, since at its core, smart city is not one-way initiative by the government, but it is massive interaction between numbers of stakeholders, components, services and technologies [20]. Smart city loop shown in **Figure 3** shown simple illustration of the interaction [21].

To be fully benefit from new possibilities brought by smart city, it is important to get the stakeholders involved in smart city loop cooperate in a coordinated way [20,21]. Further, research shows that coordination between the involved stakeholders in smart city ecosystem is the key to realize smart city benefit and to ensure long-term sustainability of smart city initiative [8,10,11,12,20,22].

To provide direction in addressing coordination issue smart city ecosystem, this paper proposes several research questions to be further elaborate. These research questions are the following:

**Question 1:** Who are the stakeholders involved in the ecosystem of IoT based services in smart city context, including their roles and interests?

This questions is proposed to define clearly all the stakeholders involved in the in the ecosystem of IoT based services in smart city context, particularly in relation with big data flows. Clear definitions of involved stakeholders including their roles and interests in the ecosystem are important to get comprehensive coordination architecture. Desired outcome in answering this question are list of all involved stakeholders in the ecosystem, roles of each stakeholder in the ecosystem, interests of each stakeholder in the ecosystem.

**Question 2:** How the coordination architecture between stakeholders should be arrange, such that satisfy the interests of each stakeholders?

This question is proposed to define coordination architecture that address relationship structure between stakeholders incorporating their interests in the ecosystem of IoT based services in smart city context. To be able to orchestrate the stakeholders in the ecosystem properly and ensure long-term sustainability of the ecosystem [22], the coordination architecture should encompassed business model that beneficial for each of the stakeholders, in which including functions performed by each stakeholder, services offered, funding structure, and business terms between involved stakeholders [22].

Desired outcome in answering this question is to have coordination architecture of stakeholders in the ecosystem of IoT based services in smart city context, encompassing business model of the ecosystem that beneficial for each stakeholder.

**Question 3:** What are the performance criteria that can evaluate sustainability of the IoT based services ecosystem in smart city context based on defined coordination architecture?

This question is proposed to define performance criteria of the coordination architecture developed in previous question. The goal of the performance criteria is to measure the impact of the coordination architecture with certain configured parameter, to sustainability of IoT based services ecosystem in smart city context. Desired outcome in answering this question is to have performance criteria that reflect the impact of the coordination architecture to sustainability of IoT based services ecosystem in smart city context.

**Question 4:** How to evaluate the coordination architecture, such that can achieve certain level of defined performance criteria?

This question is proposed to defined methodology to evaluate the coordination architecture to achieve certain level of performance criteria defined in previous question. Simulation modeling using system dynamics or agent based model (ABM) technique is proposed methodology to be further explored to evaluate the coordination architecture. Desired outcome in answering this question is to have a simulation model developed using system dynamics or ABM technique that can be parameterize to evaluate the coordination architecture to achieve certain level of performance criteria.

Elaborating on the above research questions hopefully can provide direction in addressing coordination challenges that has become significant impediment in realizing sustainable IoT based smart city ecosystem.

## 4. Conclusion

This paper proposes several research questions as recommended direction for further research in addressing challenges in smart city ecosystem, especially in coordination domain as the significant impediment in realizing sustainable IoT based smart city ecosystem. Four research questions mentioned in this paper, ultimately leads to the needs of architectural perspective of IoT based smart city ecosystem that can represent relationship of involved stakeholders in the ecosystem. Further, this architectural perspective can be the based to develop simulation model that can be parameterize to evaluate the performance of the ecosystem.

## References

[1] United Nations 2014 World Urbanization Prospects (New York: Department of Economic and Social Affairs, United Nations)

[2] Doug Washburn and Sindhu Usman 2010 Helping CIOs Understand “Smart City” Initiatives (Cambridge: Forrester)

[3] Evangelos Theodoridis, Georgios Mylonas, and Ioannis Chatzigiannakis 2013 Developing an IoT Smart City Framework, Fourth International Conference on Information, Intelligence, Systems and Applications (IISA) (Piraeus, Greece)

[4] Taewoo Nam and Theresa A Pardo 2011 Conceptualizing Smart City with Dimensions of Technology, People, and Institutions, Proc. of the 12th Annual Int. Conf. on Digital Government Research (Washington DC)

[5] Hans Schaffers et al. 2011 Smart Cities and the Future Internet: Towards Cooperation Frameworks for Open Innovation (The Future Internet)

[6] Jayavardhana Gubbi, Rajkumar Buyya, Slaven Marusic, and Marimuthu Palaniswamia 2013 Internet of Things (IoT): A Vision, Architectural Elements, and Future Directions (Future Generation Computer Systems vol. 29, no. 10) pp. 1645–1660

[7] Ignasi Vilajosana et al. 2013 Bootstrapping Smart Cities through a Self-Sustainable Model Based on Big Data Flows (IEEE Communications Magazine vol. 51, no. 6), pp. 128-134

[8] Smart Santander Team. 2016 (SmartSantander) http://www.smartsantander.eu/index.php/testbeds/item/132-santander-summary

[9] Andrea Zanella and Lorenzo Vangelista 2014 Internet of Things for Smart Cities (IEEE Internet of Things Journal, vol. 1, no. 1)

[10] Hafedh Chourabi et al. 2012 Understanding Smart Cities: An Integrative Framework, 45th Hawaii Int. Conf. on System Sciences (Maui) pp. 2289-2297.

[11] Paul Brous and Marijn Janssen 2015 A Systematic Review of Impediments Blocking Internet of Things Adoption by Governments, 14th IFIP WG 6.11 Conference on e-Business, e-Services, and e-Society (Delft) vol. 9373 pp. 81-94.

[12] Andrew Whitmore, Anurag Agarwal, and Li Da Xu 2014 The Internet of Things—A survey of topics and trends (Information Systems Frontiers vol. 17, no. 2)

[13] Jan Markendahl and Andrés Laya 2013 Business Challenges For Internet Of Things: Findings From E-Home Care, Smart Access Control, Smart Cities And Homes, 29th IMP-conf. (Atlanta, Georgia)

[14] Jiong Jin and Marimuthu Palaniswami 2014 An Information Framework of Creating a Smart City through Internet of Things (IEEE Internet of Things Journal vol. 1, no. 2) pp. 112-121

[15] Rob Kitchin 2014 The real-time city? Big data and smart urbanism (GeoJournal vol. 79, no. 1) pp. 1-14, 2014

[16] Martin Strohbach, Holger Ziekow, Vangelis Gazis, and Navot Akiva 2015 Towards a Big Data Analytics Framework for IoT and Smart City Applications, Modeling and Processing for Next-Generation Big-Data Technologies, vol. 4, 1st ed., Xhafa Fatos et al. (Springer) pp. 257-282

[17] M. Mazhar Rathore, Awais Ahmad, Anand Paul, and Seungmin Rho 2016 Urban planning and building smart cities based on the Internet of Things using Big Data analytics (Computer Networks vol. 101) pp. 63-80

[18] Ibrahim Abaker Targio Hashem et al. 2016 The Role of Big Data in Smart City (International Journal of Information Management vol. 36, no. 5) pp. 748-758

[19] Donghee Shin 2014 A socio-technical framework for Internet-of-Things design: A human-centered design for the Internet of Things (Telematics and Informatics vol. 31, no. 4)

[20] Johannes M Schleicher, Michael Vögler, Schahram Dustdar, and Christian Inzinger 2016 Enabling a Smart City Application Ecosystem (IEEE Internet Computing vol. 20, no. 2)

[21] Sotiris Zygiaris 2012 Smart City Reference Model: Assisting Planners to Conceptualize the Building of Smart City Innovation Ecosystems (Journal of the Knowledge Economy vol. 2, no. 2)

[22] C. Harrison et al. 2010 Foundations for Smarter Cities (IBM Journal of Research and Development vol. 54, no. 4), pp. 1-16